Structural Steel Drafting

DAVID MACLAUGHLIN

Delmar Publishers

an International Thomson Publishing company I(T)P

Albany • Bonn • Boston • Cincinnati • Detroit • London • Madrid
Melbourne • Mexico City • New York • Pacific Grove • Paris • San Francisco
Singapore • Tokyo • Toronto • Washington

NOTICE TO THE READER

Cover photo courtesy of Digital Stock. Contact us at 1-800-545-4514.

Delmar Staff
Publisher: Alar Elken
Acquisitions Editor: John Anderson
Developmental Editor: Michelle Ruelos Cannistraci
Production Coordinator: Toni Bolognino
Art and Design Coordinator: Cheri Plasse
Editorial Assistant: John Fisher

COPYRIGHT © 1998
By Delmar Publishers
a division of International Thomson Publishing Inc.

The ITP logo is a trademark under license.

Printed in the United States of America

For more information, contact:

Online Services

Delmar Online
To access a wide variety of Delmar products and services on the World Wide Web, point your browser to:
    http://www.delmar.com
    or email: info@delmar.com

thomson.com
To access International Thomson Publishing's home site for information on more than 34 publishers and 20,000 products, point your browser to:
    http://www.thomson.com
    or email: findit@kiosk.thomson.com

A service of **I**(**T**)**P**®

Delmar Publishers
3 Columbia Circle, Box 15015
Albany, New York 12212-5015

International Thomson Publishing Europe
Berkshire House 168-173
High Holborn
London, WCI1V 7AA
England

Thomas Nelson Australia
102 Dodds Street
South Melbourne, 3205
Victoria, Australia

Nelson Canada
1120 Birchmont Road
Scarborough, Ontario
Canada, M1K 5G4

International Thomson Editores
Campos Eliseos 385, Piso 7
Col Polanco

11560 Mexico D F Mexico

International Thomson Publishing GmbH
Konigswinterer Strasse 418
53227 Bonn
Germany

International Thomson Publishing Asia
221 Henderson Road
#05-10 Henderson Building
Singapore 0315

International Thomson Publishing—Japan
Hirakawacho Kyowa Building, 3F
2-2-1 Hirakawacho
Chiyoda-ku, Tokyo 102
Japan

1 2 3 4 5 6 7 8 9 10 XXX 03 02 01 00 99 98 97

Library of Congress Cataloging-in-Publication Data

MacLaughlin, David (David C.)
    Structural steel drafting/David MacLaughlin
        p.   cm.
    Includes index.
    ISBN 0–8273–7313–9
    1. Structural drawing.   2. Building, Iron and steel—Details-
Drawings.   3. Architecture—Designs and plans—Working drawings.
I. Title.
T355.M25   1997
624.1′ 82′ 0223—dc21   97–10389
                        CIP

Contents

iv   Contents

P reface

Structural steel is a very important part of architecture. The primary purpose of *Structural Steel Drafting* is to provide students at both the associate and bachelor's degree levels with a fundamental and practical knowledge of how structural steel is used to construct support frames for modern commercial and industrial buildings. Special emphasis is placed on how structural drafters in both structural design and fabrication offices prepare the working drawings required to help transform the architect's vision into reality.

The topics and illustrations presented in this text have been carefully chosen for students enrolled in their initial structural steel drafting and/or design course as part of an architectural, construction, or civil engineering related curriculum. An important and unique feature of *Structural Steel Drafting* is that it ties together in one book the relationship and interdependence between the types of drawings made in design offices and those prepared in structural fabrication offices. With this in mind, many practical examples of drawings produced by structural drafters in either career setting are provided throughout the text. The examples illustrate both manual and computer-aided (CAD) drafting methods to reflect the use of both techniques throughout the industry. The author believes that, although the production of both structural design and fabrication drawings will continue to move increasingly toward CAD, the computer, no matter how sophisticated, will always be a tool designed to enable drafters and designers who already know what they are doing to work more efficiently. Thus, it is strongly recommended that students of structural drafting, especially in the beginning, learn by the proven, methodical, manual method, which requires that they (the students), not the machine, do all the thinking. For this reason, considerable time is taken to acquaint the student with tables and reference information found in the American Institute of Steel Construction's *Manual of Steel Construction* and *Detailing for Steel Construction* as well as *Standard Specifications Load Tables for Steel Joists and Joist Girders* published by the Steel Joist Institute.

Examples of typical design calculations presented in this text demonstrate the type of routine technical mathematics the entry-level structural drafter can expect to encounter on the job. They are presented using the Allowable Stress Design (ASD) method because the author believes ASD is most suitable for illustrating to the beginning structural drafting and/or design student "where the numbers are coming from" when determining the required sizes for structural components. The merits of Load and Resistance Factor Design (LRFD) are discussed. Because *Structural Steel Drafting* is first and foremost a drafting book, complex structural design theory has been purposely avoided. It is assumed, however, that the student has had some exposure to basic drafting, algebra, trigonometry, structural and analysis, and/or strength of materials coursework.

The primary objective of this book is to broadly cover the subject of structural steel drafting, emphasizing the process of preparing structural steel design and fabrication drawings for commercial and industrial building applications. To that end, the student is introduced to a wide variety of practical drafting examples and assignments that a structural steel design or detail drafter might expect to encounter in an on-the-job situation.

To whatever extent possible, it is also strongly recommended that students supplement the material in *Structural Steel Drafting* by observing the erection of steel-frame structures in their own communities. Developing an awareness of how the components of a steel-frame structure arrive at the job site and are erected to become the support framework for a modern commercial or industrial building cannot be overstressed for those training for a career in the rewarding fields of structural steel drafting, design, and construction.

David C. MacLaughlin

Acknowledgments

Charles Carter, American Institute of Steel Construction

John H. Crigler, PE, Bishop State Community College, Mobile, AL

Reynold S. Davenport, Sandhills Community College, Pinehurst, NC

James Dye, Houston, TX

J. Ingram, Alexander, AR

Bill Lawson, Rock Valley College, Rockford, IL

Donald Liou, UNCC Department of Engineering Technology, Charlotte, NC

Fithugh Miller, Pensacola Junior College, Pensacola, FL

Raymond J. Nolan, PE, Middlesex County College, Edison, NJ

Edward E. Talley, Pinson, AL

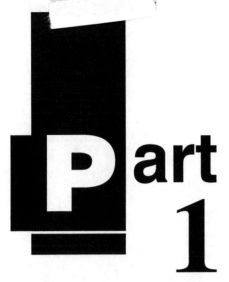

Part 1

Structural Steel Design Drawings for Steel Construction

Chapter 1

Steel: An Economical Choice for Commercial and Industrial Buildings

OBJECTIVES

Upon completion of this chapter, you should be able to:

- understand the history and economy of steel-frame construction.
- differentiate between fast-track and conventional, or linear, construction.
- know why fast-track scheduling is usually more economical than linear scheduling.
- identify the two design methods currently used to design structural steel support systems for commercial and industrial buildings: allowable stress design (ASD) and load resistance factor design (LRFD).

1.1 INTRODUCTION

American architecture has historically evinced greatness in form, function, size, beauty, and innovation. The results have been as breathtaking as they are obvious. How many people have shielded their eyes from the sun as they stared up at the reflective glass facade of a tall office tower in one of America's major cities? Who could help but be awed by the sheer size and beauty of the nation's largest retail center, the 2.6-million square-foot Mall of America in Bloomington, Minnesota? It would be a mistake to think that such architectural wonders are rare in modern-day America.

The truth is that, in recent years, the skylines of countless cities and towns all over America have experienced tremendous change. Extremely dramatic, graceful, and exciting buildings of every imaginable size and description have become almost commonplace in cities such as New York, Chicago, Atlanta, Dallas, Detroit, Pittsburgh, and others too numerous to mention. And, whether low-rise or high-rise, whether granite, brick, reflective glass, or insulated panel, the overwhelming majority of these showcase structures are evidence of both architectural achievement and structural engineering largely made possible by the flexibility, strength, and cost-effectiveness of *structural steel*.

1.2 STEEL-FRAME CONSTRUCTION

Steel-frame construction, in which a skeleton framework of structural steel supports the walls, floors, and roofs, is commonly used today in commercial and industrial buildings—not only in small one- or two-story office buildings and shopping malls, but even in such mammoth structures as the World Trade Center in New York and the Sears Tower in Chicago. In fact, steel-frame construction has made possible the modern multistory structures we so admire today.

As recently as the late nineteenth century, the height of buildings in the business districts of American cities was usually limited to four to six stories. That was because even though cast iron columns and wrought iron beams could be used for interior framing, the exterior walls of multistory buildings were still being constructed of heavy loadbearing masonry designed to support themselves as well as the adjacent floor and roof loads.

With the acceptance of the passenger elevator in the late 1870s, taller buildings became more feasible, and by the early 1890s, buildings of ten stories and more were not uncommon. However, as the structures grew in height, the loadbearing walls, of necessity, became ever thicker. The economical limits of loadbearing wall construction were reached in 1891 with Chicago's sixteen-story Monadnock Block, whose external loadbearing walls were an astounding six feet thick.

Ironically, the first steel-frame structure, the nine-story Home Insurance Building, had been erected seven years before the Monadnock Block was built. Chicago architect William LeBaron Jenny and civil engineer George Whitney had designed an innovative support system for the Home Insurance Building. The entire structure—floors, roof, and walls—was borne on a skeleton structural framework of beams, girders, and columns. F.A. Randall's book, *History of the Development of Building Construction in Chicago*, contains the following quote from the July 25, 1896, issue of *Engineering Record:*

> The principle of carrying the entire structure on a carefully balanced and braced metal frame, protected from fire, is precisely what Mr. William LeBaron Jenny worked out. No one anticipated him in it, and he deserves the entire credit belonging to the engineering feat which he was the first to accomplish.

And so the structural steel skeleton frame was born. Structural steel had made possible a new way to support structures, and the potential was truly mind-boggling. No longer dependent upon heavy loadbearing walls, tall buildings suddenly became technically very feasible. Structural steel had quite simply revolutionized building construction, not only reducing load-carrying members (such as beams or columns) to minimum sizes, which made possible longer spans and lighter weights, but also creating a clear distinction between the structural and nonstructural components of a building. Because exterior walls were no longer required to carry loads, the exterior facade could be a *curtain wall* of glass, brick, or insulated metal building panels, with the main function of protecting the interior of the building from outside elements. Led by the vision of architects such as Mies Van Der Rohe, I. M. Pei, Helmut Jahn, and others, the use of structural steel frameworks to support beautiful and functional buildings of every size, shape, and description has become commonplace throughout the world.

1.3   THE ECONOMY OF STEEL-FRAME CONSTRUCTION

From the very beginning, structural steel-frame construction proved to be the most economical support system for multistory high-rise structures. But does this economy carry over into low-rise commercial and industrial facilities? The answer is yes. Up until about 1970, cost studies found that low-rise commercial structures such as office buildings, schools, and department stores with superstructures of conventionally formed concrete were often as economical as steel-frame construction. But in the early 1970s, rising labor costs made structural steel-framing systems even more economical because they could be prefabricated in the shop and then quickly erected by relatively small field crews. By contrast, cast-in-place concrete superstructures required more labor. When comparing the two methods, it was common to find savings of $1.50 to $2.00 per square foot or more by using structural steel. For example, in 1975 the cost-per-square-foot of floors and roofs framed in structural steel was an average of 12 percent cheaper than the cost of reinforced concrete. And by the 1980s, the cost savings had widened to an average of more than 20 percent.

To the present, structural steel framing systems continue to be economical for virtually all types of commercial and industrial building structures. The two primary reasons are: (1) increased speed of production made possible through a construction method called fast-track scheduling, and (2) lower cost of materials due to the implementation of a design method known as load resistance factor design.

1.4   FAST-TRACK SCHEDULING

Certainly one of the major advantages of structural steel framing systems is speed of construction. Even on smaller, low-rise projects, steel framing systems can often be installed in 75 percent of the time required for cast-in-place concrete. Thus, a shopping mall or office building that might take 400 calendar days to complete using concrete framing could be completed in 300 days if the structural system were fabricated in steel.

How is this possible? The answer lies in *fast-track scheduling,* also known as *phased design and construction.* Fast-track scheduling is more economical because various phases of the construction process can be overlapped or condensed, thus shortening the overall time required. This is an extremely important consideration in multi-million-dollar construction projects because the owner or developer must repay money borrowed for construction and the project cannot begin to pay off until the property is leased. How can the fast-track method save time and money over conventional *linear construction?* A brief comparison of the two methods will quickly illustrate the financial advantages of fast-track scheduling for multi-million-dollar commercial and industrial projects.

Linear Construction

In linear design and construction, the phases of work for a building project are arranged end-to-end, and each

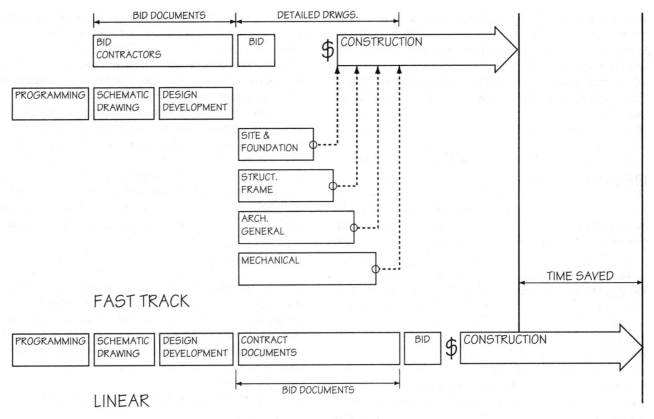

Fig. 1–1   Fast-track diagram

phase must be essentially completed before the next can begin. For example, the bidding package must be developed before construction can start. First the completed architectural, structural, and mechanical drawings must be sent out to contractors and subcontractors, each of whom then comes up with a price (*bid*) specifying the exact charge for doing the work. After that, the owner reviews the bids and selects the contractors. All of this takes time, during which the owner already has large sums of money invested in the building site, is paying interest on a substantial loan, and has not even been able to start the actual construction phase.

Fast-Track Construction

The primary advantage of fast-track construction is that many activities can take place simultaneously, saving both time and money in the end because the construction period is held to a minimum. For example, in fast-track construction, portions of the structural steel contract may have been awarded several weeks prior to the general contract. Thus, the first structural steel is being fabricated away from the job site while or before the foundations are being dug, and by the time the foundation is completed, the steel is at the job site, ready for erection. All this happens well before final

decisions about some of the upper stories or roof framing have been made. With the structural steel and steel deck in place at the earliest possible date, the mechanical trades can get an earlier start installing piping, ductwork, and electrical conduit. Such final decisions as the placement of partitions, lighting, and finishes are made at the appropriate times without holding up construction of the building shell or exterior walls. This keeps design options open longer and many times results in better design solutions. Figure 1–1 illustrates how overlapping various phases of design and construction using the fast-track method can result in substantial savings of time and money. Such savings can often determine whether or not to actually go ahead with a project.

Like any other design-build method, fast-track scheduling is not perfect. Problems can arise when construction starts before the owner knows the total cost of the building. For example, the owner or developer might overspend on the earlier stages of the project and have less money available for later contracts. Design changes can also become very expensive if they affect structural or mechanical components already in place or already being fabricated. But generally, for all of its potential problems, a steel-frame structure put up with fast-track scheduling can be completed at the earliest possible date, saving the owner or developer as much as hundreds of thousands of dollars a month in interest charges while at the same time bringing in earlier rent receipts. The cost/time analysis of construction financing has time and again led to the selection of a steel-frame superstructure for many of today's most modern and imaginative buildings.

1.5 LOAD RESISTANCE FACTOR DESIGN (LRFD)

In recent years, many structural steel skeleton frame buildings have become even more economical as a result of a new design procedure known as *load resistance factor design (LRFD)*. Although it seems unlikely that LRFD will replace the traditional *allowable stress design* method (*ASD*) for at least another decade, it has already proven to be a rational design method that can result in a more economical structure under certain conditions.

Perhaps the best way to discuss the economy of LRFD is to compare it to the ASD method.

In allowable stress design, loads on a structural member (beam, girder, column, etc.) are analyzed, and the member selected is one with a cross-sectional area large enough so that the actual stresses produced by the load do not exceed allowable stresses specified in the AISC Specification. These *allowable* or safe stresses are percentages of either the yield stress (F_y) or the ultimate stress (F_u) of a particular grade of structural steel. For example, for grade A36 steel, historically the most common grade, the F_y value is 36 ksi and the F_u is 58 ksi. For ASTM A572 grade 50 steels, which are becoming more common because of higher strength at an often minimal cost difference, the F_y value of the most frequently used steel is 50 ksi and the F_u is 65 to 70 ksi. The AISC Specification lists the *allowable* bending stress in a normally loaded, compact, braced structural member as 0.66 F_y and the allowable shear stress as 0.40 F_y. Thus, when designing a beam of A36 steel, for example, the allowable bending stress is 0.66 × 36 ksi, or 24 ksi, and the allowable shear stress is 0.40 × 36 ksi, or 14.5 ksi. These and other allowable stress values are found in the AISC Specification. When designing structural members using the ASD method, engineers usually select the most economical structural shape, which keeps the computed or actual stresses within the allowable limits for tension, shear, bending, and so on.

Load resistance factor design, also called the *strength method,* uses a probability-based rationale in which working loads are multiplied by load factors. The load factors are then applied individually for various loads such as dead load, live load, wind load, and so on. The resulting *factored loads* are used to design the structure. LRFD design methods are widely used in the design of mid-rise and high-rise buildings.

The purpose of load factors in the LRFD procedure is to increase the loads to compensate for uncertainties in estimating their magnitudes. The value of load factors varies. For example, the load factor used for dead loads is smaller than that used for live loads because the weights of gravity dead loads, such as concrete floors, concrete block, brick, steel deck, steel partitions, and so on, can be much more accurately estimated than the weights of live loads such as human occupants, stored material, and furniture, generally specified by building codes in pounds per square foot. This is especially true when attempting to estimate the greatest possible combinations of live loads likely to occur simultaneously. Determination of a critical design load using a combination of typical LRFD load factors might be written:

U (critical design or ultimate load) = 1.2D (dead load) + 1.6L (live load) + 0.5S (snow load)

The *resistance factor* part of LRFD is essentially an attempt by the designer to allow for inconsistencies or uncertainties in design theory and construction practices, including steel production and fabrication. This is done by multiplying the strength of a structural member by a resistance factor (ϕ). The values of resistance factors for various structural members, which are typically less than 1.0, are found in the AISC Specification and vary depending upon the type of structural member being designed. For example, ϕ is 0.90 for bending or shear in structural beams, but 0.85 for columns.

When using the LRFD method, the designer essentially multiplies various loads by load factors that are usually greater than 1.0 and compares these factored loads to the factored resistance in which the theoretical strength of the structural member has been reduced by an appropriate resistance factor. The fundamental concept of the LRFD method is that the sum of the loads multiplied by their appropriate load factors must be equal to or less than the member's theoretical strength (or resistance) multiplied by the appropriate resistance factor.

Σ (loads × load factors) ≤ resistance × resistance factor

Much of the economy of LRFD derives from the fact that a smaller safety factor can be used for dead loads because they are more accurately determined, while in ASD the same safety factor is used for both dead and live loads. As a result, the weight of structural steel for LRFD-designed structures, especially in floor systems when live loads are small compared to dead loads, is more cost-effective than similar steel systems designed by the ASD method. For example, a study reported in the November 1991 issue of *Modern Steel Construction* found that using LRFD for typical office floor beams spanning 30 to 46 feet resulted in beams one to two sizes lighter than those obtained using ASD procedures, and this was true for both grade A36 and A572 grade 50 high-strength steel. The same issue reported that builders of the Newport Office Tower, which is one of the tallest buildings in New Jersey and was designed using the LRFD method, realized a $2 million savings in steel costs, primarily by reducing beam and girder sizes.

Based on studies made as recently as 1994 by Thornton-Tomasetti Engineers of New York City, which show that LRFD can often save an average of 5 percent of total steel tonnage on mid- and high-rise projects, it seems inevitable that LRFD procedures will be more widely accepted with the passage of time. However, in all fairness, it must be pointed out that steel weight economies between buildings designed by the ASD and LRFD methods depend heavily upon the ratio of live loads to dead loads, and in structures with high dead-load to live-load ratios, almost no cost savings result with LRFD compared to ASD.

1.6   SUMMARY

This chapter has attempted to point out not only that structural steel framing systems have resulted in the design and construction of many of America's most exciting and imaginative structures, but also that structural steel has been and will most certainly continue to be an economical choice for future commercial and industrial buildings. For designers and drafters in structural design and fabrication offices, or for people training for such employment, the best is yet to come due to the economy, flexibility, and strength of structural steel.

STUDY QUESTIONS

1. What is the relationship between the acceptance of the passenger elevator in the late 1870s and structural steel-frame construction?

2. The economical limits of loadbearing wall construction were reached in 1891 in Chicago with the construction of the Monadnock Block. How many stories high was the Monadnock Block?

3. At their thickest part, how thick were the loadbearing exterior walls of the Monadnock Block?

4. Name the Chicago architect and engineer credited with designing the first steel-frame structure.

5. When was the first steel-frame structure erected?

6. Name an advantage of steel-frame structures over traditional loadbearing wall structures.

7. Why have steel-frame superstructures proven more economical than conventionally formed concrete structures, especially since the 1970s?

8. Fast-track construction is generally considered an economical way to design and erect steel-frame buildings. How is this economy achieved?

9. Two structural steel design methods discussed in this chapter are the ASD and LRFD methods. What do the letters ASD and LRFD stand for?

10. In ASD design, the allowable (or safe) stresses are percentages of either the _____ stress or the _____ stress of a particular grade of structural steel.

11. In the LRFD procedure, what is the purpose of load factors?

12. Name three examples of dead loads.

13. Name three examples of live loads.

14. In the LRFD method of design, why are the load factors for live loads a larger number than those for dead loads?

15. For people working as or training to become structural drafters and/or designers, the future looks very bright due to the _____ , _____ , and _____ of structural steel.

Chapter 2

The World of Structural Steel

OBJECTIVES

Upon completion of this chapter, you should be able to:

- understand steel as a structural material, including its availability in various grades and shapes.
- properly use the *Manual of Steel Construction* published by the American Institute of Steel Construction, and know why this important publication is so widely used as a reference for designing structural steel buildings and preparing structural steel design drawings.
- describe the characteristics of open-web steel joists and the Steel Joist Institute load tables.

2.1 INTRODUCTION

The primary day-to-day responsibility of the structural drafter in any engineering office is to produce drawings that depict the structural framework of a building. Structural drawings must show the framing plans and details completely enough that the ironworkers can understand what they are to do. Thus, drafting is really a language the structural drafting student must master in order to make meaningful drawings.

Before attempting to make structural drawings, the successful drafter should have a basic knowledge of structural steel. He or she must know the various strength grades of structural steel and the common structural steel rolled shapes. A familiarity with the *Manual of Steel Construction* is also a prerequisite for a competent structural drafter or designer, as is an understanding of open-web steel joists and their uses in commercial construction. This chapter will provide the structural drafting student with this essential information.

2.2 STEEL AS A STRUCTURAL MATERIAL

Steel is a man-made material consisting primarily of iron (about 98 percent) and small quantities of carbon, silicon, manganese, sulphur, and other elements. Historically in

the United States, molten iron was converted into steel by either open-hearth, basic oxygen, or electric arc furnaces. Huge ladles of fluid steel were poured into *ingots,* rectangular shapes with rounded corners that were immediately stored and reheated in underground furnaces called soaking pits. The soft, white-hot ingots were then passed between heavy rollers in the primary rolling mills where they were converted into semifinished products called blooms, billets, and slabs. Next the blooms, billets, and slabs were sent to other secondary rolling mills to be transformed into structural shapes such as pipe, tube, bar, rod, and wire.

In recent years, modern technological advances have highly automated steel-making. The continuous slab casting process has eliminated both the conventional ingot operation and the primary rolling operation by producing a continuous length of steel that can be cut into slabs. These slabs are then cut into various lengths for further processing at the secondary rolling mills, which still produce the various structural shapes. Figure 2–1 illustrates how the continuous casting process is accomplished by the **curved mold process** at Bethlehem Steel Corporation.

Grades of Structural Steel

Structural steel is produced in a variety of grades and standard rolled shapes suitable for a wide spectrum of conditions encountered by architects and engineers throughout the United States and the world. Because the chemical com-

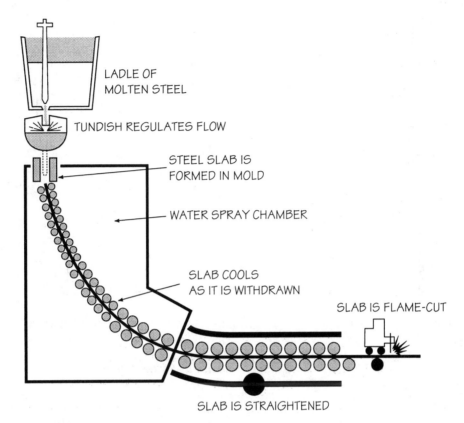

LADLE OF
MOLTEN STEEL

TUNDISH REGULATES FLOW

STEEL SLAB IS
FORMED IN MOLD

WATER SPRAY CHAMBER

SLAB COOLS
AS IT IS WITHDRAWN

SLAB IS FLAME-CUT

SLAB IS STRAIGHTENED

**Figure 2–1   Illustration of the continuous slab casting process at Bethlehem Steel Corporation
(Courtesy of Bethlehem Steel)**

position of steel is directly related to its physical properties, steel producers are constantly striving to develop new grades of steel with such qualities as higher strength, greater corrosion resistance, and better welding capabilities, while still keeping the cost reasonable. As a result of this research, several different grades of structural steel are available to the design professions, each having a unique combination of properties and economy appropriate for a specific application.

Carbon steels are the most economical grades, but they also have the lowest strength or yield point, which is usually measured in kips per square inch of cross section (ksi). (In structural engineering a kip is 1,000 pounds.) Higher-strength, low-alloy steels are stronger than carbon steels, but they may be more expensive. For example, A588 *self-weathering steel* is a high-strength, low-alloy steel with about four to six times the corrosion resistance of all-purpose carbon steel. It is not necessary to paint self-weathering steel because the rust that forms on its surface acts as a cover, like paint, preventing deeper corrosion. This grade of steel is considerably more expensive (about 30 percent more) than carbon steel, but life-cycle cost analysis often shows weathering steel to be more economical than painted carbon steel because it does not require painting or maintenance. The various grades of structural steel are identified by specifications established by the American

Society of Testing and Materials, commonly referred to as the ASTM.

ASTM A36. An all-purpose carbon steel, is the most widely used structural steel for commercial and industrial building construction. This grade of steel, which has a yield stress level of 36 ksi, has excellent welding and machining characteristics, thus is ideal for making welded and/or bolted connections. A36 is also the most economical of all structural steels on a cost-per-pound basis. At least 50 percent of the structural steel presently used for commercial and industrial building construction in the United States is ASTM A36.

ASTM A572. A higher-strength steel, is another popular grade of structural steel used in building construction. Like A36, this high-strength, low-alloy steel may be bolted or welded. It is available in four minumum stress levels: 42 ksi, 50 ksi, 60 ksi, and 65 ksi grades. Since A572 is stronger than A36, using it often results in lower costs because heavier loads may be carried at longer spans by lighter beams. This translates into cost savings because fewer footings are required, and erection time can be cut down.

ASTM A588. A corrosion-resistant, high-strength, low-alloy steel with a yield stress level of 50 ksi, is a self-

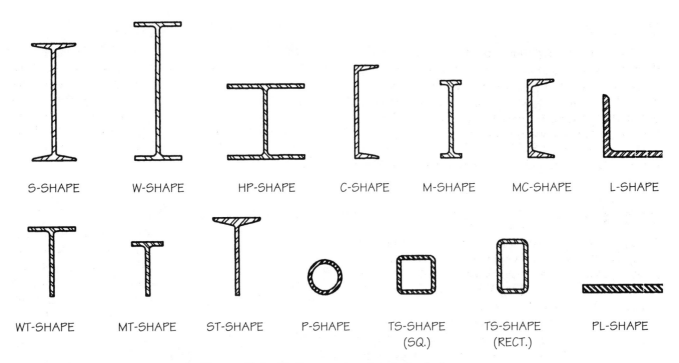

S-SHAPE   W-SHAPE   HP-SHAPE   C-SHAPE   M-SHAPE   MC-SHAPE   L-SHAPE

WT-SHAPE   MT-SHAPE   ST-SHAPE   P-SHAPE   TS-SHAPE (SQ.)   TS-SHAPE (RECT.)   PL-SHAPE

Figure 2–2 Common structural steel shapes

weathering grade of steel. The "weathering" characteristic is brought about because this type of steel contains a small amount of copper that, when exposed to the atmosphere, oxidizes to produce a thin film of reddish-brown rust, called a *patina,* on the surface. This film of rust, which is only a few thousandths of an inch thick, prevents further oxidation and eliminates the need for paint. However, this steel does not wear well if subjected to saltwater, continually submerged in water, or exposed to very dry desert conditions because wet and dry cycles are needed in order for the protective coating of rust to form.

ASTM A514. Is an extremely strong quenched and tempered alloy steel with minimum yield stresses in the 90 to 100 ksi range, but it is produced only in plates and bars.

ASTM A242. Is a corrosion-resistant, high-strength, low-alloy steel available with minimum yield stresses of 42, 46, and 50 ksi.

ASTM A529. Is a carbon steel with a minimum yield stress level of 42 ksi, but its availability is limited to certain shapes of plates and bars one-half-inch thick or less.

A discussion of structural steel shapes would be incomplete without mentioning ASTM A7, a grade of carbon steel now obsolete but used very frequently during the 1940s and 1950s when a great many connections were riveted. This steel, which had a minimum yield stress of 33 ksi, was excellent for bolted and riveted connections, but it did not have good welding characteristics, especially when members were more than one inch in thickness. With the advent

of more advanced and economical welding techniques, especially shop welding, it became necessary to develop grades of structural steel with better welding properties. This led to the development and use of A36 steel, which is stronger than the old A7 and has excellent welding and machining properties. However, structural steel drafters and designers should be aware of grade A7 steel because many structures constructed of this grade of steel during and after World War II are still in use. Thus, if an older structure is being remodeled, a load that required an 18″-deep wide-flange beam of A7 steel in the 1950s can now be supported by a smaller and much lighter W-shape of high-strength A572 steel.

2.3 COMMON STRUCTURAL STEEL ROLLED SHAPES

Having looked at the various grades of modern structural steel available to the structural drafter and designer, we will now discuss the common shapes or sections in which structural steel is produced. It is important to know these shapes because the design and erection of structural steel frames to support commercial and industrial buildings is essentially a matter of developing the most economical assembly of these standard rolled shapes.

Structural steel shapes are designated by the shapes of their cross-sections, and these designations are used to indicate structural steel members (beams, columns, etc.) on both design and shop drawings. Figure 2–2 illustrates many of the most commonly used structural steel rolled shapes.

S-Shape

The S-shape, commonly called American Standard Beam or I-beam, is a rolled section with two parallel flanges connected by a web. These sections were the first beam sections rolled in America, but they are no longer widely used in building construction. S-shape beams have relatively narrow flanges whose inner surfaces are sloped at a pitch of 2 to 12. The designation S15 × 42.9 would indicate an S section exactly 15″ deep and weighing 42.9 pounds per foot of length.

W-Shape

The W-shape is the most commonly used structural shape. The W-shape has all but replaced the S-shape in the construction of commercial and industrial buildings because it is very efficient and economical due in large part to its design. W-shapes have large moments of inertia around their principal axes, making them ideal for flexure as, for example, a beam supporting floor loads. Also, most of the areas of W-shapes are in their flanges, which are more laterally stable because they are thicker and wider than S-shape flanges. The inner and outer surfaces of the top and bottom flanges of a W-shape, unlike the S-shape, are essentially parallel. The top and bottom flanges are connected by a thin web designed to provide resistance to shear. Like the S-shape, the W-shape is designated by its depth and weight per linear foot, but the depth designation is usually approximate. For example, while a W16 × 40 is actually 16″ deep and weighs 40 pounds per foot of length, a W16 × 26 is actually 15.69″ deep, and a W16 × 89 is actually 16.75″ deep. The variation in actual depth of W-shapes is a feature structural drafters and designers must always keep in mind when designing or drawing structural steel support systems.

HP-Shape

The HP-shape is similar to the W-shape except that it is much heavier and the web and flange thicknesses are equal. Also, with HP-shapes, the width of the flange and the depth of the section are approximately equal. HP-shapes are made with very thick webs and flanges to resist the impact of pile-driving hammers because they are used primarily in pile foundations and only occasionally as building columns.

C-Shape

The C-shape, or American Standard Channel, consists of a web similar to a W-shape and two tapering parallel flanges. However, with C-shapes, the flanges extend on only one side of the web, making them ideal to use as stringers for steel stairs or for framing floor openings and stairwells. The flanges are also very narrow. The flange of a C15 × 40, which is exactly 15″ deep and weighs 40 pounds per linear foot, is only 3 1/2″ wide.

M-Shape and MC-Shape

M-shapes are essentially lightweight W-shapes, and MC shapes are miscellaneous channel sections that cannot be classified as American Standard Channels. For example, an MC12 × 10.6 is a very light channel shape often used for stair treads in industrial buildings. M- and MC-shapes are often not as readily available as other shapes, so the drafter or designer should make sure they can be obtained before specifying their use.

L-Shape

The L-shape is a rolled steel section called an angle. The angle is shaped like an L with horizontal and vertical "legs" at right angles (90 °) to each other. The inner and outer surfaces of each leg are parallel, and an angle section may have legs of either equal or unequal length. L-shapes are designated first by the length of each leg and then by the leg thickness. For example, and L-shape designated as L4 × 4 × 1/2 indicates both legs are 4″ long and 1/2″ thick. The designation L5 × 3 × 1/2 indicates an unequal-leg angle with the long leg 5″ long, the short leg 3″ long, and the thickness of each leg 1/2″. When specifying an unequal-leg angle, the longer leg is always specified first. Angles are commonly used for framing connections, cross-bracing, and constructing lintels over doors and windows.

WT-, MT-, and ST-Shapes

Called *structural tees,* these shapes are made by splitting a W-, M-, or S-shape longitudinally, usually at mid-depth. Thus, the designation WT9 × 25 indicates a structural tee cut from a W18 × 50 with a stem depth from tip of stem to the top of the flange surface of 9″ and a weight of 25 pounds per linear foot. Structural tee shapes are sometimes used as lintels and often as the top and bottom chords of prefabricated trusses.

P- and TS-Shapes

Steel pipe and tube steel (TS-shapes) are widely used as columns, although rectangular TS-shapes are also used as beams. Pipe is designated as Pipe 3 Std., which would mean a standard 3″ (nominal diameter) steel pipe. The TS shapes are rolled in square and rectangular shapes. The designation TS4 × 4 × .375 indicates a square structural tube 4″ × 4″ with a wall thickness of .375″. A designation of ST10 × 4 × .250 would indicate a structural tube section 10″ wide on one side and 4″ wide on the other, with a wall thickness of .250″. For one-story structures such as offices or restaurants, 3″ steel pipe and TS4 × 4 tubes make excellent columns because they can be easily hidden within interior and exterior walls.

PL-Shapes and Bars

Plates and bars are rectangular shapes .250″ or more in thickness. Bars are usually classified as 6″ or less in width, and plates as 8″ or more in width. The American Institute of Steel Construction recommends specifying plate size first by the thickness and then by the width. Bar size should be expressed by the width followed by the thickness. Thus, a 16″-square × 1/2″-thick column baseplate would be specified as PL 1/2 × 16 × 16, while a beam-bearing plate made of flat bar stock would be designated as 6 × 1/2 × 10.

2.4 THE *MANUAL OF STEEL CONSTRUCTION*

The *Manual of Steel Construction,* published by the American Institute of Steel Construction (AISC), is the most widely used source of information for designing and drafting steel-framed buildings. A thorough acquaintance with this invaluable handbook is absolutely necessary for anyone employed in a structural design or fabrication office. The *manual* is essentially a reference book that gives detailed information on how to make design calculations and design or shop drawings in structural steel. Divided into six parts, the manual contains charts, tables, and other drafting and design aids that will be discussed throughout this text.

The AISC now publishes two versions of its manual: one based on ASD (Allowable Stress Design) and another based on LRFD (Load Resistance Factor Design). For this text, we have chosen to use illustrations from the ninth edition of the *ASD Manual of Steel Construction.* Although both the ASD and LRFD methods are used in design offices, ASD is the most prevalent at present and is sufficient for the routine calculations performed by entry-level drafters, especially on smaller projects. Also, familiarity with ASD design methods provides an excellent background from which to progress on to LRFD concepts.

The following discussion of the organization of the *Manual of Steel Construction* is intended as an overview of the wealth of material available in either the ASD or the LRFD edition of this important book.

Part 1

Part 1 of the manual contains tables of dimensions and properties for all the standard structural steel shapes supplied by the industry. Tables designated W-Shapes Dimensions, Channels American Standard Dimensions, etc., list all the dimensional information necessary for making design or shop drawings of structural steel systems. Tables designated W-Shapes Properties or Channels American Standard Properties, etc., list properties such as section modulus, moment of inertia, and radius of gyration, which are important factors in designing and selecting structural components capable of resisting the forces that act on a structural support system.

In Part 1 of the manual, tables giving the dimensions and properties of structural shapes are placed on facing pages. Thus, a drafter or designer tentatively selecting a W18 × 50 beam to carry a given load may quickly check the manual to find that this structural W-shape is 18″ deep and has a web thickness of 3/8″, with top and bottom flanges each 7 1/2″ wide and 9/16″ thick. Scanning across to the table of properties on the adjacent page, the designer can see that the section modulus of a W18 × 50 is 88.9 in³ and the moment of inertia is 800 in⁴.

Appendix A of this text contains selected Allowable Stress Design Selection Tables and Steel Section Tables for use with design problems in subsequent chapters of this book. However, any structural drafting or design student should also have access to the latest *Manual of Steel Construction.* Table 2–1 shows one of the Dimension and Properties Tables for W-shapes found in Part 1 of the *AISC Manual,* which has similar tables for M-shapes, S-shapes, HP-shapes, American Standard and Miscellaneous Channels (C and MC-shapes), angles (L-shapes), structural tees, combination sections, steel pipe, and structural tubing.

Table 2–1 shows a few key features of the dimension table. First, notice that the W-shape dimensions are given in both decimal and fractional forms. The fractional dimensions are accurate enough for making structural drawings, while the decimal dimensions are to be used with the hand calculator or computer when performing design calculations.

For instance, in designing a beam for vertical shear, it is important to know the actual unit shear stress on the web of the beam because, according to the AISC specification, the vertical shear stress on a beam made of A36 steel should not exceed 14,400 psi (pounds per square inch). To show how to use the dimension table, we will assume that a W18 × 50 beam has a shear-producing reaction at one end of 45,000 pounds and that we want to investigate the beam for shear.

From the dimension table, we can see that the actual depth (d) of a W18 × 50 W-shape is 17.99″, while the thickness of the web (t_w) is 0.355″. The area of the web (A_w) is the product of the actual beam depth times the web thickness. Thus, the area of the beam web can be expressed:

$$A_w = dt_w = 17.99″ \times 0.355″ = 6.38 \text{ in}^2$$

Table 2–1

Table of dimensions and properties of structural steel shapes

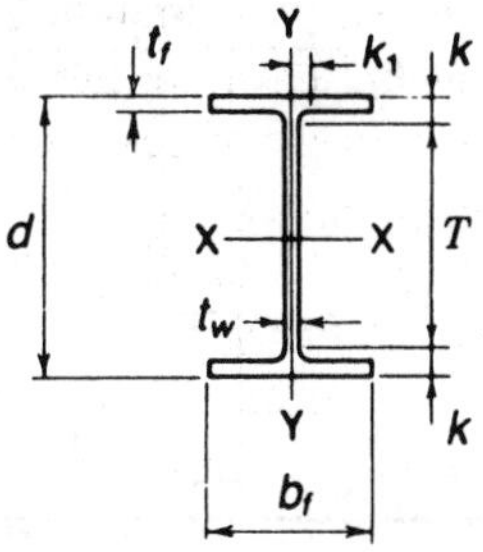

W SHAPES
Dimensions

Designation	Area A	Depth d		Web: Thickness t_w		Web: $\frac{t_w}{2}$	Flange: Width b_f		Flange: Thickness t_f		Distance: T	Distance: k	Distance: k_1
	In.^{2}	In.		In.		In.	In.		In.		In.	In.	In.
W 18×311[a]	91.5	22.32	22 3/8	1.520	1 1/2	3/4	12.005	12	2.740	2 3/4	15 1/2	3 7/16	1 3/16
×283[a]	83.2	21.85	21 7/8	1.400	1 3/8	11/16	11.890	11 7/8	2.500	2 1/2	15 1/2	3 3/16	1 3/16
×258[a]	75.9	21.46	21 1/2	1.280	1 1/4	5/8	11.770	11 3/4	2.300	2 5/16	15 1/2	3	1 1/8
×234[a]	68.8	21.06	21	1.160	1 3/16	5/8	11.650	11 5/8	2.110	2 1/8	15 1/2	2 3/4	1
×211[a]	62.1	20.67	20 5/8	1.060	1 1/16	9/16	11.555	11 1/2	1.910	1 15/16	15 1/2	2 9/16	1
×192	56.4	20.35	20 3/8	0.960	1	1/2	11.455	11 1/2	1.750	1 3/4	15 1/2	2 7/16	15/16
×175	51.3	20.04	20	0.890	7/8	7/16	11.375	11 3/8	1.590	1 9/16	15 1/2	2 1/4	7/8
×158	46.3	19.72	19 3/4	0.810	13/16	7/16	11.300	11 1/4	1.440	1 7/16	15 1/2	2 1/8	7/8
×143	42.1	19.49	19 1/2	0.730	3/4	3/8	11.220	11 1/4	1.320	1 5/16	15 1/2	2	13/16
×130	38.2	19.25	19 1/4	0.670	11/16	3/8	11.160	11 1/8	1.200	1 3/16	15 1/2	1 7/8	13/16
W 18×119	35.1	18.97	19	0.655	5/8	5/16	11.265	11 1/4	1.060	1 1/16	15 1/2	1 3/4	15/16
×106	31.1	18.73	18 3/4	0.590	9/16	5/16	11.200	11 1/4	0.940	15/16	15 1/2	1 5/8	15/16
× 97	28.5	18.59	18 5/8	0.535	9/16	5/16	11.145	11 1/8	0.870	7/8	15 1/2	1 9/16	7/8
× 86	25.3	18.39	18 3/8	0.480	1/2	1/4	11.090	11 1/8	0.770	3/4	15 1/2	1 7/16	7/8
× 76	22.3	18.21	18 1/4	0.425	7/16	1/4	11.035	11	0.680	11/16	15 1/2	1 3/8	13/16
W 18× 71	20.8	18.47	18 1/2	0.495	1/2	1/4	7.635	7 5/8	0.810	13/16	15 1/2	1 1/2	7/8
× 65	19.1	18.35	18 3/8	0.450	7/16	1/4	7.590	7 5/8	0.750	3/4	15 1/2	1 7/16	7/8
× 60	17.6	18.24	18 1/4	0.415	7/16	1/4	7.555	7 1/2	0.695	11/16	15 1/2	1 3/8	13/16
× 55	16.2	18.11	18 1/8	0.390	3/8	3/16	7.530	7 1/2	0.630	5/8	15 1/2	1 5/16	13/16
× 50	14.7	17.99	18	0.355	3/8	3/16	7.495	7 1/2	0.570	9/16	15 1/2	1 1/4	13/16
W 18× 46	13.5	18.06	18	0.360	3/8	3/16	6.060	6	0.605	5/8	15 1/2	1 1/4	13/16
× 40	11.8	17.90	17 7/8	0.315	5/16	3/16	6.015	6	0.525	1/2	15 1/2	1 3/16	13/16
× 35	10.3	17.70	17 3/4	0.300	5/16	3/16	6.000	6	0.425	7/16	15 1/2	1 1/8	3/4
W 16×100	29.4	16.97	17	0.585	9/16	5/16	10.425	10 3/8	0.985	1	13 5/8	1 11/16	15/16
× 89	26.2	16.75	16 3/4	0.525	1/2	1/4	10.365	10 3/8	0.875	7/8	13 5/8	1 9/16	7/8
× 77	22.6	16.52	16 1/2	0.455	7/16	1/4	10.295	10 1/4	0.760	3/4	13 5/8	1 7/16	7/8
× 67	19.7	16.33	16 3/8	0.395	3/8	3/16	10.235	10 1/4	0.665	11/16	13 5/8	1 3/8	13/16
W 16× 57	16.8	16.43	16 3/8	0.430	7/16	1/4	7.120	7 1/8	0.715	11/16	13 5/8	1 3/8	7/8
× 50	14.7	16.26	16 1/4	0.380	3/8	3/16	7.070	7 1/8	0.630	5/8	13 5/8	1 5/16	13/16
× 45	13.3	16.13	16 1/8	0.345	3/8	3/16	7.035	7	0.565	9/16	13 5/8	1 1/4	13/16
× 40	11.8	16.01	16	0.305	5/16	3/16	6.995	7	0.505	1/2	13 5/8	1 3/16	13/16
× 36	10.6	15.86	15 7/8	0.295	5/16	3/16	6.985	7	0.430	7/16	13 5/8	1 1/8	3/4

(Courtesy of the American Institute of Steel Construction Inc.)

Table 2–1   Continued

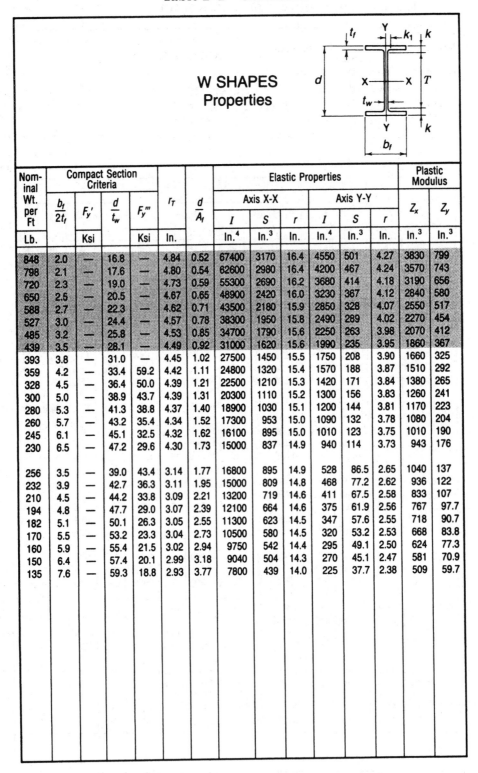

W SHAPES
Properties

Nominal Wt. per Ft	Compact Section Criteria				r_T	$\dfrac{d}{A_f}$	Elastic Properties						Plastic Modulus	
	$\dfrac{b_f}{2t_f}$	F_y'	$\dfrac{d}{t_w}$	F_y'''			Axis X-X			Axis Y-Y			Z_x	Z_y
							I	S	r	I	S	r		
Lb.		Ksi		Ksi	In.		In.^{4}	In.^{3}	In.	In.^{4}	In.^{3}	In.	In.^{3}	In.^{3}
848	2.0	—	16.8	—	4.84	0.52	67400	3170	16.4	4550	501	4.27	3830	799
798	2.1	—	17.6	—	4.80	0.54	62600	2980	16.4	4200	467	4.24	3570	743
720	2.3	—	19.0	—	4.73	0.59	55300	2690	16.2	3680	414	4.18	3190	656
650	2.5	—	20.5	—	4.67	0.65	48900	2420	16.0	3230	367	4.12	2840	580
588	2.7	—	22.3	—	4.62	0.71	43500	2180	15.9	2850	328	4.07	2550	517
527	3.0	—	24.4	—	4.57	0.78	38300	1950	15.8	2490	289	4.02	2270	454
485	3.2	—	25.8	—	4.53	0.85	34700	1790	15.6	2250	263	3.98	2070	412
439	3.5	—	28.1	—	4.49	0.92	31000	1620	15.6	1990	235	3.95	1860	367
393	3.8	—	31.0	—	4.45	1.02	27500	1450	15.5	1750	208	3.90	1660	325
359	4.2	—	33.4	59.2	4.42	1.11	24800	1320	15.4	1570	188	3.87	1510	292
328	4.5	—	36.4	50.0	4.39	1.21	22500	1210	15.3	1420	171	3.84	1380	265
300	5.0	—	38.9	43.7	4.39	1.31	20300	1110	15.2	1300	156	3.83	1260	241
280	5.3	—	41.3	38.8	4.37	1.40	18900	1030	15.1	1200	144	3.81	1170	223
260	5.7	—	43.2	35.4	4.34	1.52	17300	953	15.0	1090	132	3.78	1080	204
245	6.1	—	45.1	32.5	4.32	1.62	16100	895	15.0	1010	123	3.75	1010	190
230	6.5	—	47.2	29.6	4.30	1.73	15000	837	14.9	940	114	3.73	943	176
256	3.5	—	39.0	43.4	3.14	1.77	16800	895	14.9	528	86.5	2.65	1040	137
232	3.9	—	42.7	36.3	3.11	1.95	15000	809	14.8	468	77.2	2.62	936	122
210	4.5	—	44.2	33.8	3.09	2.21	13200	719	14.6	411	67.5	2.58	833	107
194	4.8	—	47.7	29.0	3.07	2.39	12100	664	14.6	375	61.9	2.56	767	97.7
182	5.1	—	50.1	26.3	3.05	2.55	11300	623	14.5	347	57.6	2.55	718	90.7
170	5.5	—	53.2	23.3	3.04	2.73	10500	580	14.5	320	53.2	2.53	668	83.8
160	5.9	—	55.4	21.5	3.02	2.94	9750	542	14.4	295	49.1	2.50	624	77.3
150	6.4	—	57.4	20.1	2.99	3.18	9040	504	14.3	270	45.1	2.47	581	70.9
135	7.6	—	59.3	18.8	2.93	3.77	7800	439	14.0	225	37.7	2.38	509	59.7

The actual shear stress on the web of the beam (f_v) is equal to the magnitude of the reaction (V) divided by the area of the web (A_w), as shown in the following equation:

$$f_v = V/A_w = 45,000 \text{ pounds}/6.38 \text{ sq. in.} = 7,053 \text{ psi}$$

Because the allowable shear stress on a W-shape of A36 steel is 14,400 psi, the beam is adequate for shear.

This is only one example of how the dimension table in Part 1 of the *Manual of Steel Construction* is used to perform design calculations. Another point the student might notice is that a W18 beam is not *always* 18″ deep; it might be more or less than 18″. This type of information is vitally important to structural drafters and engineers as they draw structural steel support systems that must fit together correctly at the job site.

Part 2

Part 2 of the manual is specifically intended to help the drafter/designer design and select beams, beam-bearing plates, and girders. Included are tables, charts, and other information needed to determine the most economical beam or girder for a given load. An example of one of the tables found in Part 2 of the manual is the Allowable Stress Design Selection Table for shapes used as beams. Using this table, the most economical beam for a given condition can be quickly chosen once the designer has determined either the maximum bending moment produced by a beam's loading or the required section modulus. Table 2–2 is a portion of the Allowable Stress Design Selection Table from the ninth edition of the *Manual of Steel Construction.*

In examining Table 2–2, the student should note the column on the far right, which lists the M_R (*maximum resisting moment* in kip-ft.) that a particular rolled W-shape beam can develop. This is important because most of the time, the governing factor in beam selection is to choose a beam that has a section modulus equal to or greater than the required section modulus and is capable of developing a *moment of resistance* greater than the calculated maximum bending moment. Two exceptions to this general rule might be (1) the design of roof support systems where relatively light loads may be supported on relatively long spans, often causing the beam deflection to become the governing design factor, or (2) the design of a beam to support a very heavy, superimposed load on a very short span. For this latter condition, the allowable vertical shear stress of the beam could well be the governing factor in the design and selection of the beam.

To illustrate the practical use of the Allowable Stress Design Selection Table, we will assume structural analysis has indicated that the maximum bending moment imposed on a beam by a load is 164 kip-ft. and that we will be using A36 (F_y = 36 ksi) steel.

Looking at the M_R column in Table 2–2, we see in the lower right-hand corner of the right-hand page that a W14 × 48 is capable of developing an M_R of 139 kip-ft., which is *not* adequate to resist the 164-kip-ft. bending moment. By reading up the M_R column, we find that the M_R values gradually increase. For example, at the top of the first group of W-shapes, we see that a W21 × 44 has an M_R of 162 kip-ft. This is still not equal to or greater than the 164-kip-ft. bending moment, but we are getting closer.

Continuing up the page, we see two beams grouped together, each of which has an M_R greater than our 164-kip-ft. bending moment. Obviously, either of these beams would be suitable, but the W18 × 50 is 15 pounds per linear foot lighter than the W12 × 65, thus is more economical and would normally be the beam selected. On the other hand, if space considerations should require more room to run HVAC ducts or pipe between the underside of a floor and a suspended ceiling below, the W12 × 65 would be the better choice. It should also be pointed out that the section modulus column in this table (the S_x column) lists the section modulus value of the various beams in ascending order, similar to the M_R column. Therefore, as previously stated, beams could be selected based either on the required section modulus or the maximum resisting moment.

The grouping of the W-shapes on the Allowable Stress Design Selection Table is also very important. Some groupings consist only of one or two beams, while others contain as many as eight. But in every case, the W-shape in bold print at the top of the group is always the lightest and thus the most materially economical beam in its group since structural steel is sold by the pound.

It should also be pointed out that other factors besides the weight of steel are involved in achieving true overall economy. For example, connection cost is a function of labor, and labor cost is the driving force behind project cost in today's construction industry. Thus, if selecting the lightest beam necessitates more complex connections, economy has not been achieved. Or, to go back to a previous example, if selecting a W12 × 65 rather than a W18 × 50 would leave more room for the mechanical piping and HVAC ducts, that could result in a more economical HVAC system and a more economical project overall.

Table 2–2
Allowable stress design selection table

S_x ALLOWABLE STRESS DESIGN SELECTION TABLE
For shapes used as beams

F_y = 50 ksi			S_x	Shape	Depth d	F_y'	F_y = 36 ksi		
L_c	L_u	M_R					L_c	L_u	M_R
Ft	Ft	Kip-ft	In.^{3}		In.	Ksi	Ft	Ft	Kip-ft
5.4	**5.9**	**188**	**68.4**	**W 18×40**	**17⅞**	—	**6.3**	**8.2**	**135**
9.0	22.4	183	66.7	W 10×60	10¼	—	10.6	31.1	132
6.3	**7.4**	**178**	**64.7**	**W 16×40**	**16**	—	**7.4**	**10.2**	**128**
7.2	14.1	178	64.7	W 12×50	12¼	—	8.5	19.6	128
7.2	10.4	172	62.7	W 14×43	13⅝	—	8.4	14.4	124
9.0	20.3	165	60.0	W 10×54	10⅛	63.5	10.6	28.2	119
7.2	12.8	160	58.1	W 12×45	12	—	8.5	17.7	115
4.8	**5.6**	**158**	**57.6**	**W 18×35**	**17¾**	—	**6.3**	**6.7**	**114**
6.3	6.7	155	56.5	W 16×36	15⅞	64.0	7.4	8.8	112
6.1	8.3	150	54.6	W 14×38	14⅛	—	7.1	11.5	108
9.0	18.7	150	54.6	W 10×49	10	53.0	10.6	26.0	108
7.2	11.5	143	51.9	W 12×40	12	—	8.4	16.0	103
7.2	16.4	135	49.1	W 10×45	10⅛	—	8.5	22.8	97
6.0	**7.3**	**134**	**48.6**	**W 14×34**	**14**	—	**7.1**	**10.2**	**96**
4.9	**5.2**	**130**	**47.2**	**W 16×31**	**15⅞**	—	**5.8**	**7.1**	**93**
5.9	9.1	125	45.6	W 12×35	12½	—	6.9	12.6	90
7.2	14.2	116	42.1	W 10×39	9⅞	—	8.4	19.8	83
6.0	**6.5**	**116**	**42.0**	**W 14×30**	**13⅞**	**55.3**	**7.1**	**8.7**	**83**
5.8	**7.8**	**106**	**38.6**	**W 12×30**	**12⅜**	—	**6.9**	**10.8**	**76**
4.0	**5.1**	**106**	**38.4**	**W 16×26**	**15¾**	—	**5.6**	**6.0**	**76**
4.5	**5.1**	**97**	**35.3**	**W 14×26**	**13⅞**	—	**5.3**	**7.0**	**70**
7.1	11.9	96	35.0	W 10×33	9¾	50.5	8.4	16.5	69
5.8	**6.7**	**92**	**33.4**	**W 12×26**	**12¼**	**57.9**	**6.9**	**9.4**	**66**
5.2	9.4	89	32.4	W 10×30	10½	—	6.1	13.1	64
7.2	16.3	86	31.2	W 8×35	8⅛	64.4	8.5	22.6	62
4.1	**4.7**	**80**	**29.0**	**W 14×22**	**13¾**	—	**5.3**	**5.6**	**57**
5.2	8.2	77	27.9	W 10×26	10⅜	—	6.1	11.4	55
7.2	14.5	76	27.5	W 8×31	8	50.0	8.4	20.1	54
3.6	**4.6**	**70**	**25.4**	**W 12×22**	**12¼**	—	**4.3**	**6.4**	**50**
5.9	12.6	67	24.3	W 8×28	8	—	6.9	17.5	48
5.2	**6.8**	**64**	**23.2**	**W 10×22**	**10⅛**	—	**6.1**	**9.4**	**46**
3.6	**3.8**	**59**	**21.3**	**W 12×19**	**12⅛**	—	**4.2**	**5.3**	**42**
2.6	**3.4**	**58**	**21.1**	**M 14×18**	**14**	—	**3.6**	**4.0**	**42**
5.8	10.9	57	20.9	W 8×24	7⅞	64.1	6.9	15.2	41
3.6	5.2	52	18.8	W 10×19	10¼	—	4.2	7.2	37
4.7	8.5	50	18.2	W 8×21	8¼	—	5.6	11.8	36

(Courtesy of the American Institute of Steel Construction Inc.)

Table 2–2 Continued

ALLOWABLE STRESS DESIGN SELECTION TABLE
For shapes used as beams

S_x

F_y = 50 ksi			S_x	Shape	Depth d	F'_y	F_y = 36 ksi		
L_c	L_u	M_R					L_c	L_u	M_R
Ft	Ft	Kip-ft	In.³		In.	Ksi	Ft	Ft	Kip-ft
8.1	**8.6**	**484**	**176**	**W 24× 76**	**23⅞**	—	**9.5**	**11.8**	**348**
9.3	20.2	481	175	W 16×100	17	—	11.0	28.1	347
13.1	29.2	476	173	W 14×109	14⅜	58.6	15.4	40.6	343
7.5	10.9	470	171	W 21× 83	21⅜	—	8.8	15.1	339
9.9	15.5	457	166	W 18× 86	18⅜	—	11.7	21.5	329
13.0	26.7	432	157	W 14× 99	14⅛	48.5	15.4	37.0	311
9.3	18.0	426	155	W 16× 89	16¾	—	10.9	25.0	307
7.4	**8.5**	**424**	**154**	**W 24× 68**	**23¾**	—	**9.5**	**10.2**	**305**
7.4	9.6	415	151	W 21× 73	21¼	—	8.8	13.4	299
9.9	13.7	402	146	W 18× 76	18¼	64.2	11.6	19.1	289
13.0	24.5	385	143	W 14× 90	14	40.4	15.3	34.0	283
7.4	**8.9**	**385**	**140**	**W 21× 68**	**21⅛**	—	**8.7**	**12.4**	**277**
9.2	15.8	369	134	W 16× 77	16½	—	10.9	21.9	265
5.8	**6.4**	**360**	**131**	**W 24× 62**	**23¾**	—	**7.4**	**8.1**	**259**
7.4	**8.1**	**349**	**127**	**W 21× 62**	**21**	—	**8.7**	**11.2**	**251**
6.8	11.1	349	127	W 18× 71	18½	—	8.1	15.5	251
9.1	20.2	338	123	W 14× 82	14¼	—	10.7	28.1	244
10.9	26.0	325	118	W 12× 87	12½	—	12.8	36.2	234
6.8	10.4	322	117	W 18× 65	18⅜	—	8.0	14.4	232
9.2	13.9	322	117	W 16× 67	16⅜	—	10.8	19.3	232
5.0	**6.3**	**314**	**114**	**W 24× 55**	**23⅝**	—	**7.0**	**7.5**	**226**
9.0	18.6	308	112	W 14× 74	14⅛	—	10.6	25.9	222
5.9	6.7	305	111	W 21× 57	21	—	6.9	9.4	220
6.8	9.6	297	108	W 18× 60	18¼	—	8.0	13.3	214
10.8	24.0	294	107	W 12× 79	12⅜	62.6	12.8	33.3	212
9.0	17.2	283	103	W 14× 68	14	—	10.6	23.9	204
6.7	**8.7**	**270**	**98.3**	**W 18× 55**	**18⅛**	—	**7.9**	**12.1**	**195**
10.8	21.9	268	97.4	W 12× 72	12¼	52.3	12.7	30.5	193
5.6	**6.0**	**260**	**94.5**	**W 21× 50**	**20⅞**	—	**6.9**	**7.8**	**187**
6.4	10.3	254	92.2	W 16× 57	16⅜	—	7.5	14.3	183
9.0	15.5	254	92.2	W 14× 61	13⅞	—	10.6	21.5	183
6.7	**7.9**	**244**	**88.9**	**W 18× 50**	**18**	—	**7.9**	**11.0**	**176**
10.7	20.0	238	87.9	W 12× 65	12⅛	43.0	12.7	27.7	174
4.7	**5.9**	**224**	**81.6**	**W 21× 44**	**20⅝**	—	**6.6**	**7.0**	**162**
6.3	9.1	223	81.0	W 16× 50	16¼	—	7.5	12.7	160
5.4	6.8	217	78.8	W 18× 46	18	—	6.4	9.4	156
9.0	17.5	215	78.0	W 12× 58	12¼	—	10.6	24.4	154
7.2	12.7	214	77.8	W 14× 53	13⅞	—	8.5	17.7	154
6.3	8.2	200	72.7	W 16× 45	16⅛	—	7.4	11.4	144
9.0	15.9	194	70.6	W 12× 53	12	55.9	10.6	22.0	140
7.2	11.5	193	70.3	W 14× 48	13¾	—	8.5	16.0	139

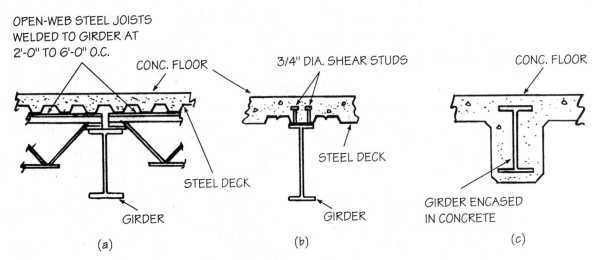

Figure 2–3 Methods of lateral beam support

Another important feature of the Allowable Stress Design Selection Table is that data for W-shapes of higher-strength steel (F_y = 50 ksi) is shown in shaded columns. In the example previously discussed, a bending moment of 164 kip-ft. could easily be resisted by a W16 × 40 using the higher-strength steel, which would be 10 pounds per foot lighter than the required W18 × 50 of A36 steel.

The Allowable Stress Design Selection Table also has two columns listed L_c and L_u, respectively. L_c and L_u refer to the required lateral support of a beam. (If the compression (top) flange is not properly supported, a beam may fail by lateral (sideways) buckling or deflection.) Most load tables in the manual assume that a beam has adequate lateral support. The L_c factor is the maximum unbraced length of the compression flange of the W-shape at which the allowable bending stress may be taken at 0.66 F_y (24 ksi for A36 steel), and the L_u factor is the maximum length at which the allowable bending stress may be taken at 0.6 F_y (22 ksi).

Figure 2–3 illustrates three common methods of providing lateral support to W-shape beams and girders. In steel joist construction, the open-web joists are welded to the top flange of the W-shape at 2′-0″ to 6′-0″ on-center as shown in Figure 2–3a. Assuming the joist spacing does not exceed L_c, concrete on steel deck that has been properly fastened to the top flange of the W-shape beam or girder below with puddle welds will generally provide the required lateral support of the top flange and thus brace any beam. Figure 2–3b illustrates composite construction in which steel studs are welded to the top flange of the W-shape beam or girder through the steel deck. Because the studs are spaced closely together along the length of the beam, continous lateral

support is provided. In Figure 2–3c, the entire beam or girder is encased in concrete, which again provides continuous lateral support. This type of construction is very often found in heavy industrial buildings such as paper mills and power plants.

Looking at the W18 × 50 previously discussed, it can be seen on the Allowable Stress Design Selection Table that L_c is listed at 7.9 ft. and L_u at 11.0 ft. Thus, if joists are being welded to the top flange of the W18 × 50 at either 2′-0″ on-center for a floor system or 4′-0″ to 6′-0″ for a roof system, the full value of 24 ksi in bending can be used.

Another helpful table in Part 2 of the manual is entitled Allowable Uniform Loads in Kips for Beams Laterally Supported. This table is illustrated in Table 2–3. Notice that, at a span of 16 feet, a W18 × 50 can support a load of 88 kips (88,000 pounds), while at a span of 24 feet, the allowable uniform load drops dramatically to less than 60 kips.

Also useful in Part 2 of the manual is the table called Allowable Moments in Beams. This table is for tentatively selecting a structural steel beam in which the unbraced length between lateral supports is greater than L_u. To illustrate how this table is used, we will assume we have already determined that the maximum bending moment due to load is 164 kip-ft. However, in this case, we will assume that the lateral supports along the beam are 12′-0″ center-to-center, and we will make a tentative beam selection using the Allowable Moments in Beams chart illustrated in Table 2–4.

To locate a point on the chart, find the 164-kip-ft. moment line on the left side of the chart and move horizontally to the right along that line until it intersects with a ver-

Table 2–3
Allowable uniform loads for beams

W 18	BEAMS W Shapes Allowable uniform loads in kips for beams laterally supported For beams laterally unsupported, see page 2-146								F_y = 36 ksi

Designation				W 18				W 18			
Wt./ft	71	65	60	55	50	46	40	35	Deflection In.
Flange Width	7⅝	7⅝	7½	7½	7½	6	6	6	
L_c	8.10	8.00	8.00	7.90	7.90	6.40	6.30	6.30	
L_u	15.5	14.4	13.3	12.1	11.0	9.40	8.20	6.70	

Span in Feet — F_y = 36 ksi

Span	71	65	60	55	50	46	40	35	Deflection
5								153	.03
6						187	162	152	.05
7	263	238	218	203	184	178	155	130	.07
8	251	232	214	195	176	156	135	114	.09
9	224	206	190	173	156	139	120	101	.11
10	201	185	171	156	141	125	108	91	.14
11	183	168	156	142	128	113	98	83	.17
12	168	154	143	130	117	104	90	76	.20
13	155	143	132	120	108	96	83	70	.23
14	144	132	122	111	101	89	77	65	.27
15	134	124	114	104	94	83	72	61	.31
16	126	116	107	97	88	78	68	57	.35
17	118	109	101	92	83	73	64	54	.39
18	112	103	95	87	78	69	60	51	.44
19	106	98	90	82	74	66	57	48	.49
20	101	93	86	78	70	62	54	46	.55
21	96	88	81	74	67	59	52	43	.60
22	91	84	78	71	64	57	49	41	.66
24	84	77	71	65	59	52	45	38	.79
26	77	71	66	60	54	48	42	35	.92
28	72	66	61	56	50	45	39	33	1.07
30	67	62	57	52	47	42	36	30	1.23
32	63	58	53	49	44	39	34	29	1.40
34	59	55	50	46	41	37	32	27	1.58
36	56	51	48	43	39	35	30	25	1.77
38	53	49	45	41	37	33	29	24	1.97
40	50	46	43	39	35	31	27	23	2.18
42	48	44	41	37	34	30	26	22	2.41
44	46	42	39	35	32	28	25	21	2.64

Properties and Reaction Values

	71	65	60	55	50	46	40	35	
S_x in.^{3}	127	117	108	98.3	88.9	78.8	68.4	57.6	For explanation of deflection, see page 2-32
V kips	132	119	109	102	92	94	81	76	
R_1 kips	44.1	38.4	33.9	30.4	26.4	26.7	22.2	20.0	
R_2 kips/in.	11.8	10.7	9.86	9.27	8.43	8.55	7.48	7.13	
R_3 kips	63.9	53.3	45.5	39.4	32.6	34.3	26.1	21.9	
R_4 kips/in.	4.96	4.05	3.45	3.18	2.67	2.61	2.04	2.20	
R kips	81	67	58	51	42	43	33	30	

Load above heavy line is limited by maximum allowable web shear.

(Courtesy of the American Institute of Steel Construction Inc.)

Table 2–4
Allowable moments in beams

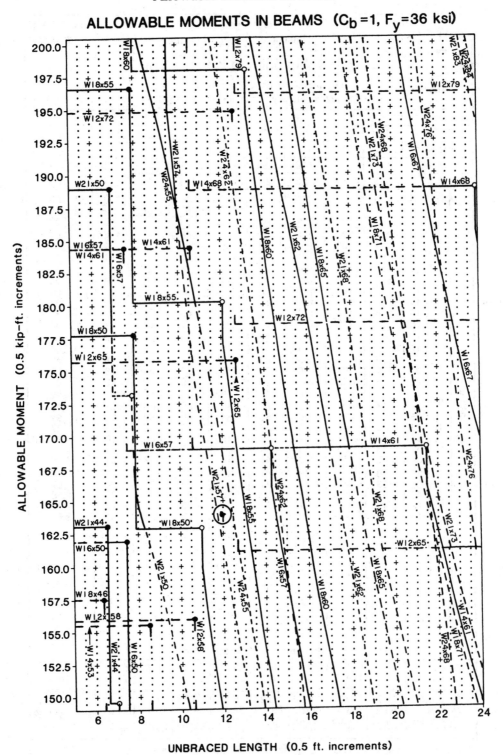

ALLOWABLE MOMENTS IN BEAMS ($C_b = 1$, $F_y = 36$ ksi)

(Courtesy of the American Institute of Steel Construction Inc.)

tical line going up from the 12′ unbraced length line at the bottom of the chart. Any beam whose graph lies above and to the right of that point will be able to resist the 164-kip-ft. bending moment. The nearest solid line graph designates the lightest or most economical shape, which in this instance turns out to be a W18 × 55. Notice that the graph line for the W18 × 50 beam, located to the left of the point, is no longer suitable to adequately resist the 164-kip-ft. bending moment now that the 12′ distance between points of lateral support exceeds the L_u distance of the W18 × 50.

Other useful charts, tables, and information in Part 2 of the *Manual of Steal Construction* include data on the design of beam-bearing plates, which will be discussed later in this text, and information on plate girders, composite design for building construction, and beam diagrams and formulas.

Part 3

Part 3 of the manual is intended to assist in designing structural steel columns and column base plates. Like Part 2 concerning beams, Part 3 contains helpful tables such as Allowable Stresses for Compression Members and several tables of Allowable Concentric Load on Columns, which are available for W- and S-shape columns, steel pipe columns, structural tubing columns, single and double angle columns, and columns made of structural tees. Table 2–5 shows the allowable concentric load in kips on standard steel pipe columns.

Part 4

Part 4 of the manual discusses the design of bolted and welded structural steel connections, which will be discussed in more detail in subsequent chapters. Again, the manual contains many helpful tables to simplify the design of standard connections.

Part 5

Part 5 of the manual is devoted to the specifications and codes relating to structural steel design, fabrication, and erection, including the RCSC (Research Council on Structural Connections) Specification for Structural Joints using ASTM A325 or A490 Bolts.

Part 6

Part 6 of the manual contains mathematical tables and data on miscellaneous subjects such as trigonometric formulas, engineering conversion factors, equivalent tables for decimals of an inch and decimals of a foot (which are very helpful for inserting feet and inch dimensions from drawings into design calculations), coefficients of linear expansion for various construction materials, and weights of building materials for use in calculating loads on structural framing systems (Table 2–6).

2.5   OPEN-WEB STEEL JOISTS

In structures such as office buildings, schools, and hotels where loads are moderate and spans between supports relatively long, it is not always economical to use the standard rolled structural W-shapes to directly support the floors or roof. This is because, if loads are very light, a standard W-shape will either be stressed exceptionally low and thus be inefficient or will be subject to unacceptable deflection at the longer spans. To solve this problem, special types of small, standard, prefabricated steel Warren trusses called *open-web steel joists* are often used to support floors and roofs. Open-web joists are generally made of light structural members such as angles, round bars, and channels. Figure 2–4 shows a detailed example of an open-web steel joist with a double-angle top chord and web and a flat bar end bearing plate.

Open-web steel joists are very economical structural members. Since the webs of these joists are *open*, they are able to span long distances with considerable less dead-load weight than a W-shape beam. Their strength is derived from the depth to which they can be fabricated, even though their chords and webs are relatively light. Another advantage of using steel joists is that the open web often makes it possible to run plumbing, electrical lines, and even small HVAC ducts directly through the web itself, resulting in a savings of floor-to-floor height and weight. The cumulative effect of this savings can be considerable in multistory buildings.

Open-web steel joists are manufactured in a variety of shapes. For instance, they may be furnished with underslung ends, square ends to facilitate column bracing or ceiling applications, or extended ends beyond the bearing plate to support overhanging roofs (Figure 2–5).

Joists used to support floors generally have parallel top and bottom chords to keep floors as level as possible. Open-web steel joists are also available with the top chord pitched 1/8″ per foot in one or two directions to facilitate roof drainage. It is common for steel joist manufacturers to provide their joists with an upward camber to compensate for deflection under load. The camber varies from

Table 2–5
Allowable concentric loads on columns

$F_y = 36$ ksi									
COLUMNS Standard steel pipe Allowable concentric loads in kips									
Nominal Dia.		12	10	8	6	5	4	3½	3
Wall Thickness		0.375	0.365	0.322	0.280	0.258	0.237	0.226	0.216
Wt./ft		49.56	40.48	28.55	18.97	14.62	10.79	9.11	7.58
F_y		36 ksi							
Effective length in ft KL with respect to radius of gyration	0	315	257	181	121	93	68	58	48
	6	303	246	171	110	83	59	48	38
	7	301	243	168	108	81	57	46	36
	8	299	241	166	106	78	54	44	34
	9	296	238	163	103	76	52	41	31
	10	293	235	161	101	73	49	38	28
	11	291	232	158	98	71	46	35	25
	12	288	229	155	95	68	43	32	22
	13	285	226	152	92	65	40	29	19
	14	282	223	149	89	61	36	25	16
	15	278	220	145	86	58	33	22	14
	16	275	216	142	82	55	29	19	12
	17	272	213	138	79	51	26	17	11
	18	268	209	135	75	47	23	15	10
	19	265	205	131	71	43	21	14	9
	20	261	201	127	67	39	19	12	
	22	254	193	119	59	32	15	10	
	24	246	185	111	51	27	13		
	25	242	180	106	47	25	12		
	26	238	176	102	43	23			
	28	229	167	93	37	20			
	30	220	158	83	32	17			
	31	216	152	78	30	16			
	32	211	148	73	29				
	34	201	137	65	25				
	36	192	127	58	23				
	37	186	120	55	21				
	38	181	115	52					
	40	171	104	47					
Properties									
Area A (in.²)		14.6	11.9	8.40	5.58	4.30	3.17	2.68	2.23
I (in.⁴)		279	161	72.6	28.1	15.2	7.23	4.79	3.02
r (in.)		4.38	3.67	2.94	2.25	1.88	1.51	1.34	1.16
B } Bending factor		0.333	0.398	0.500	0.657	0.789	0.987	1.12	1.29
$a/10^6$		41.7	23.9	10.8	4.21	2.26	1.08	0.717	0.447
Note: Heavy line indicates Kl/r of 200.									

(Courtesy of the American Institute of Steel Construction Inc.)

Table 2–6
Weights of building materials.

WEIGHTS OF BUILDING MATERIALS

Materials	Weight Lb. per Sq. Ft	Materials	Weight Lb. per Sq. Ft
CEILINGS		PARTITIONS	
Channel suspended		Clay Tile	
system	1	3 in.	17
Lathing and plastering	See Partitions	4 in.	18
Acoustical fiber tile	1	6 in.	28
		8 in.	34
FLOORS		10 in.	40
Steel Deck	See	Gypsum Block	
	Manufacturer	2 in.	9½
Concrete-Reinforced 1 in.		3 in.	10½
Stone	12½	4 in.	12½
Slag	11½	5 in.	14
Lightweight	6 to10	6 in.	18½
		Wood Studs 2 × 4	
Concrete-Plain 1 in.		12–16 in. o.c.	2
Stone	12	Steel partitions	4
Slag	11	Plaster 1 in.	
Lightweight	3 to 9	Cement	10
		Gypsum	5
Fills 1 in.		Lathing	
Gypsum	6	Metal	½
Sand	8	Gypsum Board ½ in.	2
Cinders	4		
		WALLS	
Finishes		Brick	
Terrazzo 1 in.	13	4 in.	40
Ceramic or Quarry Tile		8 in.	80
¾ in.	10	12 in.	120
Linoleum ¼ in.	1	Hollow Concrete Block	
Mastic ¾ in.	9	(Heavy Aggregate)	
Hardwood ⅞ in.	4	4 in.	30
Softwood ¾ in.	2½	6 in.	43
		8 in.	55
ROOFS		12½ in.	80
Copper or tin	1	Hollow Concrete Block	
Corrugated steel	See	(Light Aggregate)	
	Manufacturer	4 in.	21
3-ply ready roofing	1	6 in.	30
3-ply felt and gravel	5½	8 in.	38
5-ply felt and gravel	6	12 in.	55
		Clay tile	
Shingles		(Load Bearing)	
Wood	2	4 in.	25
Asphalt	3	6 in.	30
Clay tile	9 to 14	8 in.	33
Slate ¼	10	12 in.	45
		Stone 4 in.	55
Sheathing		Glass Block 4 in.	18
Wood ¾ in.	3	Windows, Glass, Frame	8
Gypsum 1 in.	4	& Sash	
		Curtain Walls	See
Insulation 1 in.			Manufacturer
Loose	½	Structural Glass 1 in.	15
Poured-in-place	2	Corrugated Cement	
Rigid	1½	Asbestos ¼ in.	3

For weights of other materials used in building construction, see pages 6-7 and 6-8

(Courtesy of the American Institute of Steel Construction Inc.)

Figure 2–4   K-series open-web steel joist (Courtesy of Bethlehem Steel)

Figure 2–5   Joist with extended toe chord (Courtesy of Bethlehem Steel)

approximately 1/4″ for a top chord length to 20′–0″ to 8 1/2″ for a top chord length of 144′–0″ (Figure 2–6).

Open-web steel joists are manufactured in three categories. The standard *K-series,* which replaced H-series joists in 1986, is available in depths from 8″ to 30″ and is recommended for spans from 8′–0″ to 60′–0″ in length. Longspan steel joists, the *LH-series* are manufactured in depths from 18″ to 48″ and may be used for spans of 25′–0″ to 96′–0″. Deep longspan joists, the *DLH-series,* are available in depths of 52″ to 72″ for spans of 89′–0″ to 144′–0″. The *KCS joist,* introduced in 1994, is a versatile K-series joist that can be specified for special loading conditions to support uniform loads plus concentrated and non-uniform loads. However, KCS joist selection requires the designer to calculate the maximum bending moment and shear stress imposed by load.

In addition to the standard categories of open-web steel joists, *joist girders,* which are open-web steel trusses, have become very popular in recent years. Joist girders are widely used to support equally spaced concentrated loads from standard open-web steel joists supporting floors or roofs. Standard joist girders are manufactured in depths of 20″ to 72″ for span lengths of 20′–0″ to 60′–0″ between columns. However, deeper depths for longer spans may be specified, with joist girder spans of up to 100′–0″ and more not uncommon.

When calling for open-web steel joists on structural design drawings, the proper convention is to list first the depth of the joist in inches, then the series of the joist, and finally a section number that relates to a relative size chord. For example, a joist specified as 24K10 would be a 24″-deep, K-series joist, with a #10 relative size chord. The chord size, denoted by the last digit, is significant for the designer when determining bridging requirements.

The method of designating joist girders is somewhat different from that used for open-web steel joists. The standard way to designate joist girders is to list first the depth of the girder in inches, followed by the number of spaces between loads, and then the kip load at each panel point. Figure 2–7, taken from an actual project, calls for a joist girder 96″ deep (96G), with 16 joist spaces at 6′–2″ center-to-center (16N) and a 9.6 kip load at each panel point (9.6K). Recommended loads for all open-web joists and joist girders are based on a maximum allowable tensile stress of 30 ksi.

2.6   STEEL JOIST INSTITUTE LOAD TABLES

When open-web joists or joist girders are to be used, the designer/drafter usually consults the most current *Standard Specifications Load Tables and Weight Tables for Steel Joists and Joist Girders,* a publication of the Steel Joist Institute (SJI). The Steel Joist Institute, formed in 1928, is a non-profit organization of steel joist manufacturers whose main function is to establish consistent steel joist standards based upon SJI specifications.

Table 2–7 shows one of the many helpful tables found in the SJI load tables, the K-Series Economy Table. The first three rows across the top of the table list the joist designation (10K1, 12K1, 8K1), the joist depth in inches (10″, 12″, 8″), and the approximate weight of the joists in pounds per linear foot (5.0, 5.0, 5.1). The lbs/ft. row is the most significant part of this table because, reading across the table from left to right, the open-web steel joists are listed progressively, not by their depths but by their weights in pounds per linear foot. Because structural steel is usually sold on a cost-per-pound basis, the joists are listed—in progressive order—by their costs, or "economy."

The far left column of the economy table shows the spans recommended for various joists, and the column under each joist lists two numbers in individual boxes that correspond to a span length. The upper number in bold print is the total recommended uniform load capacity of the joist in pounds per linear foot (dead load + live load) that the joist can safely support. This number should never be exceeded. The lower figure in each square denotes the live loads in pounds per linear foot the joist can support without exceed-

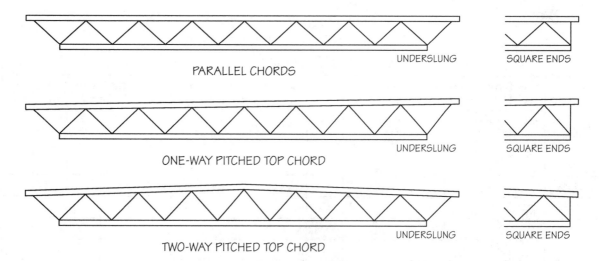

Figure 2–6   Types of open-web steel joists

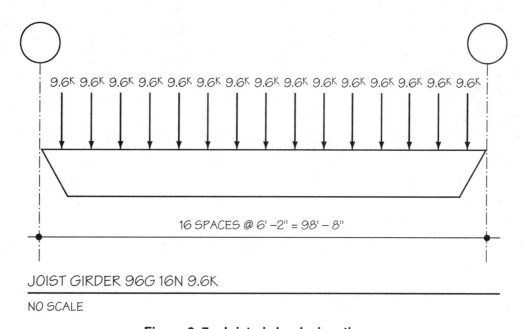

JOIST GIRDER 96G 16N 9.6K

NO SCALE

Figure 2–7   Joist girder designation

ing a deflection of 1/360 of the span. This is the maximum deflection permitted for floor loads.

For example, the economy table tells us that, at a span of 24′, a 14K4 K-series joist can take a maximum total load of 295 lbs/ft, but if the joist is to support a floor (maximum deflection 1/360 span), the maximum live load should not exceed 165 lbs/ft. Methods for determining dead loads and live loads and properly writing up the design calculations for open-web steel joists will be discussed in subsequent chapters.

Part of the selection process for steel joists includes specifying the number of rows of bridging required to stabilize the joists against lateral buckling at various spans. For this, the SJI Bridging Table is very helpful (Table 2–8).

To use the bridging table, the designer/drafter simply looks at the section number in the far left column. Reading down the column to #4, which corresponds to the last digit (4) of the 14K4 joist selected on the economy table, we see that a span of over 19′ thru 28′ requires two rows of bridging. Thus, if a 14K4 open-web steel joist had been selected

Table 2–7
K-series economy table for steel joists

K-SERIES ECONOMY TABLE

Joist Designation	10K1	12K1	8K1	14K1	16K2	12K3	14K3	16K3	18K3	14K4	20K3	16K4	12K5	18K4	16K5	20K4
Depth (in.)	10	12	8	14	16	12	14	16	18	14	20	16	12	18	16	20
Approx. Wt. (lbs./ft.)	5.0	5.0	5.1	5.2	5.5	5.7	6.0	6.3	6.6	6.7	6.7	7.0	7.1	7.2	7.5	7.6
Span (ft.)																
8			550 / 550													
9			550 / 550													
10	550 / 550		550 / 480													
11	550 / 542		532 / 377													
12	550 / 455	550 / 550	444 / 288			550 / 550							550 / 550			
13	479 / 363	550 / 510	377 / 225			550 / 510							550 / 510			
14	412 / 289	500 / 425	324 / 179	550 / 550		550 / 463	550 / 550			550 / 550			550 / 463			
15	358 / 234	434 / 344	281 / 145	511 / 475		543 / 428	550 / 507			550 / 507			550 / 434			
16	313 / 192	380 / 282	246 / 119	448 / 390	550 / 550	476 / 351	550 / 467	550 / 550		550 / 467		550 / 550	550 / 396		550 / 550	
17	277 / 159	336 / 234		395 / 324	512 / 488	420 / 291	495 / 404	550 / 526		550 / 443		550 / 526	550 / 366		550 / 526	
18	246 / 134	299 / 197		352 / 272	456 / 409	374 / 245	441 / 339	508 / 456	550 / 550	530 / 397		550 / 490	507 / 317	550 / 550	550 / 490	
19	221 / 113	268 / 167		315 / 230	408 / 347	335 / 207	395 / 287	455 / 386	514 / 494	475 / 336		547 / 452	454 / 269	550 / 523	550 / 455	
20	199 / 97	241 / 142		284 / 197	368 / 297	302 / 177	356 / 246	410 / 330	463 / 423	428 / 287	517 / 517	493 / 386	409 / 230	550 / 490	550 / 426	550 / 550
21		218 / 123		257 / 170	333 / 255	273 / 153	322 / 212	371 / 285	420 / 364	388 / 248	468 / 453	447 / 333	370 / 198	506 / 426	503 / 373	550 / 520
22		199 / 106		234 / 147	303 / 222	249 / 132	293 / 184	337 / 247	382 / 316	353 / 215	426 / 393	406 / 289	337 / 172	460 / 370	458 / 323	514 / 461
23		181 / 93		214 / 128	277 / 194	227 / 116	268 / 160	308 / 216	349 / 276	322 / 188	389 / 344	371 / 252	308 / 150	420 / 323	418 / 282	469 / 402
24		166 / 81		196 / 113	254 / 170	208 / 101	245 / 141	283 / 189	320 / 242	295 / 165	357 / 302	340 / 221	282 / 132	385 / 284	384 / 248	430 / 353
25				180 / 100	234 / 150		226 / 124	260 / 167	294 / 214	272 / 145	329 / 266	313 / 195		355 / 250	353 / 219	396 / 312
26				166 / 88	216 / 133		209 / 110	240 / 148	272 / 190	251 / 129	304 / 236	289 / 173		328 / 222	326 / 194	366 / 277
27				154 / 79	200 / 119		193 / 98	223 / 132	252 / 169	233 / 115	281 / 211	268 / 155		303 / 198	302 / 173	339 / 247
28				143 / 70	186 / 106		180 / 88	207 / 118	234 / 151	216 / 103	261 / 189	249 / 138		282 / 177	281 / 155	315 / 221
29					173 / 95			193 / 106	218 / 136		243 / 170	232 / 124		263 / 159	261 / 139	293 / 199
30					161 / 86			180 / 96	203 / 123		227 / 153	216 / 112		245 / 144	244 / 126	274 / 179
31					151 / 78			168 / 87	190 / 111		212 / 138	203 / 101		229 / 130	228 / 114	256 / 162
32					142 / 71			158 / 79	178 / 101		199 / 126	190 / 92		215 / 118	214 / 103	240 / 147
33									168 / 92		187 / 114			202 / 108		226 / 134
34									158 / 84		176 / 105			190 / 98		212 / 122
35									149 / 77		166 / 96			179 / 90		200 / 112
36									141 / 70		157 / 88			169 / 82		189 / 103
37											148 / 81					179 / 95
38											141 / 74					170 / 87
39											133 / 69					161 / 81
40											127 / 64					153 / 75

(Courtesy of Steel Joist Institute)

Table 2–8
Bridging table for K-series for open-web steel joist

			NUMBER OF ROWS OF BRIDGING** Refer to the K-Series Load Table and Specification Section 6. for required bolted diagonal bridging. Distances are Joist Span lengths – See "Definition of Span" preceeding Load Table.		
*Section Number	1 Row	2 Rows	3 Rows	4 Rows	5 Rows
#1	Up thru 16'	Over 16' thru 24'	Over 24' thru 28'		
#2	Up thru 17'	Over 17' thru 25'	Over 25' thru 32'		
#3	Up thru 18'	Over 18' thru 28'	Over 28' thru 38'	Over 38' thru 40'	
#4	Up thru 19'	Over 19' thru 28'	Over 28' thru 38'	Over 38' thru 48'	
#5	Up thru 19'	Over 19' thru 29'	Over 29' thru 39'	Over 39' thru 50'	Over 50' thru 52'
#6	Up thru 19'	Over 19' thru 29'	Over 29' thru 39'	Over 39' thru 51'	Over 51' thru 56'
#7	Up thru 20'	Over 20' thru 33'	Over 33' thru 45'	Over 45' thru 58'	Over 58' thru 60'
#8	Up thru 20'	Over 20' thru 33'	Over 33' thru 45'	Over 45' thru 58'	Over 58' thru 60'
#9	Up thru 20'	Over 20' thru 33'	Over 33' thru 46'	Over 46' thru 59'	Over 59' thru 60'
#10	Up thru 20'	Over 20' thru 37'	Over 37' thru 51'	Over 51' thru 60'	
#11	Up thru 20'	Over 20' thru 38'	Over 38' thru 53'	Over 53' thru 60'	
#12	Up thru 20'	Over 20' thru 39'	Over 39' thru 53'	Over 53' thru 60'	

* Last digit(s) of joist designation shown in Load Table
** See Section 5.11 for additional bridging required for uplift design.

(Courtesy of Steel Joist Institute)

to support a load at a span of 24', two rows of bridging would be required.

2.7 SUMMARY

This chapter has acquainted the structural drafting student with structural steel as a material, introduced some of the more common structural steel shapes used to create steel support systems, and explained the American Institute of Steel Construction, the *AISC Manual,* the Steel Joist Institute, and the SJI load tables. Open-web steel joists and joist girders have also been discussed because they are often integral parts of the support system for commercial, public, and industrial buildings. Subsequent chapters will go into more detail about designing and selecting shapes and joists for workable steel support systems and preparing the necessary design and fabrication drawings for contractors, fabricators, and ironworkers.

STUDY QUESTIONS

1. Structural steel is a man-made material consisting of about _____ percent iron.

2. In modern steel plants, steel is produced by a highly automated procedure that has eliminated both the conventional ingot and primary rolling operations. This procedure is known as the _____ process.

3. Structural steels are produced in a variety of grades. _____ steels are the most economical on a cost-per-pound basis, but they have the lowest strength.

4. In structural engineering, a _____ is 1,000 pounds.

5. Various grades of structural steel are established by the _____ , commonly referred to as the ASTM.

6. At least 50 percent of the structural steel specified for commercial and industrial building construction in the United States is ASTM grade _____ .

7. A grade of structural steel no longer produced but still supporting many commercial, public, and industrial buildings constructed during the 1940s and 1950s is grade ASTM _____ .

8. Corrosion-resistant, self-weathering steel oxidizes when exposed to the atmosphere, producing a thin, reddish-brown film of rust on its surface. This protective film is called the _____ .

9. How are S-shapes, W-shapes, and C-shapes designated on structural design and shop drawings?

10. What were the first beam sections or shapes rolled in America?

11. What is the most commonly used structural shape?

12. Depth designations for S-shapes and C-shapes indicate the _____ depth.

13. Depth designations for W-shapes usually designate the _____ depth.

14. How is the HP-shape different from the W-shape?

15. When specifying an L-shape with unequal legs, the _____ leg is always specified first.

16. What is the actual depth of a W16 × 36 structural steel rolled shape?

17. What is the actual depth of a W16 × 77 structural steel rolled shape?

18. What is the flange width of a W18 × 76 structural steel rolled shape?

19. What is the web thickness of a W18 × 71 structural steel rolled shape?

20. What is the section modulus on the x-x axis of a W16 × 57 structural steel rolled shape?

21. What maximum resisting moment can be developed in a W21 × 44 made of grade A36 steel?

22. What maximum resisting moment can be developed in a W21 × 44 made of 50 ksi steel?

23. What is the L_c distance of a W21 × 55?

24. What does the *Manual of Steel Construction* list as the maximum allowable load in kips on a W18 × 65 at a span of 24′–0″?

25. What does the *Manual of Steel Construction* list as the allowable concentrated load in kips on a 6″ standard steel pipe column with an effective length of 16′–0″?

26. What is the weight in pounds-per-square-foot of lightweight (light aggregate) concrete block?

27. What is the weight per square foot of a 3″-thick floor made of reinforced stone concrete?

28. Standard K-series open-web steel joists are available in depths from _____ to _____ for spans ranging from _____ to _____ .

29. In addition to the standard K-series open-web steel joists, steel joists are also available in the LH-series and DLH-series. What do the letters LH and DLH signify?

30. What is the live floor load in pounds-per-linear-foot that a 16K5 open-web steel joist can support at a span of 26′–0″?

31. The usual spacing of open-web steel joists for floor and roof framing systems is _____ center-to-center for floor systems and _____ to _____ center-to-center for roof systems.

Chapter 3

The Structural Drafter At Work

OBJECTIVES

Upon completion of this chapter, you should be able to:

- understand the functions and responsibility of the structural engineering team.
- name some personal qualities that are useful for a person seeking a career as a structural drafter.
- describe the organization of a typical engineering office.
- realize the career path advancement potential for structural drafters in engineering offices.

3.1 INTRODUCTION

This textbook is written primarily for the person seeking a career as a structural drafter in either a structural engineering design office or in a structural fabricator's office. In an effort to help the student better understand the career, this chapter will discuss the role of structural engineering in the design/drafting process, desirable characteristics for those training to become structural drafters, the specific technical skills needed, and the organization of a typical engineering office. What types of people are most likely to be successful in structural drafting? What skills should they have, and where will their entry-level opportunities be? What is their likely career advancement potential?

3.2 STRUCTURAL ENGINEERING

Structural engineering is a complex and exciting profession that, due to advances in technology, is currently experiencing rapid and profound changes in design methods, structural materials, and construction procedures. We say that it is a complex profession because structural engineers do much more than simply design the "bones" of a structure. They constantly strive to maintain architectural form. They calculate the shapes and sizes of steel sections required to safely support the loads. They monitor the cost of building a structure, keeping in mind the factors that can lower cost without sacrificing strength and quality.

Another concern of the structural engineer is that the structural system must be fire-resistant, able to maintain its structural integrity long enough for the occupants to evacuate. Engineers must also make sure the structural system integrates with other systems. Structural systems should not, for example, be in conflict with the plumbing systems, air handling systems, electric lighting or power systems, fire protection systems, or the movement of people within the building.

Thus, the structural engineering office or department is a vital part of the interdisciplinary design team, providing efficient structural solutions that help make the advances of modern architecture a reality.

3.3 DESIRABLE CHARACTERISTICS IN A STRUCTURAL DRAFTER

The structural drafter is a very important part of any structural design/drafting team, whether preparing design/drawings in the structural engineering office or preparing steel detail or shop drawings and erection plans in the structural fabricator's office. As a person whose primary responsibility is to produce accurate working drawings, the structural drafter needs to possess certain characteristics in order to be successful in this career. Some important qualities include:

1. The structural drafter must be reliable. As a key member of the design team, the structural drafter must be at work every day, ready to perform the required tasks in an acceptable manner. Chapter 1 discussed the importance of meeting schedules and deadlines for various phases of any building project. The drafter who is habitually not there will not only cause additional stress for the other members of the design/drafting team, but may even cause delays that could be very expensive. Such a person will not long be tolerated in most structural design/drafting departments.

2. The structural drafter must be able to concentrate. This profession requires a person to sit at a computer workstation or drafting board for long periods of time, solving technical design problems. The structural drafter must have a genuine interest in the problem and a drive to solve it as completely as possible. A nervous, flighty person who cannot sit quietly and work on a given assignment, or a person who habitually stays up partying night after night, should not even think about entering this field.

3. The structural drafter must be patient and neat. The sloppy, throw-it-all-together-quick person will not be successful in structural drafting. Working drawings for large, expensive, and complex buildings must be neat, clear, and easy to read. A drawing that is incorrect or cannot be easily understood by the tradespeople using it is virtually useless.

4. The structural drafter must have imagination and the ability to visualize. Usually the drafter is not simply copying structures that already exist, but is imagining and trying to visualize something yet to be built. For example, the drafter must be able to visualize how the component parts of a structural steel framing system will be fastened together to support the building. He or she must clearly understand building techniques like how steel beams and columns connect or how a large W-shape beam or column should fit on a footing or foundation wall.

5. The structural drafter must be organized and orderly. In any engineering office, the drafter constantly uses reference books, reference tables, calculations and drawings from previous projects, and various other sources of information. Thus, the successful drafter must be organized enough to know what references are available, where they are, how to find them, and when to use them.

6. The structural drafter must have a reasonable amount of mathematical ability. Every structural drafter or designer works with numbers daily, so at least average ability in mathematics, including a good understanding of algebra and trigonometry, is essential. There is simply no way to avoid the math requirement. This does not mean that a person must be an expert in advanced trigonometry or calculus to be a structural drafter, but it does mean drafting is not the best career choice for someone who struggled to pass general mathematics or high school algebra. Students considering the field of structural drafting should be realistic and honest with themselves about their math abilities.

3.4 ESSENTIAL SKILLS FOR A STRUCTURAL DRAFTER

In addition to having certain personal characteristics, aspiring structural drafters must master a variety of technical skills before seeking entry-level employment in most structural design/drafting offices. In today's highly competitive job market, the minimum educational requirement is usually an associate degree from a two-year technical college, although it is not uncommon for graduates of four-year college programs to begin their careers as entry-level drafters. The entry-level skills required for employment in most structural offices include:

1. The structural drafter must have a fundamental understanding of building technology and construction. The most essential part of the structural drafter's job is to produce the working drawings that tell the building contractor how the components of a building fit together. Thus, he or she must be familiar with the latest *building technology,* meaning the **processes and materials** required to assemble a building. For example, the drafter must know that structural steel is produced in a variety of grades and standard rolled shapes, as discussed in Chapter 2. He or she must be aware that steel reinforcing bars have deformed surfaces (usually ridges) to strengthen their bond with the concrete in a cast-in-place wall or footing. The drafter must also know that concrete, like structural steel, comes in a variety of strengths, that brick is a clay masonry product available in many shapes and sizes, that gypsum board is used to cover metal stud walls, and much more.

Building construction is concerned with both standard and innovative **methods** by which building materials are assembled. The structural drafter must know, for example, how steel beams are connected to concrete walls or steel columns, how walls are supported by continuous concrete footings, and how open-web steel joists are fastened to steel beams, poured concrete walls, or masonry walls.

2. The structural drafter must be able to find needed information in civil, architectural, electrical, and mechanical HVAC drawings. Because modern buildings are so technologically complex, the technical and creative skills of a diverse team of specialists are needed to create a well-designed project that efficiently and economically meets the needs of the client. On a large commercial project, the design team usually includes architectural, civil, structural, mechanical, and electrical designers and drafters.

During the conceptual phase of a project, the civil and architectural designers prepare a *site plan* indicating the orientation and location of the building on the site. They also draw architectural *floor plans* to show the sizes and locations of rooms, corridors, windows, stairs, and elevators, and *exterior elevations* to show how the outside of the building will look and tentatively establish floor and roof heights. The plans and elevations also list the basic building materials such as concrete, steel, brick, and glass.

After the preliminary conceptual drawings have been approved, the project moves to the design development phase. At this point, prints of the architectural floor plans and elevations are simultaneously issued to other specialists so that structural, electrical, plumbing, and heating and cooling systems can be developed. From the architectural floor plans, elevations, wall sections, and subsequent details, the structural drafters and designers must take the information they need to draw and design the structural support frame. The architectural drawings enable them to determine such components as: open-web steel joist spans and loads; beam, girder, and column spans and loads; window and door opening locations and sizes; and construction details. Thus, it is vitally important that the structural drafter be well-versed in reading and understanding architectural drawings.

In addition, the structural drafter should be able to read plumbing, HVAC, and electrical drawings because the effect of various subsystems on each other becomes a major concern as the building progresses. For example, it would create a real problem if a structural drafter were to show a heavy steel beam running through an elevator shaft or air duct shaft. Thus it becomes obvious that on most any construction project the "three Cs": communication, coordination, and cooperation requires that the structural drafter knows how to read and refer to drawings produced by other disciplines as he or she prepares the building's structural plans and details.

3. *The structural drafter must master the technical skills required to prepare the structural design and/or shop detail working drawings required for the construction of a building using either manual or computer-aided drafting (CAD) techniques.* The types of drawings drafters are responsible for include structural foundation plans, floor and roof framing plans, elevations, schedules, and details. The structural drafter must know such specifics as: what information should and should not be shown on a steel-framed floor or roof framing plan; how to depict a typical steel-beam-to-steel-column connection detail; how a wide-flange steel beam is anchored to a foundation wall; or how to draw a detail that will show the

ironworkers the correct way to set a column on a footing. The more skilled the structural drafter is in illustrating these concepts, the more valuable an employee he or she will be.

While the techniques of manual drafting remain very important and should not be minimized, most modern engineering offices require their structural drafters to be as competent as possible in preparing drawings using Computer-Aided Drafting (CAD) technology. With a variety of powerful software programs now available to quickly produce accurate, high-quality drawings, CAD technology is rapidly becoming the accepted, and even preferred, method of preparing structural drawings. Most of the structural programs available today operate interactively with AutoCAD, so a background in basic AutoCAD is also becoming an absolute requirement for people training to become structural drafters.

4. *The structural drafter must understand basic structural analysis and structural mechanics.* Structural design is essentially the process of creating an orderly, feasible, and economical structural support system strong enough to resist the anticipated loads in and on a building. For example, the *beams* must be designed to support the roof and floors, the *girders* to support the beams, the *columns* to support the girders, and of course, the *footings* to support the columns. While some engineering design computations can be very complex, a great many day-to-day design calculations required to develop structural layouts and details are relatively simple to solve for a drafter with a fundamental understanding of structural mechanics, simple algebra, and basic trigonometry.

3.5   ENGINEERING OFFICE ORGANIZATION

Structural drafters usually find employment either in structural engineering departments of mid- to large-sized architectural/engineering or civil engineering design firms, independent structural engineering/design consulting firms engaged in consulting services to architectural offices, or the drafting office of a structural steel fabricator. Whether working in a consulting structural engineering design office, the structural department of a multi-disciplined design firm or the structural steel fabricator's office where the main emphasis is the fabrication of structural steel, the entry-level structural drafter will be part of an organization responsible for preparing professional working drawings for a building's structural support system. Students intending to pursue careers as structural drafters should have a general idea of the type of organization they can expect to be working in, their probable responsibilities, and their prospects for career advancement. With this in mind, we will examine the career paths open to drafters in the typical structural engineering

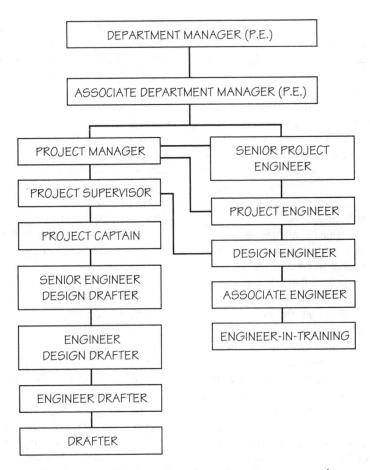

Figure 3–1 Engineering department career paths

department of a mid- to large-sized architectural/engineering design firm.

Figure 3–1 illustrates the makeup of a typical structural engineering department in a design office. Notice that the department manager and associate manager positions require registration as a P.E. (professional engineer). Notice also that the basic engineering office organization runs along two paths. The *engineering path* on the right is for graduates of a four- or five-year college or university with at least a Bachelor of Science degree in engineering. The career path on the left, the *technician path,* is followed by people with an associate degree from a two-year technical college. During the early part of a worker's career in a structural engineering office, the job descriptions and responsibilities are quite clearly defined. Thus, a person with a bachelor's degree in engineering will begin as an *engineer-in-training* and advance to the position of *associate engineer* after a few years of experience with the firm. Likewise, the two-year associate-degree graduate will begin his or her career as an entry-level *drafter,* with possible advancement after a few years of experience to *engineering drafter.*

However, as a person gains experience with a firm and advances upward through the department, job titles and descriptions tend to become less clearly defined, even to the point of crossing over from the technical to the engineering side and back again. For example, given ability, determination, and years of successful experience within a firm, it is not unusual for someone who started out as a drafter to become a design engineer, project engineer, or project manager. In fact, some two-year associate-degree people have, after years of applicable experience, taken and passed their state's professional engineering examination and gone on to become department managers. The engineering office is one place where unlimited opportunities are still available for those with genuine interest, ability, and dedication. Figures 3–2 and 3–3 illustrate in more detail the guidelines an engineering department may use to define various jobs within the organization.

Figure 3–2 shows the possible levels for two-year associate-degree technicians, with advancing job titles listed as T1, T2, T3, etc. Figure 3–3 shows the possible levels for graduates of a four-year engineering school, with job classifications listed as E1, E2, E3, and so on. To understand how

TECHNICIANS GUIDELINES	REQUIRED EDUCATION	JOB-RELATED EXPERIENCE	FINANCIAL RESPONSIBILITY	CLIENT CONTACT & MARKETING	SUPERVISION RESPONSIBILITY	SUPERVISION RECEIVED	WORK COMPLEXITY
DRAFTER T1	HIGH SCHOOL + 2 YEARS TECHNICAL COLLEGE	0 YRS.	MINOR LOSS POSSIBLE BUT NOT LIKELY	NONE	0	TOTAL UNDER WELL-DEFINED & UNIFORM PROCEDURES	SIMPLE AND ROUTINE OPERATIONS
ENGINEERING DRAFTER T2	HIGH SCHOOL + 2 YEARS TECHNICAL COLLEGE	2 YRS.	SOME LOSS POSSIBLE. PREVENTED BY ORDINARY CARE	LITTLE COMMUNICATION WITHIN WORK UNIT	0	WORK PERFORMED UNDER DIRECT AND DETAILED INSTRUCTION	SEMI-ROUTINE REQUIRING SOME IMPROVISATION ON ISOLATED PROBLEMS
ENGINEERING DESIGN DRAFTER T3	HIGH SCHOOL + 2 YEARS TECHNICAL COLLEGE	4–6 YRS.	MODERATE LOSS POSSIBLE. PREVENTED BY CONSIDERABLE CARE	MINIMAL COMMUNICATION ALONG ESTABLISHED LINES WITH USUAL COURTESY	1–3	WORK PERFORMED UNDER READILY AVAILABLE SUPERVISION	MODERATELY COMPLEX WITH ORIGINAL THINKING REQUIRED OCCASIONALLY
SENIOR ENGINEERING DESIGN DRAFTER T4	HIGH SCHOOL + 2 YEARS TECHNICAL COLLEGE	6–8 YRS.	CONSIDERABLE LOSS POSSIBLE. PREVENTED BY HIGH DEGREE OF CARE & ATTENTION	MODERATE CONTACT REQUIRING TACT AND WORKING KNOWLEDGE OF COMPANY POLICIES	4–9	DIRECTION AS TO ASSIGNMENTS & GENERAL METHODS DETERMINED BY INCUMBENT	COMPLEX WORK REQUIRING A HIGH DEGREE OF ORIGINAL THINKING
JOB OR PROJECT CAPTAIN T5	HIGH SCHOOL + 2 YEARS TECHNICAL COLLEGE	9–12 YRS.	SUBSTANTIAL LOSS POSSIBLE. PREVENTED BY SUSTAINED HIGH DEGREE OF CARE & ATTENTION	FREQUENT CONTACT REQUIRING RESOURCEFULNESS & TACT	6–12	INCUMBENT DETERMINES METHODS AND DELIVERS FINAL PRODUCT	HIGHLY COMPLEX REQUIRING SUBSTANTIAL AMOUNT OF ORIGINAL THINKING
PROJECT SUPERVISOR T6	HIGH SCHOOL + 2 YEARS TECHNICAL COLLEGE	12–15 YRS.	SUBSTANTIAL LOSS POSSIBLE. PREVENTED BY SUSTAINED HIGH DEGREE OF CARE & ATTENTION	POSSIBLE EXTENSIVE CONTACT REQUIRING RESOURCEFULNESS & TACT	10–18	INCUMBENT DETERMINES JOB & PERSONNEL METHODS	ORIGINAL THINKING FOR BOTH JOBS & PERSONNEL

Figure 3–2   Drafting technician guidelines

the classifications work, notice that the office guidelines for technicians specify minimal to no client contact for entry-level drafters. At the T1 and T2 levels, the work complexity is generally simple, routine, and performed under direct supervision, making financial loss to the company very unlikely. Then, as the drafter gains experience, his or her work gradually becomes more complex, involving more client contact and a greater chance for financial loss to the firm. But even up to the *senior engineering design drafter* classification (T4), the drafter is quite closely supervised.

On the engineering side, entry-level employees also work under supervised conditions in the beginning stages of their careers, but they usually advance into more responsible positions faster than drafters do. Notice, for example, that a graduate engineer may advance to a *design engineer* classification (E3) after four years with the firm, while promotion to the *senior engineering design drafter* position (T4) may require up to eight years of on-the-job experience. At upper-level positions such as *project managers* and *project engineers,* the work complexity, responsibilities, and financial

ENGINEERING GUIDELINES	REQUIRED EDUCATION	REQUIRED REGISTRATION	JOB-RELATED EXPERIENCE	FINANCIAL RESPONSIBILITY TO FIRM	CLIENT CONTACT & MARKETING RESPONSIBILITY	SUPERVISION RESPONSIBILITY	SUPERVISION RECEIVED	WORK COMPLEXITY	JOB MAGNITUDE AS LEAD ENGINEER
ENGINEER-IN-TRAINING E1	4-YEAR COLLEGE DEGREE	E.I.T.	0 YRS.	SOME LOSS IS POSSIBLE. ORDINARY CARE WILL PREVENT	MINIMAL ALONG ESTABLISHED PATTERNS. NORMAL COURTESY	1–3	WORK PERFORMED UNDER READILY AVAILABLE SUPERVISION	SIMPLE & ROUTINE OPERATIONS	1–2 MILLION
ASSOCIATE ENGINEER E2	4-YEAR COLLEGE DEGREE	E.I.T.	2 YRS.	MODERATE LOSS POSSIBLE. PREVENTED BY CONSIDERABLE CARE & ATTENTION	MODERATE REQUIRING TACT & WORKING KNOWLEDGE OF POLICIES	1–3	DIRECTION & GENERAL METHODS WITH PERIODIC REVIEWS	SOMEWHAT COMPLEX WITH SOME ADAPTATION REQUIRED	2–5 MILLION
DESIGN ENGINEER E3	4-YEAR COLLEGE DEGREE	P.E.	4 YRS.	CONSIDERABLE LOSS POSSIBLE. PREVENTED BY A HIGH DEGREE OF CARE & ATTENTION	FREQUENT CONTACT WITH PEOPLE OF HIGH POSITION	4–9	DIRECTION & GENERAL METHODS WITH PERIODIC REVIEWS	MODERATELY COMPLEX REQUIRING SOME ORIGINAL THINKING	5–15 MILLION
PROJECT ENGINEER E4	4-YEAR COLLEGE DEGREE	P.E.	6 YRS.	SUBSTANTIAL LOSS POSSIBLE. PREVENTED BY A HIGH DEGREE OF CARE & ATTENTION	EXTENSIVE CONTACT REQUIRING RESOURCEFULNESS & TACT	4–9	GENERAL DIRECTION AS TO BROAD ASSIGNMENTS. INCUMBENT DETERMINES METHODS	COMPLEX REQUIRING HIGH DEGREE OF ORIGINAL THINKING	20–60 MILLION
SENIOR PROJECT ENGINEER E5	4-YEAR COLLEGE DEGREE	P.E.	9 YRS.	SUBSTANTIAL LOSS POSSIBLE. PREVENTED BY A HIGH DEGREE OF CARE & ATTENTION	EXTENSIVE CONTACT REQUIRING RESOURCEFULNESS & TACT	6–15	INCUMBENT DETERMINES METHODS & DELIVERS PRODUCT	HIGHLY COMPLEX W/ SUBSTANTIAL AMOUNT OF ORIGINAL THINKING	60–100 MILLION
ASSOCIATE DEPARTMENT MANAGER E6	4-YEAR COLLEGE DEGREE	P.E.	9 TO 12 YRS.	SUBSTANTIAL LOSS POSSIBLE. PREVENTED BY A HIGH DEGREE OF CARE & ATTENTION	EXTENSIVE CONTACT REQUIRING RESOURCEFULNESS & TACT	10–24	INCUMBENT DETERMINES METHODS BOTH JOB & PERSONNEL	ORIGINAL THINKING FOR BOTH JOBS & PERSONNEL	UNLIMITED
DEPARTMENT MANAGER E7	4-YEAR COLLEGE DEGREE	P.E.	12 YRS.	SUBSTANTIAL LOSS POSSIBLE. PREVENTED BY A HIGH DEGREE OF CARE & ATTENTION	EXTENSIVE CONTACT REQUIRING RESOURCEFULNESS & TACT	25–75	DETERMINES AND EXECUTES MAJOR PROBLEMS	ORIGINAL THINKING FOR BOTH JOBS & PERSONNEL	UNLIMITED

Figure 3–3   Engineering Guidelines

compensation tend to even out because the dedicated and experienced senior engineers and designers eventually become the key people in most organizations.

The structural drafter often feels a bit apprehensive and may have many questions after accepting entry-level employment in an engineering firm. How do I get my work assignments? What do I actually do? How much help will be available? While answers to these questions vary depending on the size of the firm and the complexity of its projects, we will try to describe here the procedures followed in a typical structural engineering office.

The entry-level drafter's work is usually assigned by the associate department manager or, in a large office, by a technical manager responsible specifically for assigning work to the drafters within the department. Before assigning work, the manager checks schedules to determine approxi-

mately when the drafter will be finished with one project and ready to start another. Then the manager tells the drafter what he or she can expect to be working on next. Sometimes the drafter may have a choice of assignment, depending on his or her experience and the number of active jobs.

Generally, before the drafter becomes involved, the engineer for the project has met with the project manager, project architect, and other team members and has completed the structural analysis and some preliminary design work. After the drafter is assigned to the project, he or she often attends informal design meetings with engineers and architects where technical details are worked out. How will beams anchor to walls? How can special connections be made to columns that will conform to the steel framing plans? How will the curtain walls fasten to the steel support frame? After these questions have been answered, the drafter, supervised by the engineer, can start the working drawings.

Working drawings show the location, dimensions, and details of the structures to be built. In a steel-framed building, these drawings generally include the required steel framing plans along with appropriate details. Together with the specifications, these drawings should provide all the information required to fabricate and erect the structural steel support frame.

Framing plans and details are drawn by the drafter, usually on a CAD system, with help as required from senior design drafters and/or structural engineers. *Check prints* are then made and carefully reviewed. It is extremely important at this point to check the structural drawings against other components of the structure. Will something you have done change something already completed? If so, what adjustments can be made? After necessary changes are made, check prints are run again. Several reviews may be made, not only by the drafters and engineers assigned to the project, but also by quality-control advisers and other engineers and technicians. After all the structural drawings pass inspection, they are delivered to the project manager for "prints run for issue." The originals are then filed, and the structural drafting technician is ready for another assignment.

3.6   SUMMARY

This chapter has tried to give the structural drafting student insight into the desirable characteristics and essential skills required to enter this exciting field. It has also attempted to describe the organization of a typical engineering office and to explain the career paths available within such an organization. Subsequent chapters will explain in more depth the specific skills needed to obtain employment as an entry-level structural drafter.

STUDY QUESTIONS

1. Why is it important that the structural drafter be a reliable person?

2. Why does the structural drafter need an aptitude and an appreciation for patience and neatness?

3. Is a knowledge of calculus required for structural drafters?

4. Why is it important for the structural drafter to be able to visualize?

5. Why is it helpful for the structural drafter to be organized and orderly?

6. It is important for the structural drafter to be familiar with both *building technology* and *building construction.* What do these two terms mean?

7. Why is it unlikely that an entry-level structural drafter's mistake would cause substantial loss to his or her employer?

8. Why is a background in basic AutoCAD rapidly becoming a requirement for people training to become structural drafters?

9. Is it possible in most offices for a structural drafter with a two-year associate degree to advance to the position of project captain?

10. Why is it so important for the structural drafter to be able to read and understand architectural drawings?

Chapter 4

Reading Architectural Drawings for Steel-Framed Buildings

OBJECTIVES

Upon completion of this chapter, you should be able to:

- explain why it is important for the structural drafter or designer to be able to competently read architectural working drawings.
- recognize the basic symbols and abbreviations used on architectural drawings for commercial buildings.
- interpret architectural floor plans and exterior elevations from the point of view of a structural drafter or designer.
- understand the importance of architectural building sections, architectural wall sections, and details for structural drafters and designers.

4.1 INTRODUCTION

In building design, structure must follow architecture. As we noted in Chapter 3, structural drafters and designers receive most of their information from architectural drawings. This doesn't mean the architectural drawings have to be completed down to the last detail before the designers and drafters begin their work, but certainly a great deal of the architectural phase, must already be tentatively finished and accepted by the owner. For example, before the structural drafters and designers can begin their phase of the project, they must know such information as: types and sizes of rooms; widths of corridors; locations and sizes of door openings, window openings, elevator shafts, and stairwells; and, of course, the various construction materials the architect intends to use. It is very important for the structural

drafters and designers to know whether the outside walls of the building—the facade—are to be constructed of glass, insulated panels, or brick and concrete block. And, in order to determine structural loads, they must know whether interior walls are to be built of light-gage metal frame or concrete block, and whether the floor is to be constructed of 3"-deep cast-in-place concrete over metal deck or 8"-deep precast concrete hollow core planks.

Obviously, structural drafters and designers must know what to look for on architectural drawings and where to find specific information. To be sure, one chapter on reading architectural drawings is no substitute for a whole course in architectural drafting or construction. However, the material presented here should help the structural drafting student pick the required information off architectural drawings well enough to complete his or her assignments.

ARCHITECTURAL MATERIALS

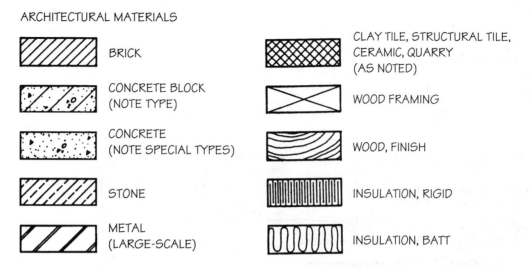

Figure 4–1   A typical architectural materials legend

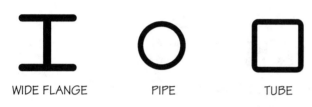

WIDE FLANGE   PIPE   TUBE

Figure 4–2   Structural steel column symbols

4.2 ARCHITECTURAL SYMBOLS AND ABBREVIATIONS

One of the most basic prerequisites of reading architectural drawings is to understand how various items are shown or referenced. The introductory student needs to know how materials such as brick, concrete block, cast-in-place concrete, metal stud walls, rigid insulation, and other building materials are identified on architectural drawings.

Architects use *materials symbols* to indicate what a wall is to be made of, be it brick, concrete block, metal stud, a combination of brick, or concrete block and metal stud. Many times the material symbols, also called cross-hatching, is shown right on the drawing, but another method is to place on the architectural plan *wall material numbers* that correspond to a wall type detail on another architectural sheet.

Figure 4–1 is an example of an architectural materials legend showing some of the standard hatching symbols for various construction materials. This information is very

valuable to the structural drafter or designer because it helps him or her determine the weights of walls, floors, and other building components prior to analyzing the overall structural requirements.

Architectural floor plans of commercial and industrial buildings also show *structural steel symbols* that indicate components of the structural framework, especially steel columns. On architectural floor plans, structural steel columns are shown by the symbols illustrated in Figure 4–2.

Figure 4–3 is an excellent example of the use of materials symbols to show the composition of a wall. The outside of the wall is brick, followed by an air space and rigid insulation. The inside of the wall is wood sheathing over metal stud frame that is filled with batt insulation and covered by gypsum board. Notice also that the W-shape column is covered with spray-on fireproofing, a common practice.

Besides materials symbols, architectural drawings contain other symbols, including reference numbers for rooms and doors, column grid reference numbers and letters, wall material type numbers, and reference drawing numbers for building sections, wall sections, and details. Some typical architectural symbols are shown in Figure 4–4.

When making working drawings, drafters often use standard abbreviations to save time and conserve space. For example, construction joints on exterior elevations or enlarged details are usually abbreviated C.J. When showing footing, floor, or top of steel elevations, the word elevation is usually designated by the abbreviation EL. Figure 4–5 is a partial list of abbreviations commonly used on architectural and structural drawings.

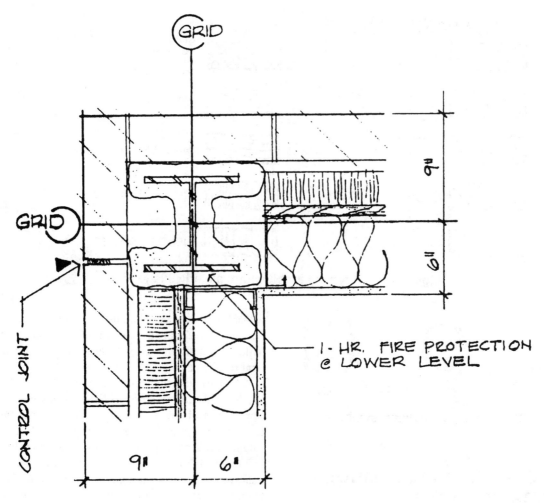

Figure 4–3   An exterior corner detail

4.3   FLOOR PLANS

On building projects, the architectural floor plan is one of the most important references for structural drafters and designers. A floor plan is a view looking down at the layout of a building as if the building were cut horizontally about three or four feet above the floor. It shows the overall size of the building as well as the sizes and locations of major corners, doors and windows, rooms and corridors, stairs, and elevators.

Floor plans are usually drawn at a scale of 1/8″ = 1′–0″ for commercial, public, and industrial buildings, although small buildings such as offices or fast-food restaurants might be drawn at a scale of 1/4″ = 1′–0″. In order to read floor plans, drafters need to know the materials symbols, architectural symbols, and abbreviations previously discussed. They also need to understand several other attributes of floor plans: the structural grid; the dimension lines, or strings; the conventions used for identifying rooms; detail drawings; and wall type numbers.

The Structural Grid

The *structural grid* is the center-to-center dimension of the columns that support the building. The center lines of all the columns make up the grid system. On a floor plan, these center lines, or grid lines, are marked with numbers reading horizontally from left to right and with letters reading vertically from top to bottom. On all drawings, the columns are identified by the grid lines that intersect at their centers. For example, a steel column located at the intersection of grid line ① and grid line Ⓑ would be column ①Ⓑ. An example of these grid line numbers can be seen at the top and left side of the floor plan shown in Figure 4–6.

The structural grid is one of the first concerns of architects, engineers, structural designers and drafters because it is the base from which the most economical steel framing

ARCHITECTURAL SYMBOLS

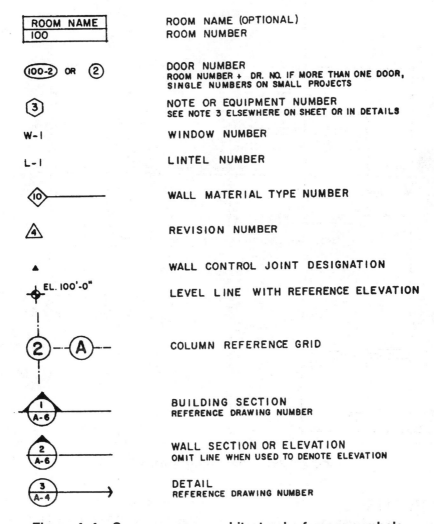

Figure 4–4   Some common architectural reference symbols

system can be designed and constructed and also a reference from which the walls are built.

Dimension Strings

The method of placing dimensions on architectural floor plans is fairly standard. Dimension lines are commonly referred to as *strings*. Most are indicated outside the building, but it is common to locate the dimension strings of interior partitions inside the building. Concrete walls are dimensioned to the outside face of the wall, and wall widths such as 8″ or 12″ are also dimensioned. Stud walls, whether wood or metal, are dimensioned to the center of the wall.

A floor plan usually shows three or four rows of dimension strings on the exterior of the structure (see Figure 4–6).

The inside string closest to the building is very important to structural drafters and designers because it shows the sizes and locations of masonry openings for doors and windows, which will often require structural steel *lintels* to support the wall above. The lintels are usually made of steel angles, plates, light W-shapes, or various combinations of these materials.

The floor plan in Figure 4–6 represents a motel. Notice that the guest rooms have 4′–0″-wide masonry openings for windows along the west wall and a 3′–4″-wide opening for an exterior door on the north wall. The center dimension string locates major corners of the building, or if the wall is straight, the entire length of a side of the building would show the overall length dimension. The outer string of

ABBREVIATIONS

& and ɩ	AND	CONST.	CONSTRUCTION	F( )	FRAME TYPE
∠	ANGLE	CONT.	CONTINUOUS	FBR.	FACE BRICK
@	AT	CONTR.	CONTRACTOR	F.C.U.	FAN COIL UNIT
℄	CENTERLINE	COORD.	COORDINATE	F.D.	FLOOR DRAIN
#	POUND OR NUMBER	CORR.	CORRIDOR	FDN.	FOUNDATION
A.A.P.	ALARM ANNUNCIATOR PANEL	CPT( )	CARPET TYPE	F.E.( )	FIRE EXTINGUISHER TYPE
A.B.	ANCHOR BOLT	C.R.	CEILING REGISTER	F.E.C.( )	FIRE EXTINGUISHER CABINET TYI
ACOUS.	ACOUSTICAL	C.R.G.	CEILING RELIEF GRILLE	FIN.	FINISH
A.C.U.	AIR CONDITIONING UNIT	C.R.U.	CONDENSATE RETURN UNIT	FL.	FLOOR
ADD.	ADDENDUM	CT( )	CERAMIC TILE TYPE	FLASH.	FLASHING
ADJ.	ADJUSTABLE/ADJUST	CU. FT.	CUBIC FEET	FLEX.	FLEXIBLE
A.F.F.	ABOVE FINISH FLOOR	C.U.H.	CABINET UNIT HEATER	FLG.	FLANGE
A.H.U.	AIR HANDLING UNIT	C.W.	COLD WATER	F. PTN.( )	FOLDING PARTITION TYPE
ALT.	ALTERNATE	C.W.( )	CURTAIN WALL TYPE	FT.	FOOT (FEET)
ALUM.	ALUMINUM	C.Y.	CUBIC YARD	FTG.	FOOTING
AM( )	ACOUSTICAL MATERIAL TYPE			FURR.	FURRING
AMP	AMPERE	D.	DEEP OR DEPTH		
AP( )	ACCESS PANEL TYPE	DEPT.	DEPARTMENT	G.	GAS
APPROX.	APPROXIMATE	DIA.	DIAMETER	GA.	GAUGE
ARCH.	ARCHITECT/ARCHITECTURAL	DIAG.	DIAGONAL	GALV.	GALVANIZED
AVE.	AVENUE	DIFF.	DIFFUSER	G.B.	GRAB BAR
		DIM.	DIMENSION	G.C.	GENERAL CONTRACTOR
B.F.E.	BOTTOM FOOTING ELEVATION	D.I.P.	DUCTILE IRON PIPE	GEN.	GENERATOR
BIT.	BITUMINOUS	DISP.	DISPENSER	GENL.	GENERAL
BLDG.	BUILDING	DIV.	DIVISION	G.F.I.	GROUND FAULT INTERRUPTER
BLK.	BLOCK	D.L.	DEAD LOAD	GL( )	GLASS TYPE
BLKG.	BLOCKING	DMPR.	DAMPER	G.P.M.	GALLONS PER MINUTE
B. LT. ( )	BORROWED LIGHT TYPE	DN.	DOWN	GYB( )	GYPSUM BOARD TYPE
B.M.	BENCH MARK	DRWG.	DRAWING	GYP.	GYPSUM
B.O.	BY OWNER				
BOT.	BOTTOM	E.	EXISTING	H.	HIGH OR HEIGHT
BRG.	BEARING	EA.	EACH	H.B.	HOSE BIBB
BSMT.	BASEMENT	E.C.	ELECTRIC CABINET	HDR.	HEADER
B.U.R.	BUILT-UP ROOF	E.E.	EACH END	HDW.	HARDWARE
		E.F.	EACH FACE	H.M.	HOLLOW METAL
CAB.	CABINET	E.F.	EXHAUST FAN	HORZ.	HORIZONTAL
CAP.	CAPACITY	EL.	ELEVATION	H.P.	HORSE POWER
C.B.	CATCH BASIN	ELEC.	ELECTRICAL	HTG.	HEATING
CBL( )	CONCRETE BLOCK TYPE	ELEC. UG.	ELECTRICAL UNDERGROUND	HTR.	HEATER
C/C	CENTER TO CENTER	EMERG.	EMERGENCY	H.V.A.C.	HEATING, VENT, AIR CONDITIONII
C.G.	CORNER GUARD	ELEV.	ELEVATOR	H.W.	HOT WATER
C.I.	CAST IRON	EQ.	EQUAL	HWD( )	HARDWOOD TYPE
C.I.P.	CAST IRON PIPE or CAST-IN-PLACE	EQUIP.	EQUIPMENT	HYD.	HYDRANT
C.J.	CONTROL JOINT	E.W.	EACH WAY		
CL.	CLASS	E.W.C.( )	ELECTRIC WATER COOLER TYPE	I.D.	INSIDE DIAMETER
CLG.	CEILING	E.W.E.F.	EACH WAY EACH FACE	IN.	INCHES
C.O.	CLEAN OUT	EXIST.	EXISTING	INSUL( )	INSULATION TYPE
COL.	COLUMN	EXP.	EXPANSION	INT.	INTERIOR
CONC.	CONCRETE	EXP. JT.	EXPANSION JOINT	INV.	INVERT
CON( )	CONCRETE FINISH TYPE	EXT.	EXTERIOR	I.P.	IRON PIPE
CONF.	CONFERENCE				
CONN.	CONNECT/-ED/-ION				

Figure 4–5   Standard architectural abbreviations

dimensions for a steel-framed building is particularly important to structural drafters and designers because it indicates the distance between columns. This relates directly to the steel framing plan because the distance between columns will determine the span lengths of structural steel beams and joists. For ease of fabrication and field erection, it is economical for the distances between columns to be as uniform as possible. The most common spacings between columns for commercial buildings range from 16′–0″ to 24′–0″. However, when using high-strength steel, spans of 30′–0″ and more are not unusual. With joist girders, spacings of 40′–0″ to 50′–0″ are not uncommon.

Reference Numbers

Figure 4–6 also illustrates the use of reference numbers on a floor plan. Notice that all rooms, corridors, and stairs are identified and numbered, and that the numbers are "100" numbers such as 106, 107, and 108. This tells us the drawing is a plan view at the first-floor level. Rooms on the second floor would be "200" numbers (201, 202, 203, etc.) and the third-floor rooms would be "300" numbers (301, 302, 303, etc.).

Details

Another point to note on Figure 4–6 is that most of the room wall locations are dimensioned, but not all. It is difficult to dimension small areas such as bathrooms and stairs on architectural floor plans at a scale of 1/8″ = 1′–0″. Thus, bathrooms such as 106A, 107A, 108A, and 109A are not dimensioned. Instead, bathroom 109A has been encircled with dashed lines and referenced to detail $\frac{6}{A14}$, which will be a separate drawing on a scale sufficient to show all the necessary dimensions. Likewise, stairs 1001 has been encircled and identified with the designation $\frac{1}{A15}$. This indicates that the stairs will be drawn to a larger scale and dimensioned on a subsequent drawing sheet. The stairs will be Detail 1 on architectural drawing sheet A15, and the bathroom will be Detail 6 on sheet A14. Design drawings for commercial

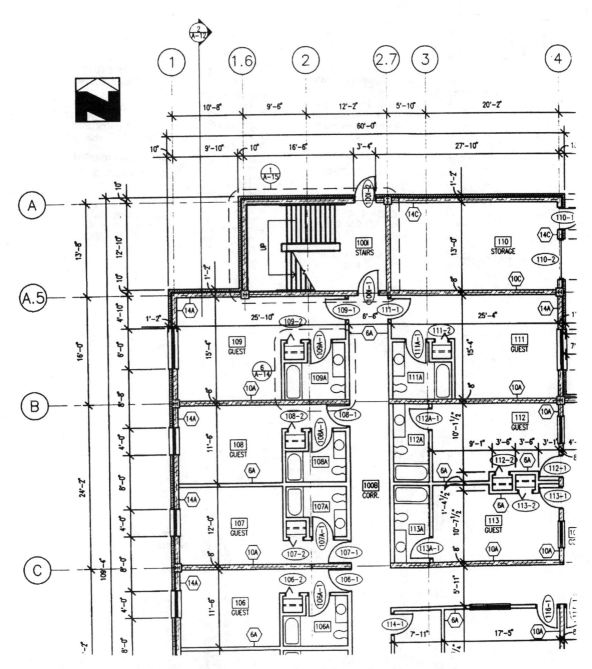

Figure 4–6   An architectural floor plan for a motel

building projects usually identify all the architectural drawings with the prefix "A," the structural drawings with the prefix "S," the plumbing drawings with the prefix "P," and the heating and ventilation drawings with the prefix "HV."

Wall Type Numbers

On a floor plan, all walls, both interior and exterior, are identified by wall type numbers. These numbers refer to larger details that will show the material composition of the wall. For example, exterior walls shown in Figure 4–6 are identified as 14A or 14C, and interior walls are identified as 6A or 10A. This information is important for the structural drafters and designers in determining the weights of walls and thus the resulting loads on the structural steel framing system. Details of selected wall types called for in Figure 4–6 are shown in Figure 4–7.

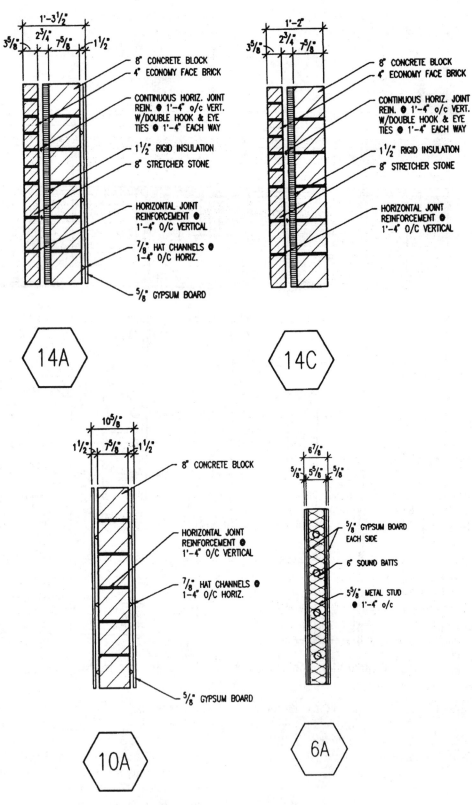

Figure 4–7   Examples of wall types

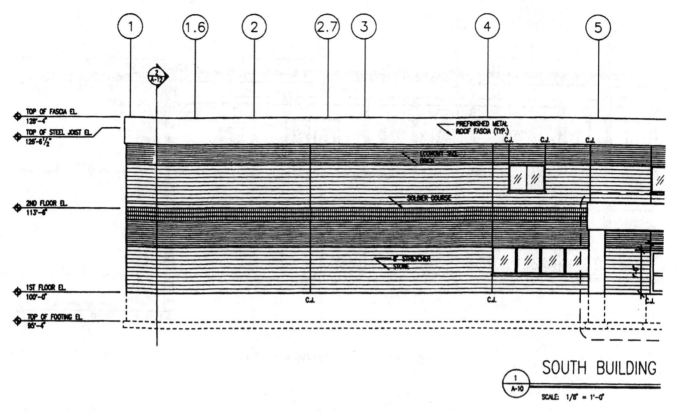

Figure 4–8   An Exterior Elevation

4.4   EXTERIOR ELEVATIONS

Exterior elevations are generally drawn at the same scale as floor plans. They show the exterior facade of the building as it will look upon completion. Exterior elevations show a building's proportions because they clearly indicate the length and height of walls, exterior building materials, window patterns, door locations, and control joint locations. Walls and footings below grade are shown as dashed lines. Exterior elevations are always referenced by points of the compass. For example, the "north elevation" is a view of the north exterior wall of a building as it will look when the building is completed.

Exterior elevations indicate footing and floor elevations, thus enabling structural drafters and designers to determine wall heights and weights. Top of steel, open-web joists, or metal deck elevations are also given on exterior elevations. In addition, elevation views show the location of section cuts, which will be drawn as larger-scale details on other architectural sheets.

4.5   BUILDING SECTIONS

Building sections are views cut through the entire building at a vertical plane perpendicular to the floor. They are usually drawn at the same scale as floor plans and exterior elevations. Typically, a set of architectural drawings will show at least one *longitudinal section,* which is a view cut through the length of the building, and one *transverse section,* which is a view cut through the width of the building.

Building sections are not dimensioned. Like exterior elevations, they show footing, floor, and top of roof, deck, or joist elevations, but the main purpose of building sections is to give a sense of space within the structure, such as ceiling or wall heights or the space between the bottom of an open-web joist and the acoustical tile ceiling as illustrated in Figure 4-9.

A building section does not have to be a straight line cut through the building but can jog back and forth to show pertinent information such as an auditorium's high ceiling and pitched floor or the varying depths of a swimming pool. And

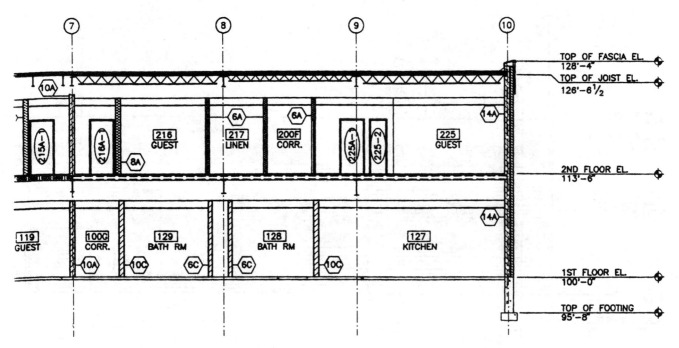

Figure 4–9 A building section

4.6 WALL SECTIONS AND DETAILS

Architectural wall sections and enlarged details are extremely important because they show most clearly how the materials are tied together to construct a building. Small-scale floor plans, exterior elevations, and building sections cannot adequately show how windows are mounted in exterior walls, how steel plate and angle lintels are put together to support the masonry wall above a steel door frame, how to locate a wide-flange structural steel column inside the corner of an exterior masonry wall (as shown in Figure 4–3, p.38), or how metal studs are fastened to the structural support frame at the roof of a building. Also, as previously mentioned, enlarged architectural details are needed to show the dimensions of small areas such as stairways and bathrooms.

Wall sections are usually drawn at scales of 1/2″ or 3/4″= 1′–0″, depending upon the amount of detail the drafter

desires to show. Occasionally, they are even drawn at a scale of 1″ = 1′–0″. Figure 4–10 is an exterior wall section for a small office building. The wall is made up of face brick over steel stud walls filled with batt insulation. Open-web steel joists set at a relatively steep pitch are supported by a wide-flange steel beam along grid line Ⓐ . Part of the roof is composed of several inches of rigid insulation over metal deck. Elevations are given for top of masonry, top and bottom of window, finished floor, recessed slab, and several brick courses.

Figure 4–11 is a wall section from a greenhouse. The lower portion is a 12″ wall made of exterior face brick, 1″ of rigid insulation in a 2″ air space, and a 6″ CMU (concrete masonry unit) interior wall. The upper portion of the wall and roof is an acrylic glazing system supported by structural steel tubes and channels. Notice how the wall section shown in Figure 4–11 refers to enlarged glazing details at the eave and exterior sill, which are shown in more detail in Figure 4–12.

Large-scale architectural details like those shown in Figure 4–12 are important because they clarify the more general specifications shown on plans and sections. Key parts of the building, such as plate and angle lintels over

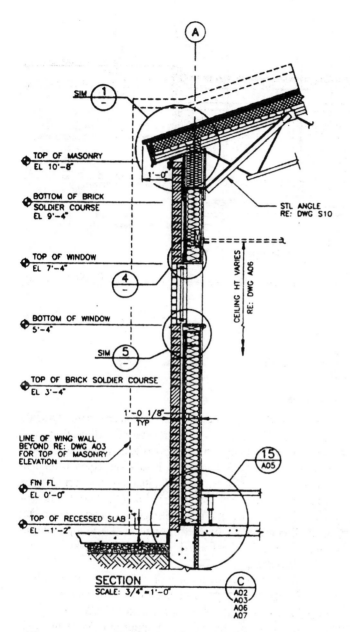

Figure 4–10   Exterior wall section from a small office building

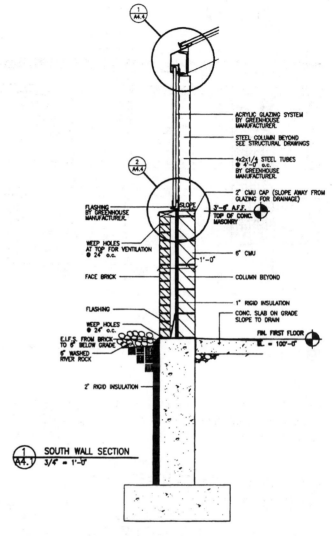

Figure 4–11   Wall section from a greenhouse

door and window openings, window sills, jambs and heads, stair sections and landings, intersections of floors and walls, and roof edge details, are drawn at scales of 1″, 1 1/2″, and sometimes even 3″ = 1′–0″ to show exactly how the materials are to be assembled. Examples of large-scale architec-

tural details are the penthouse roof edge detail in Figure 4–13 and the door head detail in Figure 4–14, both of which include items of interest to structural drafters.

As previously mentioned, enlarged stair details and enlarged bathroom details are also part of architectural working drawings. These details make it possible to show a clearer picture of how areas like stairs, bathrooms, shower rooms, are to be built. Enlarged room details are usually drawn at a scale of 1/4″ or 3/8″ =1′–0″. Figures 4–15 and 4–16 are enlarged details of First Floor Stair 1001 and a typical bathroom from the motel floor plan in Figure 4–16.

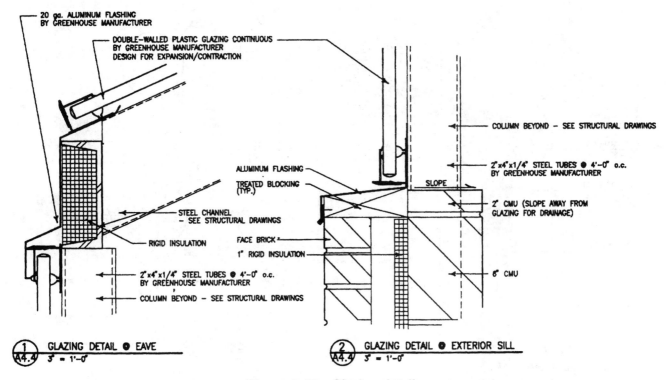

Figure 4–12   Glazing details

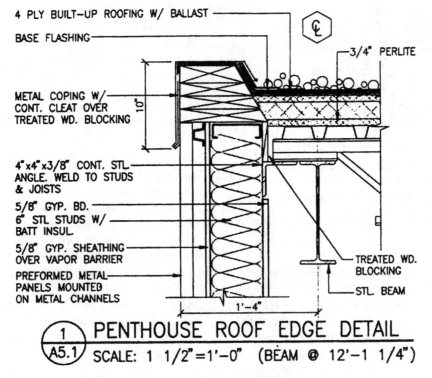

Figure 4–13   Penthouse roof edge detail

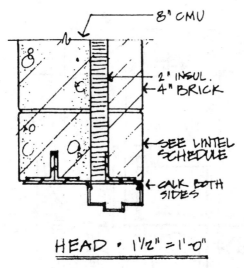

Figure 4–14   Door head detail

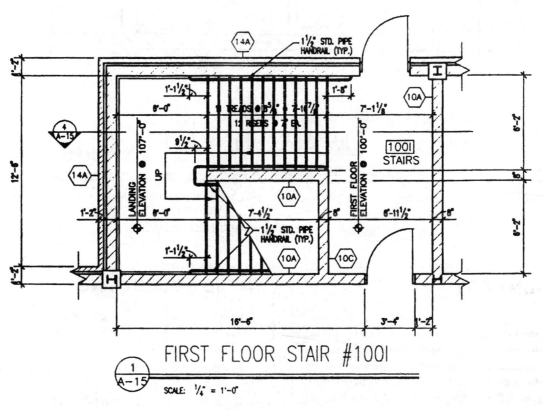

Figure 4–15   Enlarged stair detail

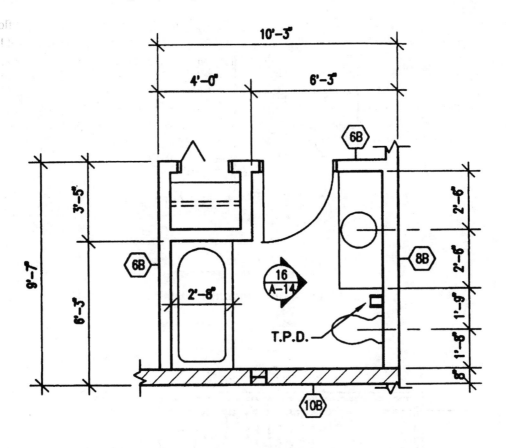

ENLARGED BATHROOM

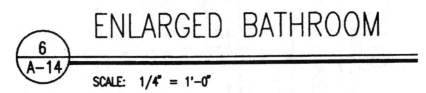

SCALE: 1/4" = 1'-0"

Figure 4–16   Enlarged bathroom detail

4.7   SUMMARY

Much of the information needed to design and draw structural steel framing systems for commercial, public, and industrial buildings can be found on the architectural drawings. Architectural drawings were discussed in this chapter primarily from the point of view of structural designers and drafters. The drafter or designer needs to know what type of information to look for in a set of architectural working drawings and where to find specific data. This chapter has not discussed all the drawings usually found in a set of architectural plans. Drawings such as the reflected ceiling plan, site plans, room finish schedules, and door schedules have not been discussed because their impact on the structural steel support system would be minimal at best.

STUDY QUESTIONS

1. When designing the structural steel support system for a building, where do structural drafters obtain most of their information?

2. Sketch the usual architectural materials symbol for concrete block.

3. Sketch the usual architectural materials symbol for batt insulation.

4. Architectural floor plans of commercial steel-framed buildings use symbols to indicate structural steel columns. Sketch the symbols for steel wide-flange (W-shape), steel pipe, and steel tube columns, and identify each.

5. On architectural floor plans, the center-to-center dimensions between structural steel columns make up the _____ .

6. Architectural drawings often abbreviate words to save time and space. What are the standard abbreviations for the words "anchor bolt," "column," and "building"?

7. Architectural drawings show the material composition of walls. Is this information of interest to the structural drafter and designer? If so, why?

8. An architectural floor plan is a view looking down at the layout of a building as if the building were cut horizontally about _____ above the floor.

9. List four items shown on architectural floor plans.

10. What is the purpose of a structural steel lintel?

11. On architectural floor plans, several dimension line strings usually appear on the exterior of the structure. Why is the inside string (closest to the building itself) important to structural drafters and designers?

12. The most common spacings between columns for commercial buildings range between _____ and _____ .

13. Why are architectural wall sections and details important to structural drafters and designers?

14. What is the purpose of drawing enlarged room details?

Chapter 5

An Overview of Basic Structural Steel Design Calculations

OBJECTIVES

Upon completion of this chapter, you should be able to:

- understand basic structural steel design considerations and terminology.
- identify structural steel component parts (joists, beams, girders, columns, etc.) on structural steel framing plans.
- perform some of the basic calculations often used by structural drafters and designers.

5.1 INTRODUCTION

The structural design of a building involves selecting and arranging the structural components into an orderly and economical system capable of supporting all loads and forces while maintaining the structure's architectural integrity. Good structural design is inseparable from architectural design, and although the structure does not dictate the architectural plan, experienced architects realize that it is critical to the success of their projects. Thus, from the inception of the design process, they must take into account structural concerns such as types of floor systems and placement of columns. Architects usually consult structural engineers early in the planning phase of a project for suggestions as to possible structural schemes. Engineers base their recommendations on experience, judgment, and a few calculations, while keeping in mind the building's architectural, structural, and mechanical/electrical requirements.

As the project moves through the planning phase into the stages where the structural designs are drawn, design calculations are performed to make sure the structural system is adequate and efficient. These calculations, which define all the anticipated loads on the structure, document

the structural design itself. In other words, the calculations justify why a certain open-web steel joist supporting a concrete floor must be a 16K4, why a W21 × 50 was selected to support several steel joists, or why a W8 × 31 column is the best choice to support the beams and girders framing into it. The design calculations also determine why a W10 × 49 steel column must have a 12″ × 12″ × 1″ – thick baseplate at its bottom to transfer the column load to a concrete footing, or why a wide-flange steel beam at the top of a concrete wall must have a beam bearing plate 6″ wide, 1′–2″ long, and ¼″ thick to transfer the beam reaction to the wall. In other words, the structural design calculations are the very backbone of both the structural system and the structural drawings.

In structural design offices, the structural engineer designs structural support systems making sure they are safe, economical, and adequate for their task. Structural drafters working under the supervision of the engineer then produce accurate, clear, and complete working drawings based upon the engineer's sketches and calculations. However, most structural engineers appreciate a drafter who can visualize the mechanics of the situation and has the initiative and ability to make routine design calculations and deci-

Figure 5–1 Modern structural steel-frame construction (Courtesy of Bethlehem Steel)

sions as the work progresses. Also, many of the day-to-day structural problems are quite simple in solution and can be competently handled by drafters with a working knowledge of algebra, basic trigonometry, and fundamental structural mechanics.

This chapter will review some of the everyday structural steel design calculations that are performed by structural drafters under the supervision of design engineers in most structural offices. It will also show how appropriate AISC and SJI tables are used in engineering offices. Because of the usual approximations made in structural steel design, beginning with the assumed dead and live loads, calculations will generally only be carried to two or three significant digits.

5.2 BASIC STRUCTURAL DESIGN CONSIDERATIONS AND TERMINOLOGY

Figure 5–1 illustrates the structural steel support system for a commercial building. The design of such a system usually begins with a sketch showing a tentative *framing plan,* a layout of the structural elements that will support a floor or roof. This sketch, which may be nothing more than a rough freehand drawing, shows only the proposed locations of the various structural steel members and does not list their sizes. After a structural system compatible with the architectural plan has been drawn, the designer determines the loads to be carried on any square foot of floor or roof area.

Loads

As you may recall from Chapter 1, the two fundamental types of loads are dead loads and live loads. *Dead loads* are the weights of all the materials permanently supported by the structure, such as concrete floor slabs, metal deck, partitions, beams, and open-web joists. *Live loads* are weights resulting from the use and occupancy of the building. Examples of live loads include furniture, people, and stored material such as book stacks in a library or rolls of paper in a warehouse. Live loads are usually expressed in pounds per square foot (psf), and recommended live load values are available from various local, state, and national codes. Table 5–1 lists some recommended live load values for various types of buildings.

Bays

Wherever possible, structural steel framing plans divide the structural support system into *bays,* which are usually square or rectangular areas defined by the dimensions between any four columns. For economy, as many bays as possible should be the same or nearly the same size, but again, for architectural reasons, that is not always possible. If architectural requirements necessitate some random arrangements of beams and columns, the structural engineers and drafters must deal with those situations as they arise. This may mean custom designing and fabricating certain structural members at a significantly higher cost. Figure 5–2 illustrates typical bays for the common beam-and-girder type of structural support system.

In this type of beam-and-girder construction, wide-flange beams or open-web steel joists support the floor or roof directly. The reactions of the beams or joists become concentrated loads on the girders, which might be either W-shapes or joist girders. The girders and beams that fasten directly to either the web or flange face of a W-shape column then become loads on the column to which they are connected. Columns in turn carry the loads of all the beams and girders connected to them down to the concrete footings and foundation walls. The purpose of the footings and foundation walls is to spread the total column load over a large enough area that the allowable bearing pressure of the soil beneath the footings will not be exceeded.

Spandrels

Figure 5–3 is a partial framing plan showing the components of a structural steel framework for a building like the one in Figure 5–1. Notice that three distinctly different types of bays are identified: *corner bays, exterior bays,* and *interior bays.* Beams and girders along the outside of the structure are called *spandrels,* and they are important for two reasons. First, spandrel beams and girders are located

Table 5–1
Typical recommended live loads for buildings

Type of Building	Live Load (psf)
Apartment house	
Corridors	80 psf
Apartments	40 psf
Public rooms	100 psf
Assembly halls	
Fixed seats	60 psf
Moveable seats	100 psf
Dance floors, restaurant serving and dining areas	100 psf
Business buildings	
Offices	50 + 20 psf partition allowance
Lobbies	100 psf
Corridors	100 psf
Schools	
Classroom/offices	50 + 20 psf partition allowance
Floors (open-plan schools	75 psf
Corridors, gymnasiums, cafeteria areas	100 psf
Storage warehouse	
Light	125 psf
Heavy	250 psf
Miscellaneous (applies to all occupancies above)	
Stairways, corridors, vestibules, lobbies	
A. In residential and institutional buildings	80 psf
B. In all other buildings	100 psf
C. Restrooms and toilets in public places	50 psf

only on the corner bays and exterior bays, so they sustain different loads than the beams and girders of interior bays. For example, the spandrel girders along grid line Ⓐ attach to floor beams or joists on only one side. However, the interior girders on grid lines Ⓑ and Ⓒ are connected to beams or joists on both sides, thus they could be carrying twice the floor or roof load of the spandrels. On the other hand, the spandrel girders along grid line Ⓐ and the spandrel beams along grid line ① would probably be supporting exterior curtain walls approximately 13′ in height along their entire length, which might add to their load, depending on the composition of the walls.

Another consideration when designing the spandrel beams and girders and their connections is the effect of lateral wind loading on tall buildings. Even in the preliminary phase of structural design work, the designer must think very methodically and constantly try to visualize the effect of the loads on each part of the entire support system. To do

a few quick calculations for a "typical" interior girder and then specify that all the other girders should be the same for the sake of economy could have very expensive, if not catastrophic, consequences.

The structural drafter also must consider how the component parts of the structural support frame will physically fit together. For example, the two most commonly used types of structural steel floor support systems in commercial buildings are the *open-web joist system* and *composite construction*.

Open-Web Joist Systems

With the *open-web joist system,* two different materials, concrete and steel, act independently of each other to carry the imposed loads because the concrete is poured over a metal decking, thus no bond forms between the bottom of the concrete slab and the tops of the open-web steel joists. As previously discussed, the joists carry the floor loads, the

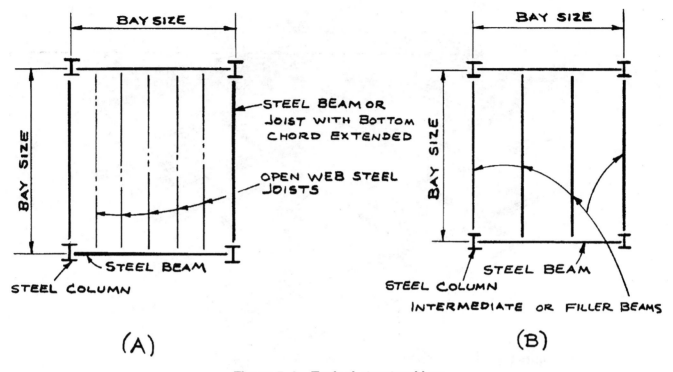

Figure 5–2   Typical structural bay

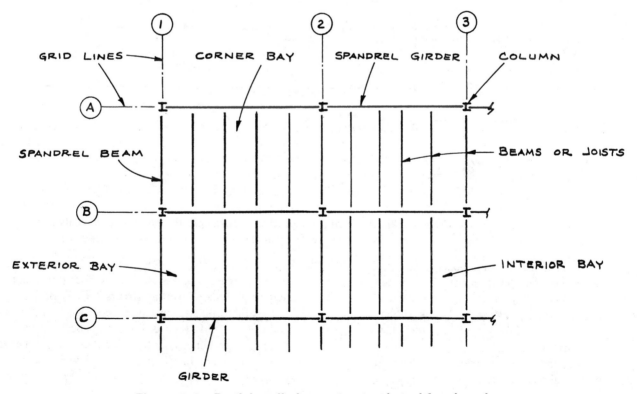

Figure 5–3   Partial preliminary structural steel framing plan

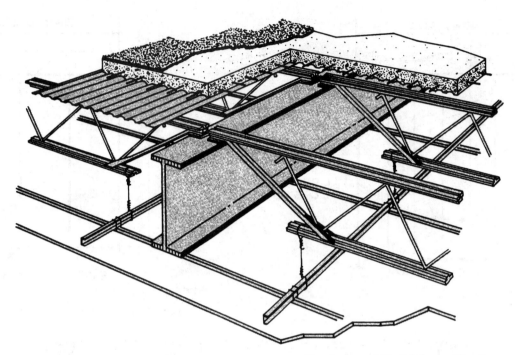

Figure 5–4   Typical open-web joist system (Courtesy of Bethlehem Steel)

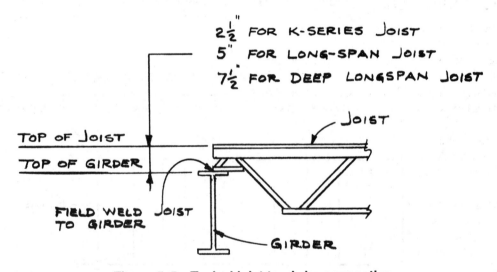

Figure 5–5   Typical joist to girder connection

reactions of the joists become concentrated loads on the W-shape girders, and the reactions of the girders become concentrated loads at the structural steel columns. With this type of system, the tops of the steel joists will obviously be higher than the tops of the girders supporting them, as Figure 5–4 illustrates.

The difference between top-of-joist elevation and top-of-supporting-girder elevation is 2 ½″ for K-series joists, 5″ for long-span joists, and 7 ½″ for deep long-span joists as shown in Figure 5–5. It must be remembered that, not only are the tops of the supporting girders 2 ½″, 5″, or 7 ½″ below the tops of the joists, but any wide-flange beams running parallel to the joists, such as the spandrel beams between the columns on grid line ① or the wide-flange beams between the columns of grid line ② in Figure 5–3, would be installed at the same top-of-steel elevation as the open-web joists.

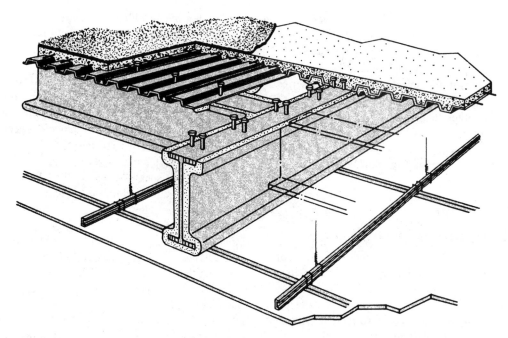

Figure 5–6 Composite construction (Courtesy of Bethlehem Steel)

Composite Construction

If wide-flange intermediate beams, instead of open-web joists, support a floor, the top-of-steel elevation for all beams and girders is the same. This type of floor framing, called *composite construction,* is also commonly used in commercial buildings. An example of composite construction is illustrated in Figure 5–6.

In composite construction, two different materials combine to form a structural member that utilizes each material to the fullest. To accomplish this, shear studs are welded to the tops of steel beams through the metal deck. Concrete is then poured onto the metal deck, and after it cures, the concrete and steel act as one unit, or compositely. The tendency of the concrete slab to slide on the beam (horizontal shear) is resisted by the shear studs, and the steel beam resists tensile stresses. Thus, both materials are utilized to their best advantage.

The result of this composite action is a more efficient structural unit because composite sections have continuous lateral support and thus much greater stiffness than a noncomposite beam of equal depth, loads, and span length. Also, better economy is possible because of longer beam and girder spans (up to 30 feet and more) and fewer columns. A typical composite steel-and-metal deck floor system might consist of an 18-gage metal deck supported by beams spaced at 8- or 10-foot centers, which themselves are connected to girders spanning 30 to 32 feet. A concrete floor consisting of 2 ½″ of normal-weight concrete over the metal deck (total slab thickness of 4 ½″) would be held in place by ¾″ diameter shear studs welded to the tops of the structural steel beams through the metal forms.

5.3 THE DESIGN AND SELECTION OF OPEN-WEB STEEL JOISTS

In designing structural steel framing systems, the usual procedure is to first select the beams or joists that will directly support the floor or roof loads. Choosing open-web standard, longspan, or deep longspan steel joists is easily accomplished using the tables in the *Standard Specifications Load Tables for Steel Joists and Joist Girders* published by the Steel Joist Institute, as explained in Chapter 2. The procedure consists of two steps: (1) calculating the load (in weight per lineal foot) that will be supported by the joist, and (2) selecting the proper joist from the load tables.

Calculating the Load

Calculating the load that will be supported by the joist includes first determining the weight of all the materials that the joist will support, such as the metal deck, concrete, insulation, mechanical plumbing and piping, HVAC ductwork, and electrical conduit. These constitute the so-called dead loads on the structure. The designer then adds to that the live loads recommended by code for various floors and roofs, and makes allowance for the weight of the joist itself.

After the loads have been determined, the structural drafter or designer consults either the Standard Load Tables or the Economy Tables in the SJI *Standard Specifications and Load Tables*. As previously mentioned, the economy table is especially useful for selecting standard K-series open-web joists because it lists joists progressively according to their weight rather than their size, making it easy to find the lightest or most economical joist. At this point, we will describe the procedure for selecting an open-web K-series steel joist for a floor loading application.

General Procedure for Selecting Joists

Let us assume we are designing the structural steel floor framing for the lobby of a bank using an open-web steel joist system similar to the one shown in Figure 5–4. We will also assume that a preliminary structural steel framing plan similar to that shown in Figure 5–3 has been worked out, but no sizes have been determined for any of the joists, beams, girders, or columns. The fire-resistant floor system is to be a commercial carpet over 3″ (total thickness) of reinforced concrete on 28-gage formed sheet steel deck supported by K-series open-web steel joists spaced at 2′-0″ center-to-center. A channel suspended acoustical tile ceiling system will hang from the bottom of the joists as shown in Figure 5–4. The joists will also support various mechanical systems such as sanitary drainage and water supply pipes, electrical conduit, and HVAC ducts. We will assume the open-web steel joists are spanning 24′-0″ and design them according to AISC specifications, allowing a maximum calculated live load deflection of 1/360 span. The first step in steel joist design and selection is to determine the actual loading conditions. In this case, they are:

Live load	100 psf (from Table 5–1 or applicable code)
Commercial carpet and carpet pad	4 psf (from manufacturer)
3 inches of concrete	36 psf (figuring 12 psf per inch of thickness, see Table 2–6)
28-gage metal deck	1 psf approximately (from AISC Manual)
Channel suspended acoustical tile ceiling system	2 psf (Table 2–6)
Mechanical allowance	7 psf (for pipes, ducts, etc.)
Total	150 psf

The floor joists are spaced at 2′-0″ centers, so each joist would support a portion of the floor, or *distributive area,* halfway between itself and the nearest joist on either side. In this case, the distributive area would be equal to 1′-0″ on either side of the joist or a 2′-0″ width of floor along its entire 24′-0″ length, as diagrammed in Figure 5–7.

Figure 5–7 shows that the *total load* (dead load + live load) per lineal foot along the joist span would be 150 psf × 2 = 300 plf. It is very important that the structural drafter or designer keep in mind the difference between plf (pounds per lineal foot) and psf (pounds per square foot) because loads are calculated in psf, while joists are selected in loads of plf along the span of the joists.

Having established that the design load along the span of the joist (excluding the weight of the joist itself) is 300 plf, the designer turns to the SJI economy table for K-series open-web steel joists. Although the economy table in Table 2–7 has been briefly discussed, we will now look at it in more depth and detail.

When using the table, the drafter/designer reads down the left-hand column to find a span in feet equal to or slightly greater than the span of the joist. For example, if an actual joist span were 22′-8″, the designer would select a joist based on the loads for a joist span of 23′-0″ on the table. He or she would then read across the table to find an allowable load value approximately equal to the pounds per lineal foot that have already been determined for the joist.

Each division of the table contains two numbers for each joist. The bold black figures (top figures) list the *total* safe, uniformly distributed load-carrying capacity of the joists. If economically feasible, this figure should not be exceeded—and certainly never by more than a few pounds. The lower figures in each box show the live loads per lineal foot of joist that will produce a deflection of 1/360 span. This is the recommended allowable deflection for floor joists, so when selecting floor joists, these lower figures can be used directly.

Drawing a Load Diagram. Once the loads on the joist are known, the usual procedure is to draw a *load diagram,* or *free-body diagram,* showing all the known loads or forces that will act on the joist. Total load, live load, dead load, and joist reactions are noted separately as in Figure 5–8.

The diagram in Figure 5–8 shows that the total uniform load on the joist is 300 plf × 24′ = 7,200 pounds, or 7.2 kips. Because the loads are uniformly distributed along the full span of the joist, each reaction (R_1 and R_2) would equal $\frac{7.2 \text{ kips}}{2}$, or 3.6 kips. It is extremely important to make note of the reactions because they will become concentrated loads on the girders supporting the joists. Now everything required to select the proper joist has been documented, so we can go to the joist tables and select the open-web joist itself.

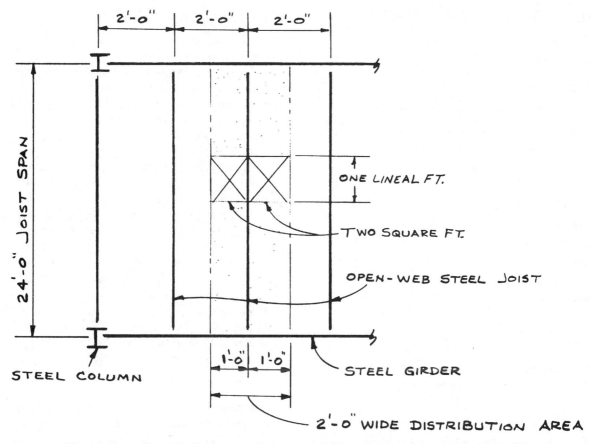

Figure 5–7   Distributive area of structural floor slab supported by one joist

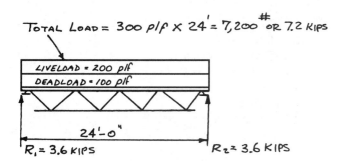

Figure 5–8   Load or free-body diagram

Selecting Floor Joists   From the economy table in Table 2–7 (p.26), we first read down the far left column (Span Ft.) to 24 and then across to $\frac{320}{242}$. These are the first numbers that equal or exceed our known 300 plf total load and 200 plf dead load. Looking up to the top of the table, we see that we are in the column for an 18K3 joist, which weighs 6.6 plf. Adding the joist weight of 6.6 plf to the 300

plf we already have yields a total load of 306.6 plf, which is clearly less than the total allowable load of 320 plf listed on the table. The 200 plf live load previously calculated is also less than the 242 plf allowable live load for this joist. Thus, the actual deflection of the joist under load will be even less than the allowable 1/360 span for floor joists.

It seems clear that the 18K3 open-web joist is the most economical (lightest-weight) joist to carry the calculated loads at a span of 24″–0′. However, the 18K3 is not the only possible selection for this loading condition. The table shows that a 16K4 with total and live load capacities of $\frac{340}{221}$ would also be acceptable for the loading conditions. Because the 16K4 is not as deep as the 18K3, it might be considered if space between the bottom of the joist and the suspended ceiling below were a concern. However, at a weight of 7 plf, the 16K4 would be heavier and thus more expensive than the 18K3, so it would not be the most economical choice. Assuming that an 18K3 is selected, one more step is necessary to complete the design procedure. We must determine the number of rows of bridging required for an 18K3 open-web steel joist at a span of 24′–0″.

OPEN-WEB STEEL JOIST DESIGN SHEET

LOADING CONDITIONS:

Live Load

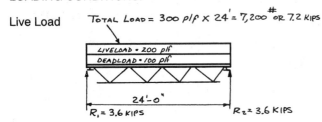

LOADING CONDITIONS: – – – – 100

Live load – – – – – – – – – – – –	psf
Carpet – – – – – – – – – – – –	4 psf
3" concrete – – – – – – – – –	36 psf
28 gage metal deck – – – – – – –	1 psf
Suspended ceiling – – – – – – –	2 psf
Mechanical – – – – – – – – – –	7 psf
Total loads – – – – – – – – –	150 psf

Total psf along joist span = 150 psf × 2 = 300 plf

Total load = 300 plf × 24' = 7,200 lbs or 7.2 kips.

$R_1 = R_2 = \dfrac{W}{2} = \dfrac{7.2 \text{ kips}}{2} = 3.6 \text{ kips}$

From Economy Table select a 18K3 joist $\dfrac{320}{242}$

Weight of 18K3 joist = 6.6 plf

Total weight of 306.6 plf < 320 plf (OK)

Total live load of 200 plf < 242 plf (OK)

Select 18K3 joist with 2 rows of bridging for a 24' span

Figure 5–9   Typical floor joist calculations

Selecting the Bridging. To select the bridging, we refer to the SJI bridging table in Table 2.8. Reading down the Section Number column, we stop at #3 because that number corresponds to the last digit of an 18K3 joist. Reading across the table, we see in the 2 Rows column that an 18K3 joist with a span of over 18' thru 28' would require two rows of bridging. Thus, the most economical floor joist for the loading condition described would be an 18K3 open-web steel joist with two rows of bridging. On the framing plan drawing, which will be discussed in detail in a subsequent chapter, the two rows of bridging would be indicated at third points of the span. Figure 5–9 shows how the floor joist calculations might be written up in an engineering office.

Selecting Roof Joists

Selecting open-web steel joists for roof systems is a slightly different procedure, first because roof loads are usually lighter than floor loads, so the distance between joists can be increased. Spacings of 4'–0" to 6'–0" are recommended, though in northern climates where snow loads must be considered, joist spacings commonly range from 4'–0" to 5'–0" center-to-center.

A second factor that makes selecting roof joists different from selecting floor joists is the type of ceiling involved. If roof joists have a suspended acoustical tile ceiling below the joist, like the one shown in Figure 5–4, (a very common condition in public and commercial buildings) a joist deflection of 1/240 of the span is allowed. A deflection of 1" for every 240" of roof joist span is 1.5 times the recommended 1" deflection for every 360" of floor joist span. Thus, the live loads in the joist table (the lower figures), which are based on 1/360 deflection, may be multiplied by 1.5 to produce 1/240 deflection. However, the upper figure in the joist tables (the total load capacity of the joist) should not be exceeded. We will now go through the procedure for selecting a typical roof joist for a commercial building.

General Procedure. For this example, we will assume the joists are spanning 28'–0" and the roof is constructed as shown in Figure 5–10. We will also assume a snow load allowance (live load) of 30 psf, with the joists to be placed 4'–0" on center. Again, as in floor joist selection, the first step is to consider the actual loading conditions, which would be:

Live load (snow)	30 psf
Gravel ballast over single-ply membrane roofing	6 psf
4" of rigid insulation	6 psf
1" perlite board	3 psf
22-gage metal deck	1.5 psf
Mechanical allowance	5.5 psf
Total	52 psf

Because the open-web steel joists are to be spaced 4'–0" on center, each joist will carry a total load of 52 psf × 4 = 208 plf, excluding the weight of the joist. Of the total load of 208 plf, the live load (snow) is 30 psf × 4 = 120 plf.

As previously discussed, to satisfy the 1/240 deflection criterion for roof loading, we find a table entry in which live load (lower figure) × 1.5 = 120 plf. To find that amount, we divide 120 by 1.5:

$$\frac{120}{1.5} = 80 \text{ plf}$$

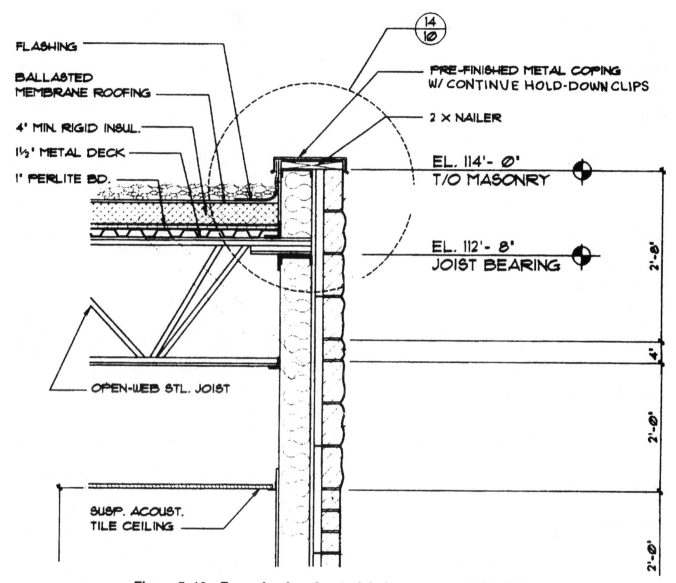

FLASHING

BALLASTED
MEMBRANE ROOFING

4' MIN. RIGID INSUL.

1½' METAL DECK

1' PERLITE BD.

OPEN-WEB STL. JOIST

SUSP. ACOUST.
TILE CEILING

14
10

PRE-FINISHED METAL COPING
W/ CONTINUE HOLD-DOWN CLIPS

2 × NAILER

EL. 114'- 0'
T/O MASONRY

EL. 112'- 8'
JOIST BEARING

2'-8'

4'

2'-0'

2'-0'

Figure 5–10   Example of roof materials for a commercial building

Thus, in the 28 Span Ft. line of the economy table, we could go as low as 80 plf for the bottom figure. We may not be able to find a joist with an 80 plf live load rating; however, the most economical joist with a lower number of 80 plf or above and an upper number not less than about 215 (to make allowance for the weight of the joist itself) should be acceptable.

Looking at the K-series Economy Table in Table 2–7, we find at the 28 Span Ft. line of the far-left column that a 14K4 is listed at $\frac{216}{103}$. We then look up to the top of the column and see that the weight of a 14K4 joist is 6.7 plf. Adding the 6.7 plf to the weight of 208 plf we have been working with, we have a total weight of 214.7 plf.

Because this is less than the maximum allowable 216 total load on the joist, the 14K4 is clearly an acceptable choice. Selecting the required bridging from Table 2–8 (p.27), we again find that a 14K4 joist at a span of 28′ would require two rows of bridging.

5.4   THE DESIGN AND SELECTION OF W-SHAPE BEAMS AND GIRDERS

After the steel joists have been designed and their reactions are known, the structural designer usually designs the beams and girders. Girders support the beams and joists and are in turn supported by the building's columns and walls. The following discussion will explain some commonly used beam and girder design procedures and conditions. In all cases, we will assume that the beams and girders are standard W-shapes.

Depth and Weight

One of the first considerations in designing wide-flange beams and girders is the approximate depth and weight of the W-shape itself, which will add an additional load, however slight, to the calculations. For many types of commercial construction, the depth and weight per lineal foot of most floor and roof beams and girders can be adequately estimated using two general rules of thumb.

First, the depth of a wide-flange beam or girder is rarely less than ½″ of depth per foot of span, and the most economical design oftens uses W-shapes having slightly more than ½″ of depth per foot of span. For example, at a span of 20′–0″, the depth of a W-shape beam or girder would usually be at least 10″, though more probably 12″ and possibly as much as 14″. And a 6″- or 8″-deep W-shape would almost never be used to 20′–0″.

Using second rule of thumb, designers can estimate the weight-per-foot-for floor or roof beams and girders under normal load as follows:

A. Assume 30 to 40 plf for spans up to 20′–0″.
B. Assume 40 to 50 plf for spans of 21′–0″ to 24′–0″.
C. Assume 50 to 60 plf for spans of 25′–0″ to 28′–0″.
D. Assume 60 to 80 plf minimum for W-shapes spanning 29′–0″ to 32′–0″. For beams spanning more than 32′–0″, W-shapes become very expensive, thus plate girders, joist girders, or trusses are quite commonly used instead.

Let us look at an example of how the two rules work in practice: If a roof or floor beam has a span of 24′–0″, we assume the beam is at least a W12, though a W14 is probably a better and more economical guess and a W16 is not out of the question. The beam probably weighs between 40 and 55 plf. Although specific calculations might indicate that a W14 × 53 or W16 × 45 should be used, you would not be far off for either the depth or weight of the beam by using these rules. Of course, when estimating beam and girder sizes and weights for various spans and loads, there is no substitute for experience gained from previous projects.

When designing structural steel rolled beams and girders, the designer must analyze the various loads and stresses acting on the member, then select a shape that can ade-

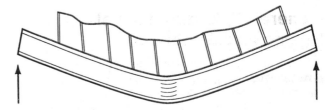

Figure 5–11   Bending failure in a beam

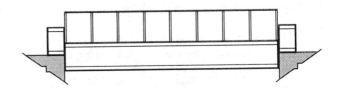

Figure 5–12   Vertical shear failure in a beam

quately resist these stresses and loads. The goal is to achieve *equilibrium*, a state of balance between the applied forces acting on the member and the internal resisting forces developed within the structural shape. For most beams and girders subjected to transverse loads, the main considerations are bending strength, shear resistance, lateral support, deflection, and web crippling.

Bending Strength

Flexure, or bending, is an indirect stress due to moments in the beam caused by the beam's loads and reactions. The bending stress in a structural steel rolled shape due to the *maximum bending moment* is usually the governing factor in beam selection. Thus, unless the rolled shape has adequate material to develop a resisting moment equal to or greater than the maximum bending moment, the beam will fail in bending. The allowable bending stress for a compact braced section of A36 steel is normally assumed to be 24 ksi. Figure 5–11 illustrates the effect and possible failure of a beam caused by bending.

Shear Resistance

Vertical shear is the concentrated force at the supports that could make the beam shear or break, allowing adjacent sections to slide past each other and causing the beam to drop between its supports. Shear failure is rarely a governing factor in steel beam design unless the beam, because of an unusually short span, is capable of sustaining a tremendously heavy load. The allowable shear for A36 steel is 14.4 ksi. Figure 5–12 illustrates beam failure due to vertical shear.

General Procedure for Selecting Beams and Girders

The following general procedure lists the usual steps followed when designing steel beams and girders.

Step 1   Compute the loads the beam must support, and draw a diagram showing the loads and their locations.

Step 2   Compute the reactions and record them on the load diagram.

Step 3   Compute the maximum bending moment and determine the required section modulus using the flexure formula

$$S = \frac{M}{F_b}$$

where M is the maximum bending moment in inch-pounds or kip-inches and F_b is the allowable bending stress in psi and ksi, respectively.

Step 4   Refer to the AISC Allowable Stress Design Selection Table (Table 2–2, p17) and select the most economical (lightest) structural shape with both a M_R (maximum resisting moment) and S_x (section modulus) equal to or greater than the actual maximum bending moment and required section modulus calculated in step 3.

Step 5   Check and compare the distance between applied loads on the load diagram with the recommended L_c and L_u distances listed in the Allowable Stress Design Selection Table (Table 2–2, p17). If the beams are compact, which is the usual condition, and if they are supported laterally at intervals not greater than L_c, they are considered adequately braced, and the full allowable bending stress (24 ksi for A36 steel) may be used. If the distance between lateral supports is greater than L_c but not greater than L_u, the allowable bending stress (F_b) is lowered to 22 ksi. When distances between points of lateral support exceed L_u, beams must be designed by procedures not covered in this textbook.

Step 6   Check the beam for shear by comparing the greatest computed or *actual shearing stress*

$$(f_v = \frac{V}{A_w})$$

with the *allowable shearing stress* (F_v) of 14.4 ksi for A36 steel. The actual shearing stress must of course be equal to or smaller than the allowable shearing stress.

Step 7   Compute the *maximum deflection* due to the design load and compare it with the *allowable deflection*. If the computed deflection exceeds the allowable deflection (usually $\frac{1}{360}$ span for floor beams and girders), try a *deeper* beam. For example, if a W16 × 40 W-shape is acceptable for everything except deflection, the deflection problem can often be solved by choosing a deeper section such as a W18 × 40 or even a W21 × 44. It should be pointed out that there is virtually an unlimited number of loading conditions in actual design situations, and deflection is directly related to load, span length, depth of section, and allowable bending stress. Thus, the experience and judgment of the structural designer play a very important part in determining how much a beam or girder will deflect under the design load. Various methods, charts, and tables are available to help the designer calculate deflections once the span length, depth of section, and allowable bending stress are known. Table 5–2, reprinted in Part 2 of the *Steel Construction Manual,* is an example of deflection coefficients for rolled beams and girders.

We will now apply these steps to design a W-shape interior girder spanning 24′–0″ between supporting columns. We will assume the girder will support open-web steel joists that frame into it from both sides at 2′–0″ centers. The reaction of each joist that becomes a load on the girder is 2.5 kips as shown in Figure 5–13. For this example, we will limit the deflection of the girder to $\frac{1}{360}$ span.

Step 1   Estimate the beam weight and make a load diagram showing the loads and their locations. Since the joists are framing into the girder from both sides, the loads on the girder at each 2′ interval will be: 2.5 kips × 2 = 5.0 kips. At a span of 24′, we will assume the beam weight is approximately 50 plf (Figure 5–14).

Step 2   Compute the reactions, and record them on the load diagram. Because the loads are symmetrical, inspection reveals that the

Table 5–2

CAMBER AND DEFLECTION
Coefficients
For beams and girders with constant cross section

Given the simple span length, the depth of a beam or girder and the design unit bending stress, the center deflection in inches may be found by multiplying the span length in feet by the tabulated coefficients given in the following table.

For the unit stress values not tabulated, the deflection can be found by the equation $0.00103448 \, (L^2 f_b / d)$ where L is the span in ft, f_b is the fiber stress in kips per sq. in. and d is the depth in inches.

The maximum fiber stresses listed in this table correspond to the allowable unit stresses as provided in Sects. F1.1 and F1.3 of the AISC ASD Specification for steels having yield points ranging between 36 ksi and 65 ksi when $F_b = 0.66 F_y$; and between 36 ksi and 100 ksi when $F_b = 0.60 F_y$.

The table values, as given, assume a uniformly distributed load. For a single load at center span, multiply these factors by 0.80; for two equal concentrated loads at third points, multiply by 1.02. Likewise, for three equal concentrated loads at quarter points multiply by 0.95.

Ratio of Depth to Span	Maximum Fiber Stress in Kips Per Sq. In.												
	10.0	22.0	24.0	25.2	27.0	28.0	30.0	33.0	36.0	39.0	42.9	54.0	60.0
1/8	.0069	.0152	.0166	.0174	.0186	.0193	.0207	.0228	.0248	.0269	.0296	.0372	.0414
1/9	.0078	.0171	.0186	.0196	.0209	.0217	.0233	.0256	.0279	.0303	.0333	.0419	.0466
1/10	.0086	.0190	.0207	.0217	.0233	.0241	.0259	.0284	.0310	.0336	.0370	.0466	.0517
1/11	.0095	.0209	.0228	.0239	.0256	.0266	.0284	.0313	.0341	.0370	.0407	.0512	.0569
1/12	.0103	.0228	.0248	.0261	.0279	.0290	.0310	.0341	.0372	.0403	.0444	.0559	.0621
1/13	.0112	.0247	.0269	.0282	.0303	.0314	.0336	.0370	.0403	.0437	.0481	.0605	.0672
1/14	.0121	.0266	.0290	.0304	.0326	.0338	.0362	.0398	.0434	.0471	.0518	.0652	.0724
1/15	.0129	.0284	.0310	.0326	.0349	.0362	.0388	.0427	.0466	.0504	.0555	.0698	.0776
1/16	.0138	.0303	.0331	.0348	.0372	.0386	.0414	.0455	.0497	.0538	.0592	.0745	.0828
1/17	.0147	.0322	.0352	.0369	.0396	.0410	.0440	.0484	.0528	.0572	.0629	.0791	.0879
1/18	.0155	.0341	.0372	.0391	.0419	.0434	.0466	.0512	.0559	.0605	.0666	.0838	.0931
1/19	.0164	.0360	.0393	.0413	.0442	.0459	.0491	.0541	.0590	.0639	.0703	.0885	.0983
1/20	.0172	.0379	.0414	.0434	.0466	.0483	.0517	.0569	.0621	.0672	.0740	.0931	.1035
1/21	.0181	.0398	.0434	.0456	.0489	.0507	.0543	.0597	.0652	.0706	.0777	.0978	.1086
1/22	.0190	.0417	.0455	.0478	.0512	.0531	.0569	.0626	.0683	.0740	.0814	.1024	.1138
1/23	.0198	.0436	.0476	.0500	.0535	.0555	.0595	.0654	.0714	.0773	.0851	.1071	.1190
1/24	.0207	.0455	.0497	.0521	.0559	.0579	.0621	.0683	.0745	.0807	.0888	.1117	.1241
1/25	.0216	.0474	.0517	.0543	.0582	.0603	.0647	.0711	.0776	.0841	.0925	.1164	.1293
1/26	.0224	.0493	.0538	.0565	.0605	.0628	.0672	.0740	.0807	.0874	.0962	.1210	.1345
1/27	.0233	.0512	.0559	.0587	.0628	.0652	.0698	.0768	.0838	.0908	.0999	.1257	.1397
1/28	.0241	.0531	.0579	.0608	.0652	.0676	.0724	.0797	.0869	.0941	.1036	.1303	.1448
1/29	.0250	.0550	.0600	.0630	.0675	.0700	.0750	.0825	.0900	.0975	.1073	.1350	.1500
1/30	.0259	.0569	.0621	.0652	.0698	.0724	.0776	.0853	.0931	.1009	.1110	.1397	.1552

Courtesy of the American Institute of Steel Construction Inc.

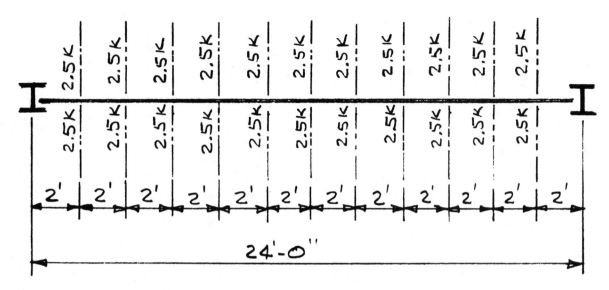

Figure 5–13 Example for girder design procedure

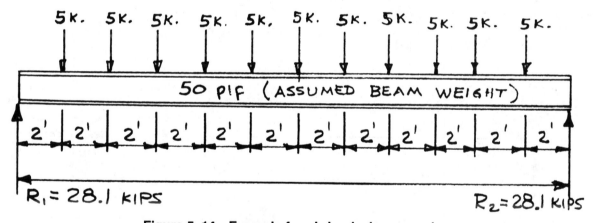

Figure 5–14 Example for girder design procedure

reactions will be equal and each reaction will equal half the sum of the loads.

Total load = (5 kips × 11) + (50 plf × 24) = 56.2 kips

$$R_1 = R_2 = \frac{W}{2} = \frac{56.2 \text{ kips}}{2} = 28.1 \text{ kips}$$

Step 3 Compute the maximum bending moment, and determine the required section modulus. Because the loads are symmetrical, the maximum bending moment will be at the center of the span. The value of the maximum bending moment will be:

M_{12} = (28.1 kips × 12′) − [(5 kips × 2′) + (5 kips × 4′) + (5 kips × 6′) + (5 kips × 8′) + (5 kips × 10′) + (.05 kips × 12′ × 6)] = 183.6 kip-ft.

$$\text{Section modulus required} = \frac{183.6 \text{ kip-ft.} \times 12''}{24 \text{ ksi}} = 91.8''^3$$

Step 4 The AISC Allowable Stress Design Selection Table (Table 2–2, p16) indicates that a W21 × 50 can produce a resisting moment (M_R) of 187 kip-ft. and has a section modulus (S_x) of $94.5''^3$. A resisting moment of 187 kip-ft. > the maximum bending moment of 183.6 kip-ft. ⃝OK

A W21 × 50 has a section modulus of $94.5''^{33}$, the required section modulus of $91.8''^3$. ⃝OK

Since the joists are framing into the girder at 2'–0" on center and the L_c is listed as 6.9' for a W21 × 50 on the Allowable Stress Design Selection Table, the 24-ksi allowable bending stress is obviously ⃝OK.

A W21 × 50 is ⃝OK for bending.

Step 5 Check for shear.

$$f_v \text{ (actual shear stress)} = \frac{V \text{ (maximum shear from reactions)}}{A_w \text{ (gross area of beam web)}}$$

$$f_v = \frac{V}{dt_w} = \frac{27.43 \text{ kips}}{(20.83 \times .380)} = 3.467 \text{ ksi}$$

The actual shear stress is 3.467 ksi; the allowable shear stress is 14.4 ksi.
The W21 × 50 is ⃝OK for vertical shear.

Step 6 Check for deflection.
Allowable deflection = 1/360
span = $\frac{24' \times 12''}{360}$ = .8"
Beam span = 24'
Beam depth = 1.75'
Ratio of depth to span = x = $\frac{24'}{1.75}$ = 13.7
(Round up to 14.)
The AISC Camber and Deflection Table shows that, at a ratio of 1/14 and a 24 ksi maximum fiber stress, the tabulated coefficient is .029
Deflection = 24' span × .029 = .696"
The actual deflection is .696"; the allowable deflection is .8". ⃝OK
Use a W21 × 50 W-shape Girder.

5.5   THE DESIGN OF BEAM BEARING PLATES

The steel members in a support system can be connected to each other and to the structure itself by a wide variety of methods, known as *framing connections*. The design and selection of framing connections has long been considered a responsibility of the structural fabricator, thus this work has traditionally been done in the fabricator's structural drafting office. For this reason, design procedures for standard framing connections will be discussed in chapter 12 of this book.

However, on every commercial building project, certain special connections or connections requiring details must be shown on the design drawings. An example might be a situation in which the end of a steel beam or girder must be supported by either a poured concrete or CMU (concrete masonry unit) wall. For this type of connection, it is often necessary to design a steel *bearing plate* that will transfer the load of the beam over a suitable area of the wall so the concrete will not be overstressed. The bearing plates are usually shipped separately to the job site and leveled and grouted in place. The beams and girders are then welded to the bearing plates during construction.

Figure 5–15 shows a beam bearing plate spreading a concentrated beam load over a CMU wall. Notice that the hollow cores in the upper three courses of concrete block have been filled with grout to distribute the beam load over a larger area of the wall and thus reduce the unit stress on the wall to an acceptable level. When designing beam bearing plates, the designer must determine the length of bearing plate parallel to the beam illustrated by N in Figure 5–16, the required width of the bearing plate parallel to the wall, illustrated by B in Figure 5–16, and the required thickness (t) of the bearing plate.

Determining Dimensions

The N dimension is often dictated by the thickness and makeup of the wall itself. For example, if a wide-flange beam were to be mounted in a beam pocket in a 12"- or 14"-thick cast-in-place concrete foundation wall, and if the wall had a 4" brick ledge on its exterior side, the N dimension might be limited to 6" or 8" unless the designer were to design a pilaster on the inside of the wall. If any event, N is usually a given, known before the bearing plate calculations are begun.

The B dimension, the width of the bearing plate parallel to the wall, is a calculated dimension. It is determined by calculating the width of plate required to make the total area of the plate large enough that the allowable bearing pressure on the wall is not exceeded. Note that dimensions N and B are usually rounded up to full inches. For example, if the required B dimension is calculated as 7.61", the designer always rounds up to the next larger inch, which in this case would be 8".

The required thickness of the bearing plate (t) is calculated using the following formula based on the cantilever of the plate under uniform pressure from the concrete beneath it.

$$t = \sqrt{\frac{3f_p n^2}{F_b}}$$

In this formula: t = thickness of the plate in inches
f_p = actual bearing pressure of the plate on the wall in ksi
F_b = allowable bending stress in the bearing plate. (27 ksi for grade A36 steel)

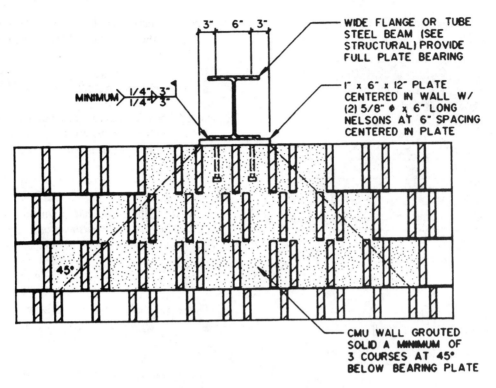

WIDE FLANGE OR TUBE STEEL BEAM (SEE STRUCTURAL) PROVIDE FULL PLATE BEARING

1" x 6" x 12" PLATE CENTERED IN WALL W/ (2) 5/8" ⌀ x 6" LONG NELSONS AT 6" SPACING CENTERED IN PLATE

MINIMUM

45°

CMU WALL GROUTED SOLID A MINIMUM OF 3 COURSES AT 45° BELOW BEARING PLATE

Figure 5–15   Beam bearing detail

$$n = \frac{B}{2} - k, \text{ in inches}$$

k = distance from the center of the beam web to the toe of the fillet in inches. The k distance for various beams is found in Part 1 of the Manual *Steel Construction.*

Because structural plates are rolled in thickness increments of 1/8″, it is common to round up to the next eighth-inch of thickness. For example, a calculated thickness of .921″ would be increased to 1″.

An important consideration in the design of beam bearing plates is the *allowable bearing pressure* on the masonry or cast-in-place concrete wall beneath the plate. These allowable bearing pressures, which are specified by code, may vary greatly depending on the loading condition and the composition of the supporting wall—that is, whether it is of solid brick, concrete block, 3,000 psi concrete, 4,000 psi concrete, or whatever. For example, the allowable bearing stress on an unreinforced concrete hollow block wall with type S mortar is 225 psi. The allowable bearing pressure on a 3,000-psi poured concrete wall is 750 psi if the bearing plate covers the full area of the wall and 1,125 psi if the bearing plate bears on one-third or less of the support area. Thus, as one of the very first steps in designing bearing plates for beams and girders, the drafter/designer must check the proper code to verify the allowable bearing pressure on the concrete or masonry wall.

General Procedure for Selecting a Bearing Plate

We will now go step-by-step through the typical procedure for designing a bearing plate for a W-shape beam to be supported by a concrete wall.

We will assume a W18 × 46 beam is supported on one of its ends by a reinforced poured concrete wall made of 3 ksi concrete. The end reaction (R) is 48 kips. We will also assume that both the beam and bearing plate are A36 steel for which the allowable bending stress in plate F_b is 27 ksi. And finally, we will assume that the length of the bearing plate parallel to the beam is 6″ and the allowable bearing pressure on the wall (F_p) is .75 ksi.

Solution

Step 1   Solve for the minimum required area of the plate. The required area of the plate is:

$$A = \frac{R}{F_p} = \frac{48 \text{ kips}}{.75 \text{ ksi}} = 64 \text{ in.}^2$$

Step 2   Solve for the required width of the plate:

$$N = 6 \text{ in, so } B = \frac{64 \text{ in.}^2}{6 \text{ in.}} = 10.66 \text{ in.}$$

(Round up to 11 in.)

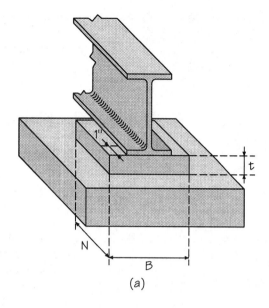

(a)

ANCHOR AS REQUIRED

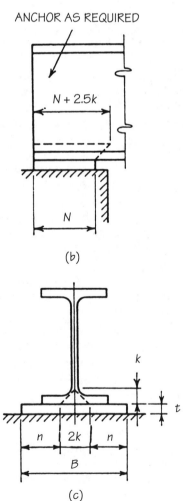

(b)

(c)

Figure 5–16 Dimensions for beam bearing plates (Figure 5–16b and c, Courtesy of the American Institute of Steel Construction Inc.)

Step 3 With the overall dimensions of the plate known, we can solve for the *actual bearing pressure* on the wall:

$$f_p = \frac{R}{A} = \frac{48 \text{ kips}}{6 \text{ in.} \times 11 \text{ in.}} = 0.727 \text{ ksi}$$

Step 4 Solve for the required thickness of the plate. To do this, we must first determine the *n* dimension as shown n Figure 5–16c.

$$n = \frac{B}{2} - k_1$$

From the Table of W-Shapes Dimensions in Part 1 of the *Manual of Steel Construction* (Table 2–1), we find the value of k_1 for a W18 × 46 to be $^{13}/_{16}$ in. (0.813):

$$n = \frac{B}{2} - k_1 = \frac{11 \text{ in.}}{2} - 0.813 \text{ in.} = 4.687 \text{ in.}$$

The required plate thickness will be:

$$t = \frac{3f_p n^2}{F_b} = \frac{3 \times .727 \times (4.687)^2}{27} = 1.33 \text{ in.}$$

The measurement 1.33″ should be rounded up to 1.375″, which is the next eight-inch increment. Thus, the required bearing plate size is 6″ × 1⅜″ × 11″. In an on-the-job situation, it might be most economical to select a plate thickness of 1½″ or even greater if that same plate thickness could be used for the majority of bearing plates on a project.

5.6 THE DESIGN OF STRUCTURAL STEEL COLUMNS

After the joists, beams, and girders have been designed and their reactions are known, the designer can design the steel columns. *Columns* may be defined as long, slender vertical compression members that support the framing system's structural beams and girders and thus transfer the loads of floors and roofs down to the footings and foundation walls. Structural steel columns are usually HP-shapes, W-shapes, P-shapes (steel pipe) and TS-shapes (square or rectangular structural tube). Because of their length and slenderness, the tendency of columns to bend or buckle under load as illustrated in Figure 5–17 becomes a major factor in structural steel column design.

The most common shapes for columns are W-shapes (such as W8 × 31, W10 × 49, and W12 × 65) TS-shapes

Figure 5–17 Bending due to column slenderness under axial load. (Courtesy of the American Institute of Steel Construction Inc.)

(particularly square structural tubes such as TS4 × 4 × .250 or TS6 × 6 × .375) and P-shapes (such as 3″ std. or 4″ x-strong steel pipe).

Designing steel columns is essentially a trail-and-error procedure because the allowable unit stress on the column depends upon the column's slenderness ratio (which will be discussed shortly) and the distribution of the cross-sectional area, while the required area depends upon the allowable stress. Other important factors are the grade of steel used, such as 36 ksi or 50 ksi, the effective unbraced length in inches, the location of the applied loads, and the effect of

end conditions or the method of end restraint at the top and bottom of the column.

We will now review some of the major considerations of steel column design.

Factors Involved in Steel Column Design

Slenderness Ratio. The *slenderness ratio* of a structural steel column is the ratio of the unsupported length of the column (*1*) in inches, to the radius of gyration of the column section (*r*), in inches, with respect to the direction of potential buckling, which may be either the strong or weak axis. The *radius of gyration* is the measure of a column's stiffness or resistance to buckling. Values for *r* for various steel shapes can be found in the Tables of Dimensions and Properties in Part 1 of the *Manual of Steel Construction*. These tables show that for W-shapes the values of *r* for the strong axis (x–x axis) are always greater than the values for the weak axis (y–y axis). For symmetrical shapes such as round steel pipe or square structural tubing, the radius of gyration is the same for both the x–x and y–y axis.

Some people believe that the least radius of gyration (r_y) is always used when determining the slenderness ratio of a column, but this is often not true. Determining the slenderness ratio for columns depends on how the loads are applied. For example, in determining the slenderness ratio for a W8 × 31 column 13 feet in height supporting only an axial load from roof beams, the least radius of gyration would govern because the column is not braced against buckling on either the x–x or y–y axis (Figure 5–18).

If however, the column was to be braced at mid-height on only the minor (y–y) axis, as shown in Figure 5–19, it would be very possible that the slenderness ratio or the tendency to buckle would be greatest with respect to the major (x–x) axis because on that axis the unsupported length would be 26′ compared to 13′ on the minor axis.

To illustrate this point, we will assume the W-shape column in Figure 5–19 is a W8 × 31 and calculate the slenderness ratio with respect to both the major and minor axes. From Part 1 of the *Manual of Steel Construction*, we would find that for a W8 × 31:

$$r_x = 3.74″ \text{ and } r_y = 2.02″$$

Because the column is supported at mid-height perpendicular to the y–y axis, the effective length for buckling

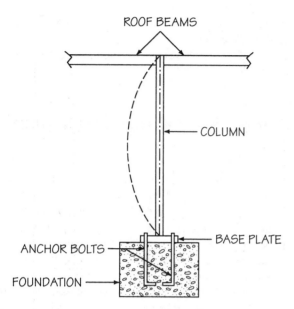

Figure 5–18 Column buckling under axial load

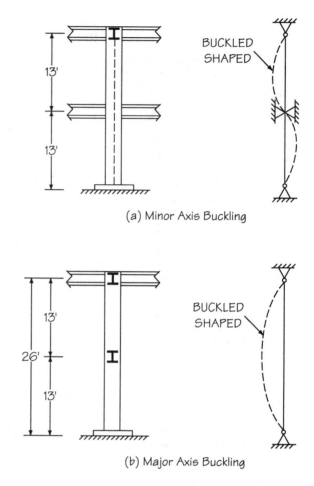

(a) Minor Axis Buckling

(b) Major Axis Buckling

Figure 5–19 Column buckling at (a) major and (b) minor axis

around that axis will be 13′. Thus, the slenderness ratio with respect to the y–y axis is:

$$\frac{1}{r_y} = \frac{13' \times 12''}{2.02''} = 77.22$$

Because the column is only supported with respect to the major (x–x) axis at the top, the unsupported length of the column with respect to that axis is 26′. Thus, the slenderness ratio with respect to that axis would be:

$$\frac{1}{r_x} = \frac{26' \times 12''}{3.47''} = 89.9$$

In this case, the maximum slenderness ratio of 89.9 calculated for the major (x–x) axis would be used to design the column.

Types of Connections. Another consideration when designing a column is the manner in which it is anchored or restrained at the top and bottom ends. For example, Figure 5–18 illustrates a very common situation whereby the column is welded at is bottom to a steel baseplate, which is in turn fastened to the foundation wall or footing by heavy steel bolts called anchor bolts. Because the holes through the baseplate for the anchor bolts are many times up to $\frac{3}{16}''$ or more oversized. This type of connection will not restrain the column from rotation. Nor will the roof beams, which are

usually also bolted to the column, keep it from rotation at the top end. This results in what is called a pinned, or pin-ended, connection. *Pin-ended connections* are suitable for most conditions in commercial and industrial construction, but other types of end restraint for columns are also possible. For example, on very tall structures where lateral wind forces may become the governing consideration for exterior columns, fixed *moment-resisting connections* are very common. Effects of varying types of end restraint become a factor, called the K factor, in calculating the effective length of a steel column.

Performing the Calculations

The *K factor* is a modifying or compensating factor that takes into consideration the effects of various types of col-

Table 5–3
Recommended K values for columns

	(a)	(b)	(c)	(d)	(e)	(f)
Buckled shaped of column is shown by dashed line						
Theoretical *K* value	0.5	0.7	1.0	1.0	2.0	2.0
Recommended design value when ideal conditions are approximated	0.65	0.80	1.2	1.0	2.10	2.0
End condition code		Rotation fixed and translation fixed				
		Rotation free and translation fixed				
		Rotation fixed and translation free				
		Rotation fixed and translation free				

Courtesy of the American Institute of Steel Construction Inc.

umn end restraint. This may result in either a reduced or magnified value for *l*. Thus, the slenderness ratio is usually expressed as K*l*, although for common pin-ended connections, the factor of K is given as 1, so the slenderness ratio is expressed as:

$$\frac{Kl}{r} = \frac{l}{r}$$

Table 5–3, reprinted from Part 5 of the *Manual of Steel Construction,* lists the theoretical K value for various types of column end restraint.

If the slenderness ratio has been determined and the main building columns do not require a $\frac{Kl}{r}$ of more than 120, the designer then refers to Part 3 of the *Manual of Steel Construction* where a table entitled Allowable Stress for Compression Members lists the allowable unit compression stress in ksi for various slenderness ratios Table 5–4.

Once the allowable unit stress for the column is known, the total allowable compression load can be easily calculated by multiplying the allowable unit stress (F_a) from the table by the cross-sectional area of the trial column. For example, assume a structural drafter/designer is supporting two roof beams with a W-shape steel column similar to the loading condition shown in Figure 5–18. Each roof beam has a reaction of 42 kips, and the unsupported length is 20′. Would a W8 × 31 be adequate?

Part 1 of the *Manual of Steel Construction* shows that the area of a W8 × 31 is 9.13 in², the r_x is 3.47 in, and the r_y is 2.02 in. Since the end restraint is pinned, the K factor will be 1.

Solution

1. The total load on the column is 42 kips × 2 = 84 kips.
2. Since the loading condition show in Figure 5–18 would allow the column to buckle in either direction, the least radius of gyration, r_y, would govern. Thus, the slenderness ratio is:

$$\frac{Kl}{r_y} = \frac{1 \times (20' \times 12'')}{2.02''} = 118.8 \text{ (Round up to 119.)}$$

Table 5–4
Allowable stresses for compression members

Table C-36
Allowable Stress
For Compression Members of 36-ksi Specified Yield Stress Steel[a]

$F_y = 36$ ksi

$\frac{Kl}{r}$	F_a (ksi)	$\frac{Kl}{r}$	F_a (ksi)	$\frac{Kl}{r}$	F_a (ksi)	$\frac{Kl}{r}$	F_a (ksi)	$\frac{Kl}{r}$	F_a (ksi)
1	21.56	41	19.11	81	15.24	121	10.14	161	5.76
2	21.52	42	19.03	82	15.13	122	9.99	162	5.69
3	21.48	43	18.95	83	15.02	123	9.85	163	5.62
4	21.44	44	18.86	84	14.90	124	9.70	164	5.55
5	21.39	45	18.78	85	14.79	125	9.55	165	5.49
6	21.35	46	18.70	86	14.67	126	9.41	166	5.42
7	21.30	47	18.61	87	14.56	127	9.26	167	5.35
8	21.25	48	18.53	88	14.44	128	9.11	168	5.29
9	21.21	49	18.44	89	14.32	129	8.97	169	5.23
10	21.16	50	18.35	90	14.20	130	8.84	170	5.17
11	21.10	51	18.26	91	14.09	131	8.70	171	5.11
12	21.05	52	18.17	92	13.97	132	8.57	172	5.05
13	21.00	53	18.08	93	13.84	133	8.44	173	4.99
14	20.95	54	17.99	94	13.72	134	8.32	174	4.93
15	20.89	55	17.90	95	13.60	135	8.19	175	4.88
16	20.83	56	17.81	96	13.48	136	8.07	176	4.82
17	20.78	57	17.71	97	13.35	137	7.96	177	4.77
18	20.72	58	17.62	98	13.23	138	7.84	178	4.71
19	20.66	59	17.53	99	13.10	139	7.73	179	4.66
20	20.60	60	17.43	100	12.98	140	7.62	180	4.61
21	20.54	61	17.33	101	12.85	141	7.51	181	4.56
22	20.48	62	17.24	102	12.72	142	7.41	182	4.51
23	20.41	63	17.14	103	12.59	143	7.30	183	4.46
24	20.35	64	17.04	104	12.47	144	7.20	184	4.41
25	20.28	65	16.94	105	12.33	145	7.10	185	4.36
26	20.22	66	16.84	106	12.20	146	7.01	186	4.32
27	20.15	67	16.74	107	12.07	147	6.91	187	4.27
28	20.08	68	16.64	108	11.94	148	6.82	188	4.23
29	20.01	69	16.53	109	11.81	149	6.73	189	4.18
30	19.94	70	16.43	110	11.67	150	6.64	190	4.14
31	19.87	71	16.33	111	11.54	151	6.55	191	4.09
32	19.80	72	16.22	112	11.40	152	6.46	192	4.05
33	19.73	73	16.12	113	11.26	153	6.38	193	4.01
34	19.65	74	16.01	114	11.13	154	6.30	194	3.97
35	19.58	75	15.90	115	10.99	155	6.22	195	3.93
36	19.50	76	15.79	116	10.85	156	6.14	196	3.89
37	19.42	77	15.69	117	10.71	157	6.06	197	3.85
38	19.35	78	15.58	118	10.57	158	5.98	198	3.81
39	19.27	79	15.47	119	10.43	159	5.91	199	3.77
40	19.19	80	15.36	120	10.28	160	5.83	200	3.73

[a]When element width-to-thickness ratio exceeds noncompact section limits of Sect. B5.1, see Appendix B5.
Note: $C_c = 126.1$

Courtesy of the American Institute of Steel Construction Inc.

3. The Allowable Stress Table shows that for a slenderness ratio of 119, the allowable unit stress on the W8 × 31 is 10.43 ksi.

4. The allowable column load is equal to the allowable unit stress multiplied by the number of square inches of steel in the W-shape. Thus, for the W8 × 31 column with an effective length of 20', the allowable load is:

$$P = AF_a = 9.13 \text{ in}^2 \times 10.43 \text{ ksi} = 95.2 \text{ kips}$$

$$\begin{array}{ccc} 84 \text{ kips} & & 95.2 \text{ kips} \quad \text{OK} \\ \text{(actual load)} & < & \text{(allowable load)} \end{array}$$

A W8 × 31 is adequate.

It should be pointed out that, prior to selecting the W8 × 31 for a trial selection, the structural drafter/designer would have consulted the *Manual of Steel Construction* and found in the Table of Allowable Concentric Loads on Columns that a W8 × 31 W-shape column of A36 steel is capable of supporting 95 kips. Thus, the calculation process previously described is mainly a way to make sure the designer did not misread the table. Most structural design offices require that calculations be performed because the time involved in doing so is minimal compared to the potential catastrophic consequences of misreading the AISC table.

Dealing with Combinations of Stresses

One more consideration should be discussed in relation to steel column design. In actual practice, steel columns are often simultaneously subjected to combinations of stresses. For example, in exterior columns of a steel framing system, the floor or roof beams might frame into the web of a W-shape structural steel column, causing an axial or concentric load on the member (Figure 5–20). At the same time, an interior girder supporting the reactions of beams or joists might frame into the flange face, delivering a moment to the column that would cause bending at the X–X axis.

Inspection reveals that the two 20-kip loads framing into the column web from either side would cause axial or concentric loads that would not develop significant bending stresses on the Y–Y axis of the column. However, the 40-kip load is an *eccentric load,* which would cause a bending stress because its connection point at the face of the flange is several inches away from the column's X–X axis, as shown in Figure 5–21.

The usual method of designing W-shape steel columns for eccentric loading is to convert the eccentric loads into an equivalent axial load by multiplying the bending moment resulting from the eccentric load by the appropriate *bending*

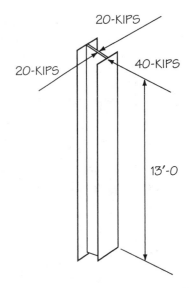

Figure 5–20   Example of eccentric loading on exterior column

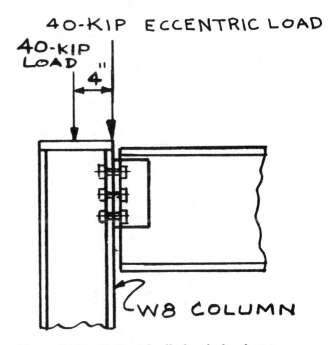

Figure 5–21   Eccentrically loaded column

factor found in Part 3 of the *Manual of Steel Construction.* This equivalent axial load is then added to the actual eccentric load and any concentric loads to come up with the total equivalent load for the column. Table 5–5, reprinted from the *Manual of Steel Construction,* illustrates the B_x and B_y bending factors for several different W8 columns.

The B_x and B_y bending factors listed at the bottom of the table are found by dividing the cross-sectional area of the shape by the section moduli of both the X–X and Y–Y axes. For example, the two bending factors for a W8 × 31 column are found by taking the area of a W8 × 31, which is 9.13 in², and dividing it by the section moduli of 27.5 and 9.27 in³, respectively:

$$B_x = \frac{A}{S_x} = \frac{9.13}{27.5} = 0.332$$

and

$$B_y = \frac{A}{S_y} = \frac{9.13}{9.27} = 0.985$$

Notice that $B_x = 0.332$ and $B_y = 0.985$ correspond to the bending factors listed in Table 5–5 for a W8 × 31 column.

Applying the Procedure

We will now select a W8 column of A36 steel for the loading condition illustrated in Figure 5–20. By inspection, we see an axial load of 40 kips created by the two 20-kip reactions of beams fastening into either side of the column web. We also see a 40-kip eccentric load being applied at the face of the flange, which would be approximately 4 inches from the X–X asis on a W8 column.

Solution

Step 1   The loads applied by the beam reactions are:

20 kips + 20 kips + 40 kips = 80 kips.

Step 2   The bending moment produced by the eccentric load is:

M = 40 kips × 4 in = 160 kip-in.

Step 3   Table 5–5 shows that the average bending factor listed for W8 × 31 through W8 × 67 columns is approximately 0.3333 for B_x. Thus, the bending moment multiplied by the average bending factor is:

160 × 0.333 = 53.28 kips

So we assume 54 kips as the equivalent axial load for the eccentric load.

Step 4   The total equivalent load on the column is then:

20 kips + 20 kips + 40 kips + 54 kips = 134 kips

Step 5   Referring again to Table 5–5, we see that, at a 13-foot unsupported height, a W8 × 31

W-shape column is adequate to support an axial load of 143 kips.

143 kips	>	134 kips	
(allowable load)		(equivalent axial load)	Ⓞ𝐊

A W8 × 31 column is sufficient to support the loading condition illustrated in Figure 5–20.

5.7   THE DESIGN OF COLUMN BASEPLATES

After the columns have been designed, the drafter/designer must select the column baseplates. *Baseplates* are steel plates, usually welded to the bottom of the column and fastened to the top of the footing or foundation wall with heavy anchor bolts as shown in Figures 7–1 through 7–9. Baseplates are necessary to distribute the column load over a sufficient area of a concrete wall or footing so that the allowable compressive strength of the concrete, which is considerably lower than that of steel, is not exceeded.

The AISC specification recommends two allowable pressures (F_p) for concrete supports. These allowable pressures are based on the compressive strength of the concrete (f_c') and the percentage of support area covered by the baseplate. If the area of the baseplate (A_1) is equal to the area of the support (A_2), the allowable compressive stress on the concrete is:

$$F_p = .35\, f_c$$

If the area of the baseplate (A1) is less than the area of the concrete support (A_2, the allowable stress is increased because the mass of concrete beyond the contact area provides considerable confinement to the directly loaded area of concrete. A common example is when a 10″ × 10″ baseplate under a structural steel column rests on a 3′–0″ square concrete footing. For this condition, the allowable bearing stress becomes:

$$F_p = .7\, f_c'$$

When fully loaded, the bearing pressure between the baseplate and the footing or foundation wall is assumed to be uniformly distributed within a rectangle of dimensions 0.95d and 0.80b, as shown by the dashed lines in Figure 5–22a. However, the reaction of the footing pushing back against the cantilevered parts of the plate outside the column tend to cause the plate to curl upward as illustrated in Figure 5–22b.

Table 5–5
Allowable Axial Loads on Columns

COLUMNS
W shapes
Allowable axial loads in kips

F_y = 36 ksi
F_y = 50 ksi

Designation		W8										
Wt./ft		67		58		48		40		35		31	
F_y		36	50	36	50	36	50	36	50	36	50	36	50
Effective length in ft KL with respect to least radius of gyration r_y	0	426	591	369	513	305	423	253	351	222	309	197	274
	6	387	525	336	455	276	375	229	310	201	272	178	241
	7	379	510	328	442	270	363	223	300	197	264	174	234
	8	370	494	320	428	263	352	218	290	191	255	170	226
	9	360	477	312	413	256	339	212	279	186	246	165	217
	10	350	459	303	397	249	326	205	268	180	236	160	208
	11	339	440	293	380	241	312	199	256	174	225	154	199
	12	328	420	283	363	233	297	192	244	168	214	149	189
	13	316	399	273	344	224	282	184	231	162	202	143	179
	14	304	378	263	325	215	266	177	217	155	190	137	168
	15	292	355	251	305	206	249	169	203	148	177	131	156
	16	279	331	240	284	196	232	160	188	141	164	124	145
	17	265	307	228	263	186	214	152	172	133	150	117	132
	18	251	281	216	240	176	195	143	156	125	136	110	119
	19	236	254	203	217	165	175	134	140	117	122	103	107
	20	221	230	190	196	154	158	124	126	109	110	95	97
	22	190	190	162	162	131	131	104	104	91	91	80	80
	24	159	159	136	136	110	110	88	88	76	76	67	67
	26	136	136	116	116	94	94	75	75	65	65	57	57
	28	117	117	100	100	81	81	64	64	56	56	49	49
	30	102	102	87	87	70	70	56	56	49	49	43	43
	32	90	90	76	76	62	62	49	49	43	43	38	38
	33	84	84	72	72	58	58	46	46	40	40	35	35
	34	79	79	68	68	55	55	44	44				
	35	75	75	64	64								

Properties												
U	2.48	2.48	2.50	2.50	2.54	2.54	2.56	2.56	2.59	2.59	2.61	2.61
P_{wo} (kips)	147	205	120	167	86	119	69	96	56	78	48	67
P_{wi} (kips/in.)	21	29	18	26	14	20	13	18	11	16	10	14
P_{wb} (kips)	744	877	533	628	257	303	187	221	120	141	93	110
P_{fb} (kips)	197	273	148	205	106	147	71	98	55	77	43	59
L_c (ft)	8.7	7.4	8.7	7.4	8.6	7.3	8.5	7.2	8.5	7.2	8.4	7.2
L_u (ft)	39.9	28.7	35.3	25.4	30.3	21.8	25.3	18.2	22.6	16.3	20.1	14.5
A (in.^{2})	19.7		17.1		14.1		11.7		10.3		9.13	
I_x (in.^{4})	272		228		184		146		127		110	
I_y (in.^{4})	88.6		75.1		60.9		49.1		42.6		37.1	
r_y (in.)	2.12		2.10		2.08		2.04		2.03		2.02	
Ratio r_x/r_y	1.75		1.74		1.74		1.73		1.73		1.72	
B_x } Bending	0.326		0.329		0.326		0.330		0.330		0.332	
B_y } factors	0.921		0.934		0.940		0.959		0.972		0.985	
$a_x/10^6$	40.6		33.9		27.4		21.7		18.9		16.4	
$a_y/10^6$	13.2		11.2		9.1		7.3		6.3		5.6	
$F'_{ex} (K_x L_x)^2/10^2$ (kips)	144		138		135		129		128		125	
$F'_{ey} (K_y L_y)^2/10^2$ (kips)	46.6		45.7		44.9		43.2		42.7		42.3	

Note: Heavy line indicates Kl/r of 200.

Courtesy of the American Institute of Steel Construction Inc.

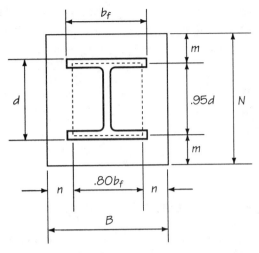

P = Total column load, kips
A_1 = B × N = Area of plate, in.^{2}
A_2 = Full cross-sectional area of concrete
          support, in.^{2}
F_b = Allowable bending stress in baseplate, ksi
F_p = Allowable bearing pressure on support, ksi
f_p = Actual bearing pressure, ksi
f_c = Compressive strength of concrete, ksi
t_p = Thickness of baseplate, in.

(a)   **(Courtesy of the American Institute of Steel Construction Inc.)**

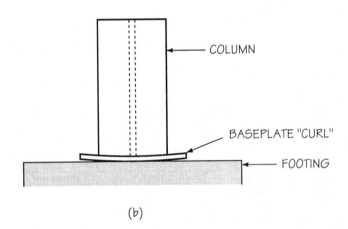

(b)

Figure 5–22   Column baseplate design consideration

To resist this tendency to curl, the AISC has developed formulas to calculate required baseplate thickness based upon *m* and *n*, the cantilevered distances of the plate beyond the rectangular dimensions of 0.95*d* and 0.80*b*. Using the larger value of *m* and *n*, the thickness of the plate is determined by either of the two formulas:

$$t_p\, 2m = \sqrt{\frac{f_p}{F_y}} \text{ and } t_p = 2n \sqrt{\frac{f_p}{F_y}}$$

The following example will demonstrate how the previously discussed equations are used to design a typical baseplate.

Sample Problem   Design a column baseplate for a W8 × 40 column to support a total column load of 180 kips. The column is to rest on a 7′–0″ × 7′–0″ square footing to be made of $f_c' = 3$ ksi concrete.

Solution

1. Since the concrete support area is considerably larger than the baseplate area, the minimum number of square inches in the baseplate could be:

$$A = \frac{P}{.7\, f_c'} = \frac{180 \text{ kips}}{.7\, (3 \text{ ksi})} = 85.7 \text{ in.}^2$$

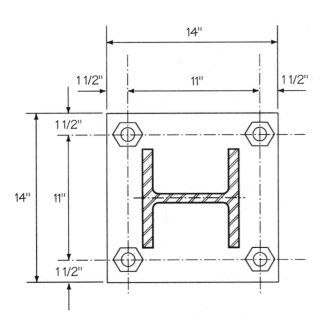

Figure 5–23   Baseplate for W8 × 40 column

2. A W8 × 40 is essentially a square column, and the column is resting on a square footing. Thus, the baseplate should be square and should equal or exceed 85.7 in.² OSHA requirements presently dictate that all column baseplates be fastened to their footings with four anchor bolts. The anchor bolts are usually, but not always, placed outside the column, as shown in Figure 5–23. The edge distance from the end of the baseplate to the center of the anchor bolt hole is usually a minimum of 1½″. Thus, the baseplate should extend a minimum of 3″ beyond the column in all directions. This means the dimensions of the baseplate, for erection purposes, should be 14″ × 14″ square. The area of the baseplate will then be 14″ × 14″ = 196 in², and the actual stress on the baseplate will be:

$$\frac{180 \text{ kips}}{196 \text{ in.}^2} = 0.9 \text{ ksi}$$

3. Based upon the dimensions of a W8 × 40 steel column centered on a 14″ × 14″ baseplate, determine the dimensions of *m* and *n*. See Figure 5–24.

$$m = 14'' - \frac{(.95 \times 8.25)}{2} = 3.085''$$

$$n = 14'' - \frac{(.80 \times 8.07)}{2} = 3.772''$$

4. Solving for the required thickness of the plate, use the largest value found in step 3 in the equation:

$$t = 2n \sqrt{\frac{f_p}{F_y}} = t_p = 2 \ (3.772) \ \sqrt{\frac{.9 \text{ ksi}}{36 \text{ ksi}}} = 1.19''$$

Because structural steel plate is rolled in ⅛″ increments, the baseplate should be 1 ¼″ (1.25″) thick.

The required size and thickness of the column baseplate as designed should be: 14″ × 1¼″ × 14″.

5.8   SUMMARY

This chapter has presented an overview of basic structural design concepts and the types of design considerations routinely encountered and performed by structural drafters under the supervision of design engineers in most structural design offices. Emphasis has been placed on the basic concepts and load conditions that arise as the component parts of a structural steel support frame are assembled. The chapter shows how the structural drafter can use design tables and/or simple algebra to help design and select a variety of structural members, including open-web steel joists to support floor or roof loads, beams and girders to support the joists, columns to support the beams and girders, and beam bearing plates and column baseplates to fasten beams to walls and columns to walls or footings.

This chapter is not intended to substitute for a course in structural analysis or structural design. Rather it has been an attempt to illustrate to the structural drafting student "where the numbers are coming from" or, in other words, methods by which experienced structural drafters, designers, and engineers use algebraic formulas to determine the required physical sizes of structural steel beams, columns, baseplates, etc., which taken together become structural steel support frames for commercial and industrial buildings. The primary purpose has been to illustrate why structural steel design calculations are the very backbone of both the structural steel support system and structural steel drawings.

STUDY QUESTIONS

1. What is the responsibility of the structural engineer in the structural design office?
2. What is the responsibility of the structural drafter in the structural design office?

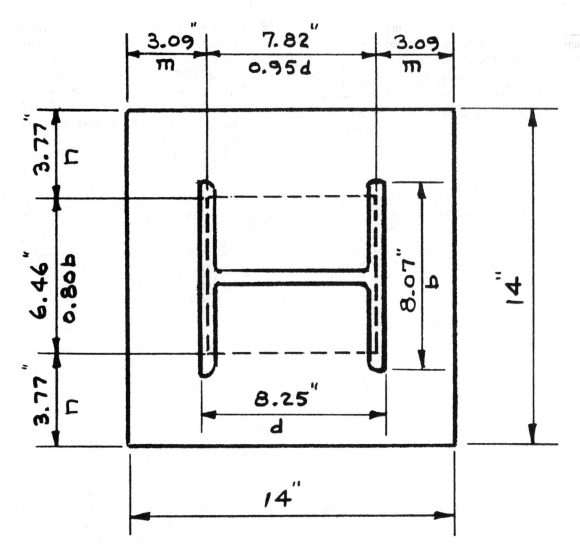

Figure 5–24   Column baseplate layout

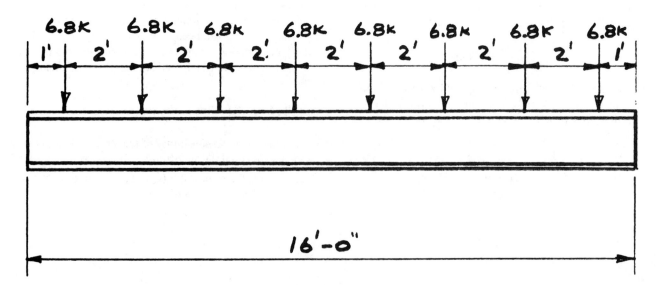

Figure 5–25

3. _____ loads are usually expressed in pounds per square foot (psf) and are recommended by various local, state, and national codes.

4. Structural _____ are usually rectangular in shape and defined by the dimensions within any four columns.

5. Beams and girders along the outside walls of a structure are called _____.

6. If a K-series open-web steel joist is supported by a wide-flange (W-shape) girder, the top of the joist will be _____ above the top of the girder.

7. Select an open-web steel floor joist for a span of 30′–0″. The total design load is 150 psf, of which the live load is 100 psf, and the joists are to be spaced 2′–0″ on center.

8. Select a wide-flange structural steel girder for the loading condition shown in Figure 5–25.

9. A W24 × 62 beam of A36 steel has an end reaction of 60 kips and is to be supported by a concrete wall of 3 ksi concrete for which the allowable bearing pressure is .75 ksi. The bearing length dimension (N) of the plate is limited to 6″, and the k_1 value of the beam is 0.938″. Design the beam bearing plate.

10. Compute the maximum allowable axial load for a W8 × 48 structural steel column of A36 steel if the unbraced height is 14 feet. Assume K = 1, the average end conditions found in most steel construction.

11. Design a column baseplate of A36 steel for a W8 × 58 column to transfer a column load of 250 kips to a large footing of 3 ksi concrete. Because the plate is to be fastened to the footing with four anchor bolts, assume the baseplate must extend beyond the column a minimum of 3″ in all directions.

Chapter
6

The Preparation of Structural Steel Design Drawings and Details

OBJECTIVES

Upon completion of this chapter, you should be able to:

- define the basic objectives of structural drafters and designers as they begin the design phase of a building.
- develop a structural grid system.
- prepare structural steel framing plans and sections using the proper single-line symbols to represent various structural steel members.
- properly draw steel details.

6.1 INTRODUCTION

Before structural steel design drawings can be developed, a basic architectural floor plan showing overall building dimensions and general layouts of rooms, corridors, doors, and windows is essential. Ideally, while designing this floor plan, the architects have considered structural space requirements such as placement of the structural steel columns and the amount of space needed between floors for the support system itself as well as for the plumbing and heating, HVAC, and electrical systems. After the preliminary floor plans have been drawn and tentative building materials and floor-to-floor elevations established, the structural design phase of the project, including preparation of the structural design drawings, can begin.

6.2 THE BASIC OBJECTIVES OF STRUCTURAL DESIGN

Structural steel design involves developing a structural steel framework that supports the building and enables it to fulfill its intended use. To that end, structural drafters and designers often consider several possible structural schemes or layouts and figure cost estimates for each, always keeping in mind that certain basic objectives must be fulfilled.

One of the most important objectives of structural drafters and designers is to ensure that the steel support frame is capable of sustaining all imposed loads and forces. Another very important objective is to design the structural steel support frame as economically as possible, taking into account not only the most efficient use of materials but also any problems that might be encountered by the steel erection crew on the job site. A third objective is to make sure the steel support system in no way alters the appearance of the building as envisioned by the architect and the client.

Once these basic objectives have been accomplished, preparation of the structural steel design drawings can begin. Usually the first step in preparing the structural steel design drawings is to make framing plans or layouts of the structural steel beams, girders, and columns that comprise the building's floor and roof support systems. These framing plans are generally built around the building's grid system.

6.3 THE STRUCTURAL GRID SYSTEM

A structural steel *grid system* simply establishes the location of the structural steel columns so that beam spans will not be uneconomically long and steel fabrication costs will be reasonable. The spacing between columns for most commercial buildings falls into the 16 ft. to 28 ft. range, with 30 ft. being about the maximum unless the designer intends to use high-strength steel or open-web joist girders. Also, it is most economical to lay out a grid system so that as many spacings as possible are the same. This does not mean all the columns in the grid system must be uniformly spaced. On any building project, there are valid reasons, many of them based on architectural requirements, to change the spacing of columns in some areas. However, a prime responsibility of structural drafters and designers is to use their best judgment in laying out a structural steel grid system that is the most efficient possible for the type of building being designed.

For the majority of structural steel framing plans, the structural grid divides the support frame into a matrix of adjacent rectangles called bays. As discussed in chapter 5, a *bay* is the dimensions between any four columns that comprise a somewhat rectangular shape. The center lines of the columns forming adjacent bays, taken together in their entirety, become the structural grid system.

Grid lines are shown on the structural framing plan as thin but clearly visible lines, easily distinguishable from the thicker, darker lines symbolizing the actual components of the structural framework (beams, girders, columns, etc.). The grid lines must be drawn to extend well beyond the exterior columns of a building because they are used as extension lines for the dimension lines that will show both the distances between columns and the overall dimensions of the building. Because the structural grid system is shown on the architectural floor plans, even the walls can be located or referenced from the grid lines. In fact, on most large commercial building projects, virtually everything on both the architectural and structural drawings is ultimately located off the grid system.

Since grid lines are very important reference lines, they must be quickly and easily identified on the architectural floor plans and structural framing plans. This is usually accomplished by placing small circles or hexagons at the end of each grid line. The circles of hexagons must be large enough that identification numbers placed within them can be easily seen. The general rule is to make the circles or hexagons at the ends of grid lines about ⅜″ in diameter, and of course, they aligned.

The usual identification system is to number the grid lines along the top of the sheet, starting from the upper left-hand corner, as ①, ②, ③, etc. Along the side of the sheet, the grid lines are lettered, as Ⓐ, Ⓑ, Ⓒ, etc. Then each structural steel column on the sheet can be identified by the intersection of the grid lines passing through it. For instance, a structural steel column intersected by grid lines Ⓑ and ② would be identified as column B2 at every floor on both the architectural floor plans and the structural framing plans. Figure 6–1 illustrates how grid lines are shown on a structural steel framing plan.

6.4 STRUCTURAL STEEL FRAMING PLANS

The *structural steel framing plan* is a view cut on an imaginary plane just above and parallel to a floor or roof elevation, showing only those items that are structural or of structural concern. The main purpose of a framing plan, as illustrated in Figure 6–1, is to show the size and location of beams, girders, open-web steel joists, columns, and other steel members. Framing plans are usually drawn at the same scale as the architectural floor plans of the building. For commercial and industrial projects, structural steel framing plans are typically drawn at a scale of ⅛″ = 1″–0″, although for smaller structures, plans drawn at a scale of ¼″ = 1′–0″ are not uncommon. The accepted practice in preparing a set of structural design drawings is to draw a structural foundation plan and framing plans for each level of the structure including the roof, and then to follow with the required sections and details as needed. Thus, a three-story office building would have (1) a foundation plan for the foundation showing the grid system column locations on footings, the overall building dimensions, etc., and (2) separate framing plans for the first floor, second floor, third floor, and roof levels showing the size and/or location of the components of the structural system.

On structural steel framing plans, all of the steel members are located off the grid system. The sizes of beams, girders, and open-web steel joists are shown. Column sizes may or may not be shown. On some framing plans, column sizes and types are noted at each location, but it is also very common to simply show the column, usually oversized for clarity, and then give all the necessary information about that column on a column schedule. Lintels are shown on the framing plan but the general practice is to give all the required lintel information on a lintel schedule. Column schedules and lintel schedules will be discussed later in this chapter. One last point to be made about the preparation of structural steel framing plans is that the interior and exterior walls of a building, if shown at all, are drawn with sharp, thin lines to clearly distinguish them from the heavy, dark lines used to denote the structural steel because on steel framing plans the building outline itself is considered a secondary feature.

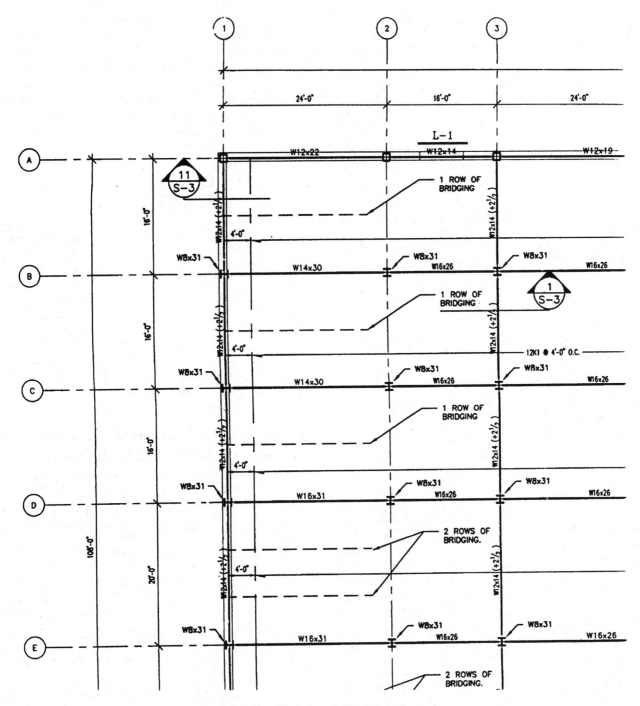

Figure 6–1   Structural steel framing plan

Symbolic Representation on Structural Steel Framing Plans

Figure 6–1 shows that structural members on steel framing plans are indicated by symbols or by heavy single lines. Symbols for structural steel columns are usually drawn oversized for clarity, and their shapes represent the end-view shapes of the members, as shown by the W-shape and structural tube columns in Figure 6–1. (The symbols for structural steel W-shape, tube, and pipe columns are also illustrated in Figure 4–2.) Beams, girders, and other structural members appear as heavy lines that represent the work-

Table 6–1

USUAL GAGES FOR ANGLES, INCHES														
Leg	8	7	6	5	4	3½	3	2½	2	1¾	1½	1⅜	1¼	1
g	4½	4	3½	3	2½	2	1¾	1⅜	1⅛	1	⅞	⅞	¾	⅝
g_1	3	2½	2¼	2										
g_2	3	3	2½	1¾										

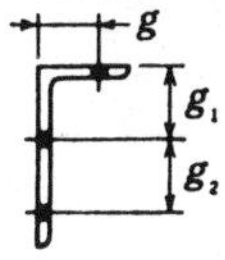

Courtesy of the American Institute of Steel Construction Inc.

ing line of the member. *A Working line* is the line to which locating dimensions are given on plan views and elevations are given on elevation views. Thus, these lines represent different parts of structural members on plans and elevations.

On structural framing plans, heavy single lines represent the center lines of the webs of W-shapes used as beams but the back of channels or angles used as beams. However, when structural steel angles or tees are used for bracing, the heavy lines indicating the working line will be locating the gage line for bolted connections and the centrodial axis or gravity line for welded connections. *Gage lines* indicate where bolt holes are to be placed, according the guidelines established by the American Institute of Steel Construction. The locations of gage lines and gravity lines for structural steel angles can be found in Part 1 of the AISC *Manual of Steel Construction.* AISC-recommended gage lines for structural steel angles are shown in Table 6–1.

On structural elevation drawings, the heavy lines locate either the top or the bottom of beams and channels, depending upon which is the important dimension to hold. For columns, the single heavy line indicates the center line of the column. When showing W-shape columns in single-line representation, a short length of the column is usually shown as double line to indicate whether the view is of the flange face or the web face of the column. On framing elevations, the heavy line again indicates the backs of angles or channels being used as columns and the working lines of angles or structural tees using as bracing.

At certain times, a heavy line may not, by itself, give a completely clear picture of what the structural drafter wants to show. This is especially true on framing plans where channels or angles are being used as beams because the orientation of the channel flange or angle leg might be on either side of the working line. For these situations, the accepted practice is to show a short portion of the structural member along the working line in double-line representation. Figures 6–2, 6–3, and 6–4 illustrate standard symbols used to represent structural steel members in plan and elevation on structural steel design drawings.

Notice on Figure 6–3 that, when drawing structural angle bracing in single-line representation, the drafter must show a short section of the angle in double line to indicate the orientation of the horizontal leg of the angle. The horizontal leg of the 6 × 4 × ¼ angle is shown turned toward the viewer. Also notice the letters LLV on the 6 × 4 × ¼ angle, which indicate *long leg vertical.* The 4 × 4 × ¼ angle indicates that the viewer is looking at the back of the angle with the horizontal leg turned away. It is important to show this when drawing diagonal cross bracing or "X" bracing because the angles must pass each other back-to-back for clearance.

Connection Representation on Structural Steel Framing Plans

When preparing a structural steel framing plan, the drafter should indicate the structural connections as completely as possible. Although this can be done on the framing plan, it is impossible to clearly show every connection because the same symbol on a framing plan may indicate two, or sometimes even three, possible connection arrangements, as shown in Figure 6–5. Thus, the structural drafter must often clarify structural connections by drawing details and or sections, which will be discussed later in this chapter.

Beam Symbols. The single-line symbols in Figure 6–5 indicate in plan view how W-shape or wide-flange beams fasten into a W-shape column. The double-line elevation views under each symbol show that the beams could either be bolted to a steel cap plate on top of the column (Figure 6–5A) or fastened to the column flanges directly with clip angle connections (Figure 6–5B). In either case, the single-line representation would be the same, so a detail would be required.

A very important point to notice in Figure 6–5 is that the single heavy lines representing the center lines of the beams stop short of touching the column. On most structural steel framing plans, the open distance between the end of the

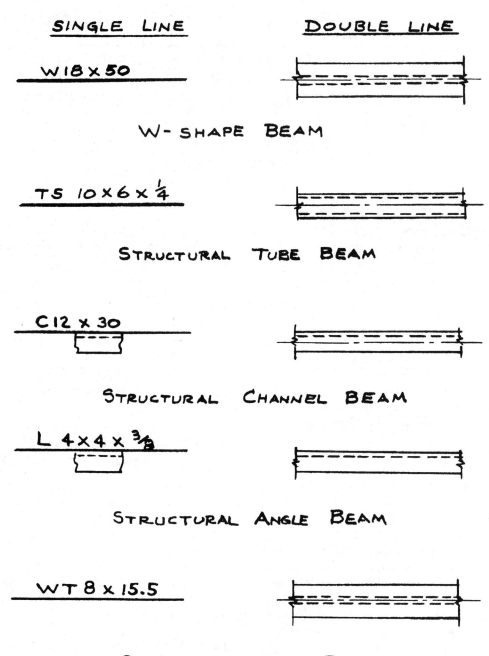

Figure 6–2 Standard structural steel symbols in plan view

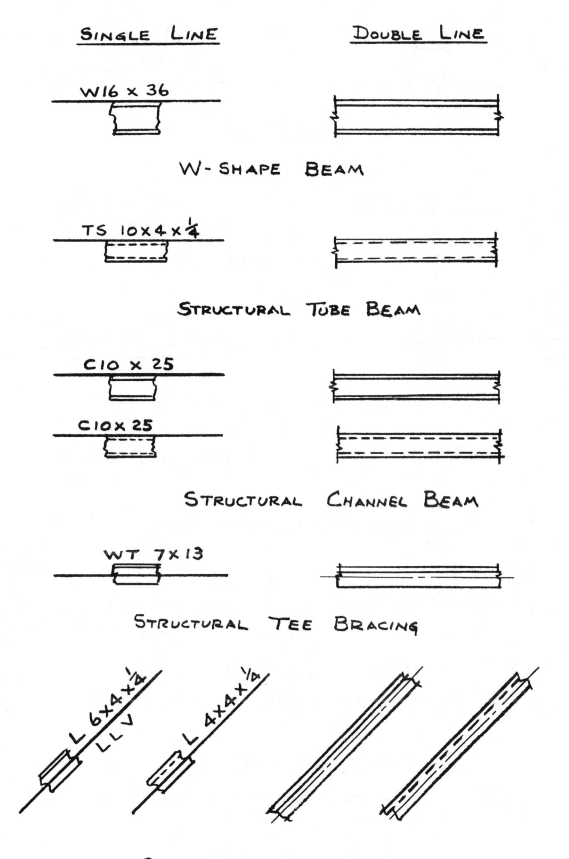

Figure 6–3   Standard structural steel symbols in elevation view

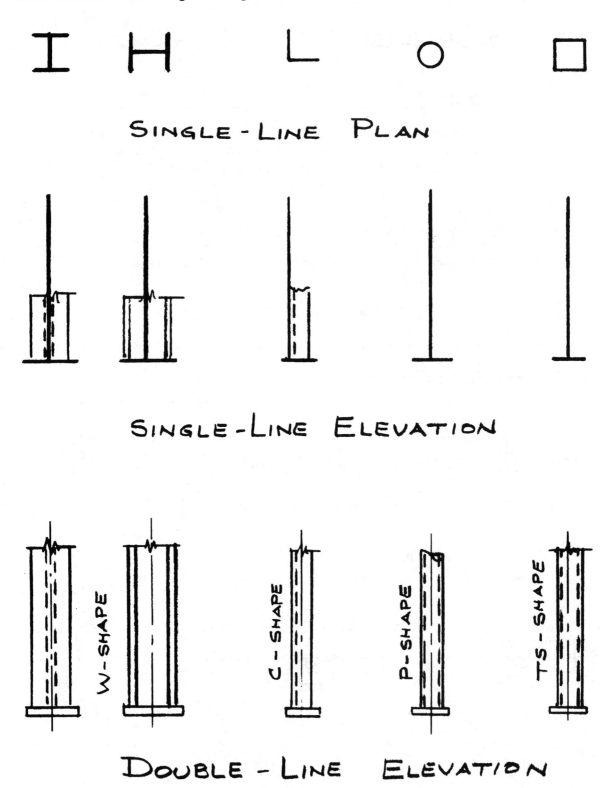

Figure 6–4   Standard structural steel column symbols in plan and elevation

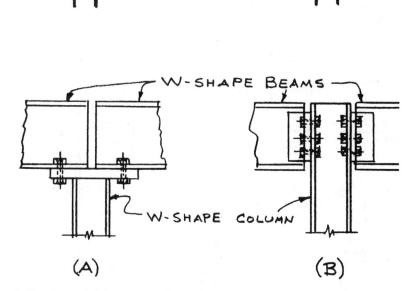

Figure 6–5   Standard structural steel column symbols in plan and elevation

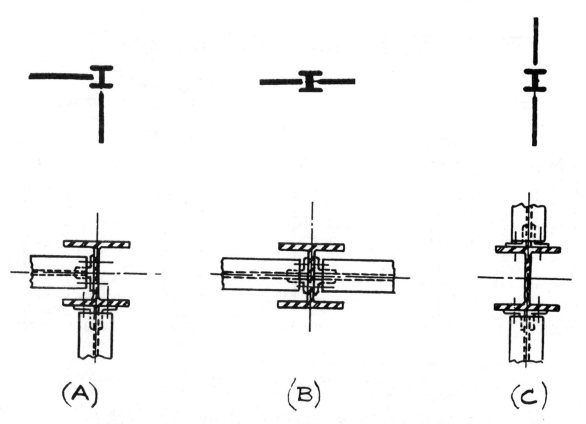

Figure 6–6   Standard structural steel beam-to-column symbols

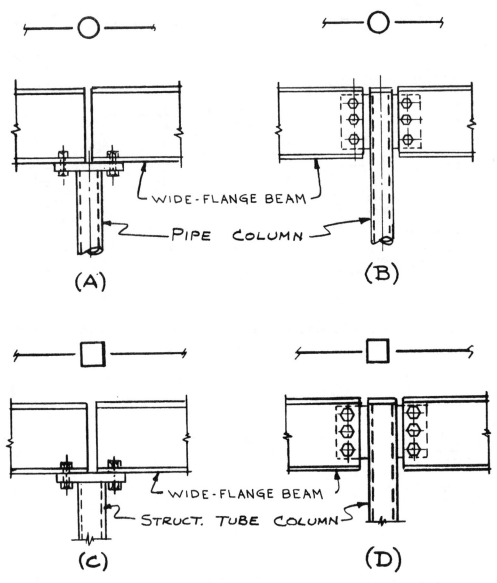

Figure 6–7   Standard structural steel beam-to-column symbols

beam and the column to which it connects will be approximately ¹⁄₁₆″. Figure 6–6 shows examples in single and double line of typical beam-to-beam connections in which the beams are connected to either the flanges or web of the column with framing angles.

Column Symbols. Beams framing into structural pipe or tube columns are also shown by symbol on framing plans. But once again, the symbol does not clearly indicate the type of connection desired. Figure 6–7 illustrates how symbols of W-shape beams connecting to pipe or tube columns might indicate that the beams could fasten either to a column cap plate or to a shear plate (sometimes called a web plate) welded to the side of the column.

Many times the drafter must indicate that a beam should extend or overhang beyond its column support. This is done

in single-line plan view through symbols like those used in Figure 6–8. This illustration shows why it is so important that the heavy lines indicating beams not touch the column in the framing situations in Figures 6–6 and 6–7.

A type of connection frequently illustrated on structural steel framing plans is that of beams connecting to beam pockets in walls. A *beam pocket* is an open space left in either a poured concrete or concrete block wall for the future installation of a structural steel beam. Another commonly used type of connection is that of beams connecting to structural steel wide-flange girders. When illustrating this type of connection, the drafter always stops the beam just short of the girder if the beams frame into the girder web. Figure 6–9A illustrates a beam-to-girder connection as it would look in single-line plan view and double-line elevation. Figure 6–9B illustrates how a beam framing into a

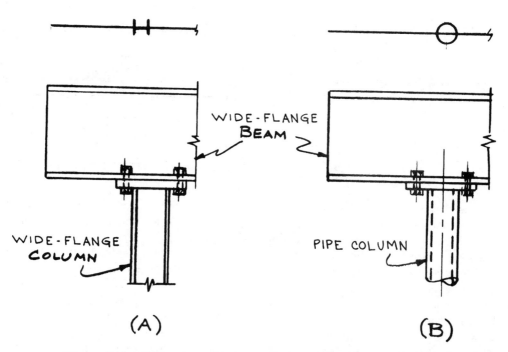

Figure 6–8   Beam-to-column symbols for overhanging beams

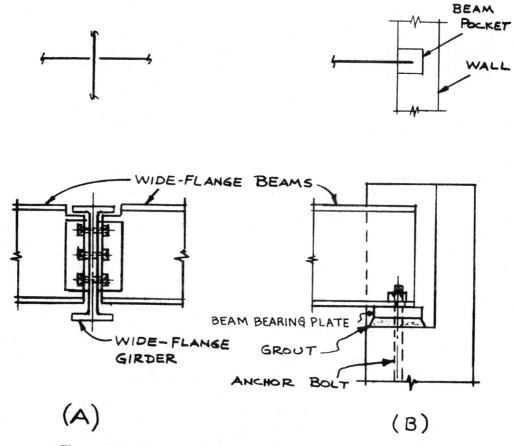

Figure 6–9   Beam-to-girder and beam-to-beam pocket symbols

Figure 6–10   Partial steel framing plan with overhanging beams

beam pocket would be shown in single-line plan view and double-line elevation. In the elevation view, notice that the beam bearing plate, anchor bolts, and grout are shown. *Grout* is a non-shrink cement-mortar material, usually between ¼″ and 1 ½″ thick, which is packed between the bottom of the beam bearing plate and the poured concrete or concrete block wall to ensure a level and solid contact between the wall and the bottom of the beam bearing plate.

When beams overhang their column supports as previously discussed, they often connect to intermediate beams that may or may not be the same depth as the overhanging beam. Figure 6–10 illustrates two W6 × 26 beams overhanging W8 × 31 columns with a shorter W16 × 26 fastened to the overhanging beams with flexible shear connections. The usual overhang is 4′–0″ to 6′–0″. In this very common situation for structural steel roof framing systems, the negative moments caused by the loads on the overhanging beams offset the maximum positive moments found between the columns. This makes the value of the positive moments less than they would otherwise be and results in a smaller, lighter, and thus more economical beam. Figure 6–11 illustrates in single-line plan view and double-line elevation view how the flexible connection between the overhanging beams and the intermediate beams can be made with either structural steel angles or splice plates.

Joist Symbols. Other very common symbols shown on structural steel framing plans are those representing open-web steel joists. There does not seem to be a common symbol presently in use for steel joists other than the fact that joists are usually shown as thinner lines than those used to represent W-shapes or structural tube. The symbols representing open-web steel joists, which indicate the center line of the joist, are sometimes a thin solid line, sometimes a thin line made up of one long and two short dashed lines, and

sometimes even thin double lines. The main point to keep in mind is that a definite contrast should be discernible between the lines representing steel joists and the lines representing the other structural steel members such as W-shapes that support the joists.

If open-web steel joists are to be supported on load-bearing walls, the walls are usually shown with lighter lines than the joists because the walls are a secondary feature. Joist sizes, spacings, and locations should always be designated on the structural steel framing plan. The required rows of joist bridging are also shown in their proper locations, usually by dashed lines, and the bridging is identified. Figure 6–12 shows one method of designating open-web steel joists on structural framing plans. Detail ⓵⁄ₛ₃ below the plan view illustrates in elevation how the structural drafter envisions the joists being set on the beams.

Detail Reference Symbols. As previously stated, the structural framing plan can never show enough information to tell the ironworker exactly how the structural steel is to be erected, thus structural details must be drawn. *Details* are larger-scale drawings that clearly illustrate such information as how structural steel beams connect to columns, how W-shape or structural tube columns connect to footings or walls, or how joists will connect to beams and/or columns. Details are usually drawn at scales of ½″ = 1′–0″ to 1 ½″ = 1′–0″. Details are cross-referenced to the structural framing plan by the use of detail symbols. The detail symbol usually consists of a heavy cutting plane line that indicates the location on the framing plan at which the detail is taken and then the detail symbol itself.

The detail symbol is usually drawn as a circle ½″ in diameter, partially enclosed in an arrowhead that indicates the direction of the detail. The detail cross-reference circle is also divided into two parts, with the number of the detail

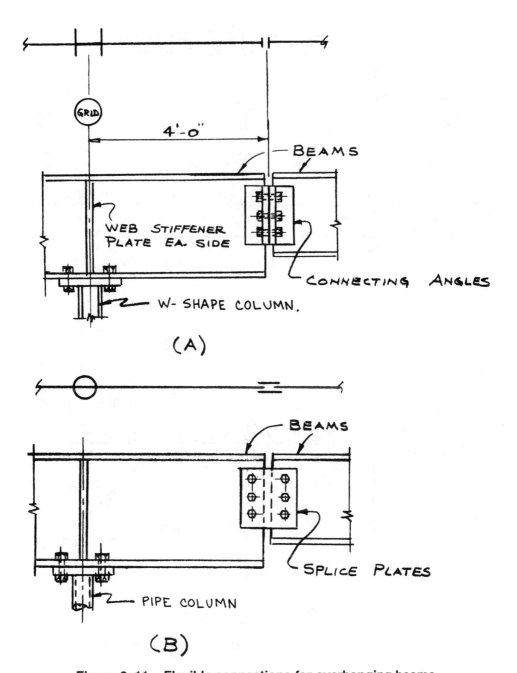

Figure 6–11   Flexible connections for overhanging beams

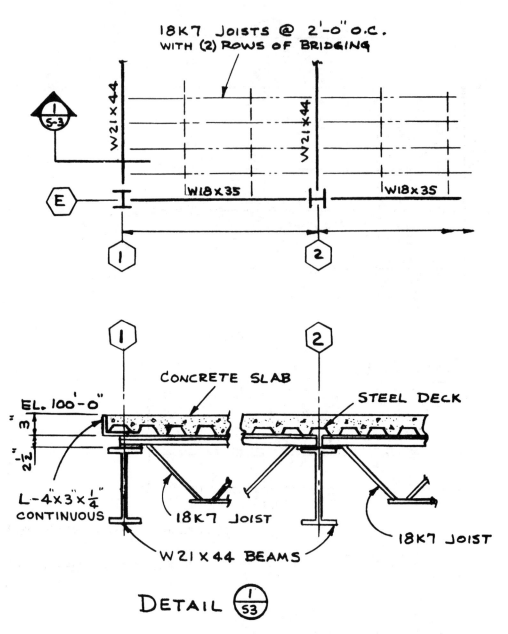

Figure 6–12   Partial structural steel framing plan and detail

in the upper half and the number of the drawing sheet where the detail can be found in the lower half. Figure 6–13 illustrates a detail reference symbol being used to show how structural steel beams are to be connected at a roof corner to each other and to the top of a structural steel tube column.

Support Angle Symbols. Another symbol commonly found on structural steel framing plans represents the *support angle*, sometimes called a shelf angle, which is used to support the end of the steel deck at the juncture of the

floor and the foundation wall. Figure 6–14 illustrates the support angle on the framing plan (top drawing) and how the support angle will be fastened to the foundation wall as well as how it will support the end of the steel deck (bottom detail drawing).

Notice in the framing plan view in Figure 6–14 that the 12K3 open-web steel joists are supported on the ends by the foundation wall itself and not supported by structural steel beams as shown in Figure 6–12. It is very common for the joists to bear directly on either the foundation walls or on

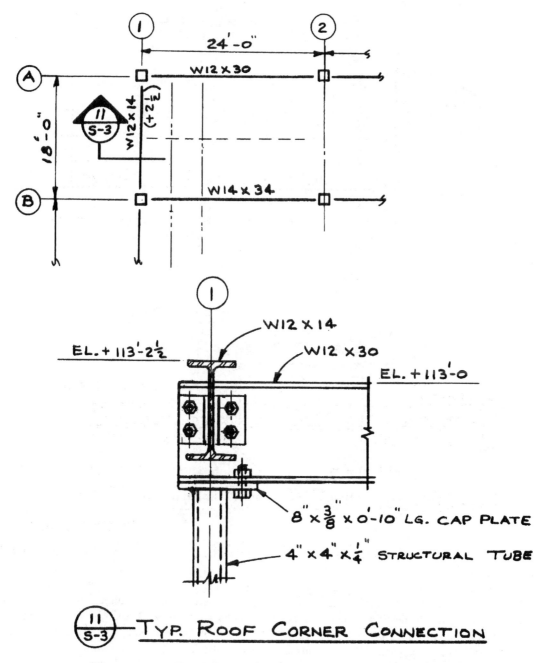

Figure 6–13   Partial structural steel framing plan and detail

masonry walls. Figures 6–15 and 6–16 illustrate how framing plans indicate open-web steel joists being supported by either cast-in-place foundation walls or masonry walls. In Figure 6–15, the deck is supported by a support angle at the foundation end wall parallel to the steel joists, while in Figure 6–16, a steel joist set very close to the masonry end wall supports the end of the steel deck.

Weld Symbols. The detail in Figure 6–15 shows the steel joist supported at the top of the foundation wall by small fillet welds at the top of a 6″-long 4″ × 3″ × ¼″ structural steel angle embedded in the poured concrete foundation wall. Weld symbols will be discussed in detail in Part 2 of this book, but for now, we will explain that Figure 6–15 indicates two small welds connecting the steel joist to its supporting

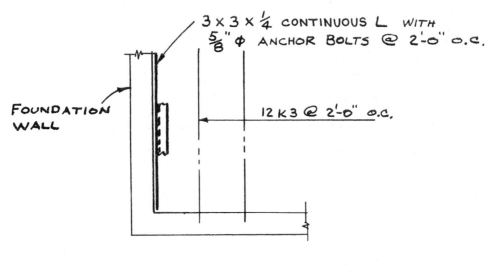

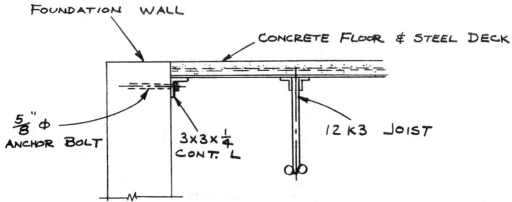

Figure 6–14   Partial structural steel framing plan and detail

angle. One weld will be on the front side and one on the back side of the joist, and each weld will be a ⅛″ fillet weld 2″ in length. The black flag on the weld symbol indicates that the weld is to be made "in the field," or on the job site as the joists are being erected. The detail also indicates the 4″ × 3″ × ¼″ angle is to be embedded into the foundation wall with the 4″ leg in the horizontal position because this is the leg the drafter wants the steel joist bearing upon and thus welded to.

Figure 6–16 shows a steel joist supported by a continuous weld plate that has been embedded in a bond beam at the top of a masonry wall. The weld plate, also called a bearing plate, has hooked anchor rods welded to its underside and wrapped around one of the #4 (½″-diameter) reinforcing steel bars embedded in the bottom of the bond beam. Notice that the elevations at both the top of the bearing plate and the top of the steel roof deck are given.

Lintel Symbols. The last type of symbol to be discussed is the symbol denoting structural steel lintels. A *lintel* is a

horizontal member or beam placed above a wall opening (such as a door or window) to carry the weight of the wall above the opening. Structural steel shapes are very commonly used as lintels; angles are suitable for spanning smaller openings, while steel beams and plates can support wide openings or openings with heavy loads. Figure 6–17 shows details of lintels supporting various types of masonry walls above openings for steel door frames.

On structural steel framing plans the symbol used to indicate a steel lintel is usually a heavy, dark line, either solid or dashed, with a lintel identification number such as L–1, L–2, L–3, etc., which identifies the type of lintel to be used over that opening. On the framing plan, the window or door opening itself is represented by thin, light lines because in this instance the building walls are considered secondary to the steel structure. An example of a lintel symbol is shown in Figure 6–1 on grid line Ⓐ between structural tube columns Ⓐ1 and Ⓐ2 .

The lintel identification numbers on the framing plan refer to how those lintels are identified on the lintel sched-

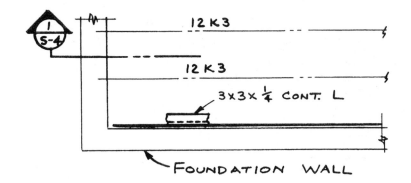

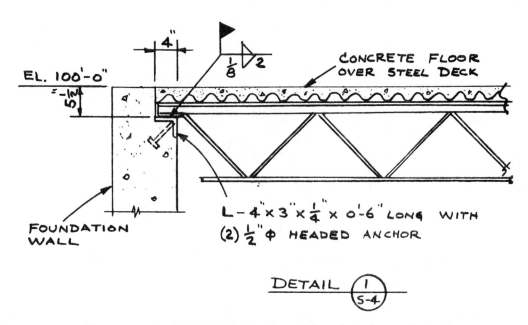

Figure 6–15   Partial structural steel framing plan and detail

ule. For each type of lintel, a *lintel schedule* shows the lintel number, a small section view of the makeup and sizes of the beams, angles, and plates from which the lintel will be constructed, the length of wall bearing required for the lintel if it is to be placed on a wall and, usually, the elevation at the bottom of the lintel after it is installed. Note that, for steel lintels, unlike steel beams, the bottom-of-steel elevation is the most important and must be held if the door and window frames under them are to fit properly. Most lintel schedules will also have a column for special remarks such as "Bear on solid masonry." An example of a lintel schedule is shown in Figure 6–18.

Now that we have discussed structural steel shapes and the symbols used to represent them on framing plans, we will examine several specific types of structural steel framing plans.

Types of Structural Steel Framing Plans

Good structural framing plans, like all construction drawings, must be easily understood, accurate, neat, orderly, complete, informative, and instructive. Thus, the structural drafter must take care not to clutter the drawing with unnecessary lines.

Foundation Plans. Usually the first plan drawn in the structural framing set is the foundation plan. The *foundation plan* is a plan view of the building's footings and foundation walls that also shows the grid system and all the structural steel columns setting on top of the walls or extending down to the continuous wall footings or independent column footings. Top- or bottom-of-footing elevations

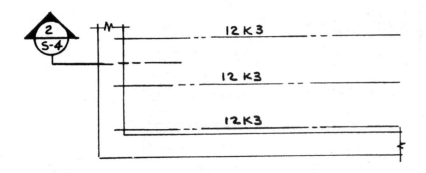

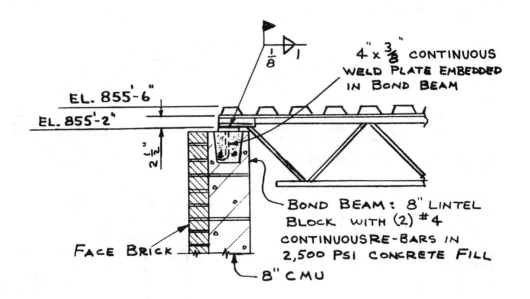

Figure 6–16 **Partial structural steel framing plan and detail**

are noted, and the independent column footings, which are usually square in shape, are identified with a reference number such as F–1, F–2, F–3, etc. These numbers refer to a *footing schedule,* which gives complete information about each type of footing, including size, depth, and size and quantity of reinforcing bars. The floor slab thickness (either slab-on-grade or basement slab) is noted on the foundation plan. Figure 6–19 illustrates how information is shown on a foundation plan.

Small Floor Framing Plan. Figure 6–20 shows a small floor framing plan. Notice that the intermediate beams in

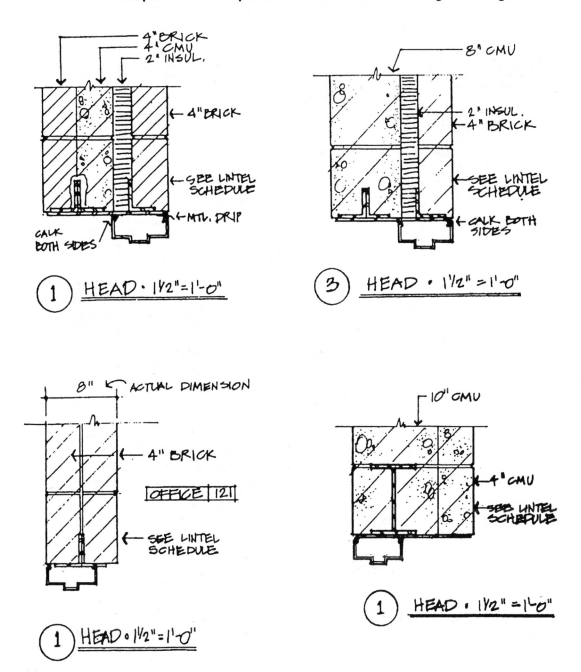

Figure 6–17   Structural steel lintels for masonry walls

LINTEL SCHEDULE						
NO.	SECTION	MEMBERS	RO	BRG EA END	BRG EL	REMARKS
L-1	I	W 8 X 10 1/4″ X 11″ PLATE		16′	108′-0′	
L-2	⌐L	2-31/2″ X 31/2″ X 1/4″ ∠s		8″	108′-0″	
L-3	⌐LL	3-4″ X 4″X 1/4″ ∠s		8″	108′-0′	
L-4	I	W 8 X 10 1/4″ X 11″ PLATE		6′	108′-0″	BEAR ON SOLID MASONRY
L-5	I	W 8 X 10 1/4″ X 7″ PLATE		8′	108′-0″	
L-6	I	W 8 X 15 1/4″ X 7″ PLATE		12′	108′-0″	BEAR ON SOLID MASONRY
L-7	⌐LL	3-3 1/2″ X 3 1/2″ X 1/4″ ∠s		8″	108′-0″	
L-8	I	W 8 X 10 1/4″ X 13″ PLATE		12′	VARIES	SEE NOTES 1, 3, 4, 5
L-9	I	W 8 X 24 1/4″ X 11″ PLATE		10′	109′-4″	BEAR ON SOLID MASONRY
L-10	I	W 8 X 28 1/4″ X 13″ PLATE		11′	112′-0″	BEAR ON SOLID MASONRY

Figure 6–18   Structural steel lintel schedule

each bay are W14 × 26 wide-flange beams spaced at 6′-0″on center rather than open-web steel joists spaced at 2′-0″ centers. The concrete floor slab depth, the welded wire fabric reinforcing for the slab, and the top of slab elevation are all given by notation. Columns are shown, but column information is not given, which indicates it can be found on a column schedule on another sheet. Notice also how all beams, girders, and columns are shown with the single-line symbolic representation previously described.

If the structural drafter were preparing a floor framing plan for composite floor construction, as illustrated in Figure 5–6, it would be prepared very much like the one in Figure 6–20 except that the quantity of shear studs required for each beam and girder would be called out as shown in Figure 6–21.

Partial First-Floor Framing Plan. Figure 6–22 is a partial first-floor framing plan for a commercial office building. This is an open-web steel joist floor framing system as illustrated in Figure 5–4. Notice that the office building has tube columns on the side walls, but all interior and end wall columns are W-shape columns. Column sizes are not given,

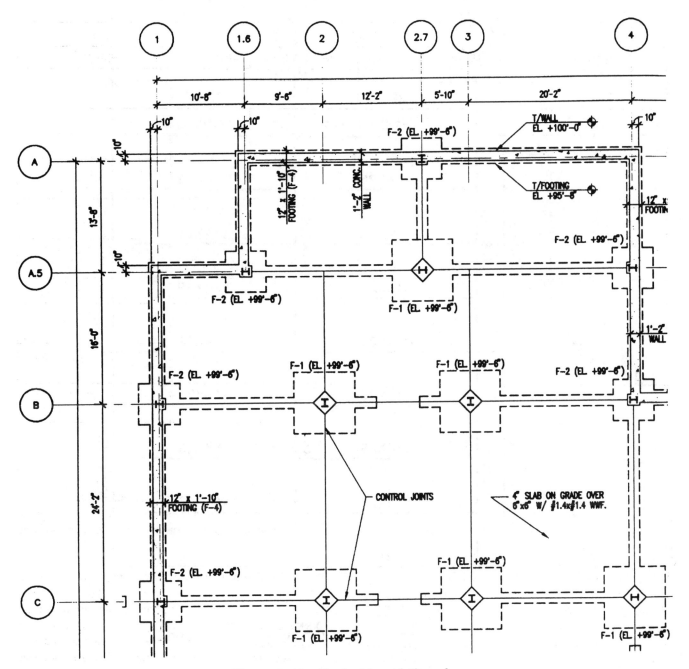

Figure 6–19   Typical foundation plan

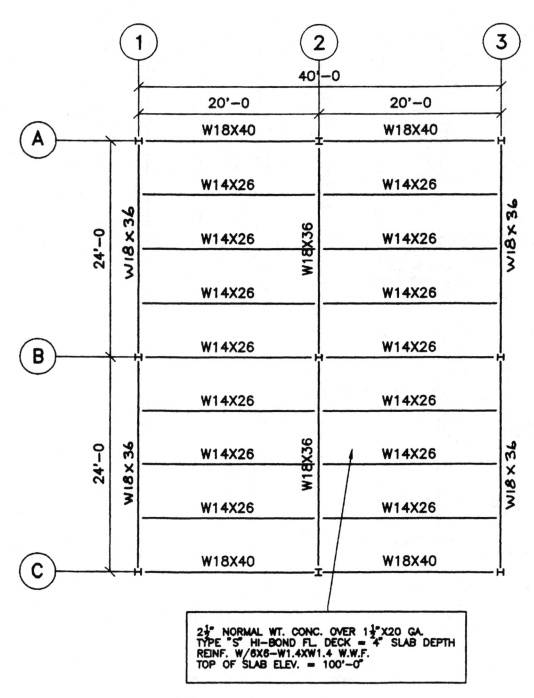

Figure 6–20   Structural steel framing plan

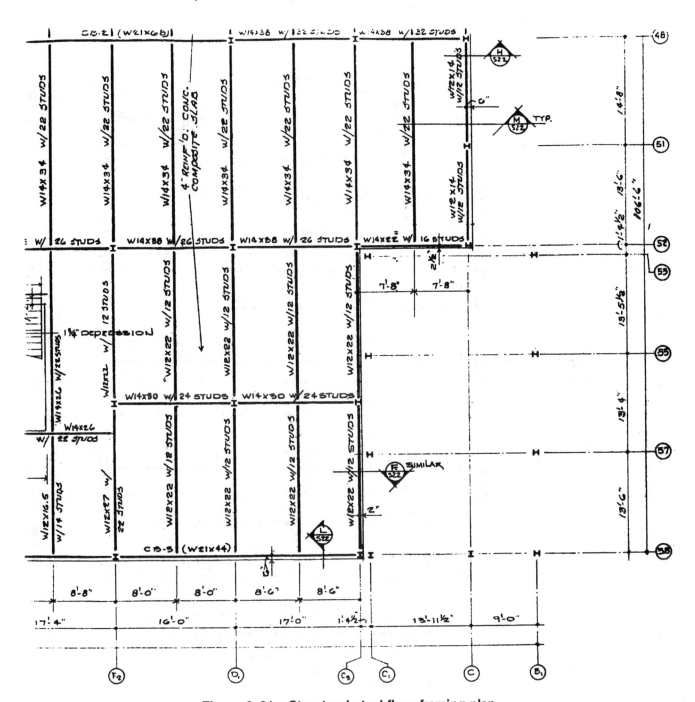

Figure 6–21   Structural steel floor framing plan

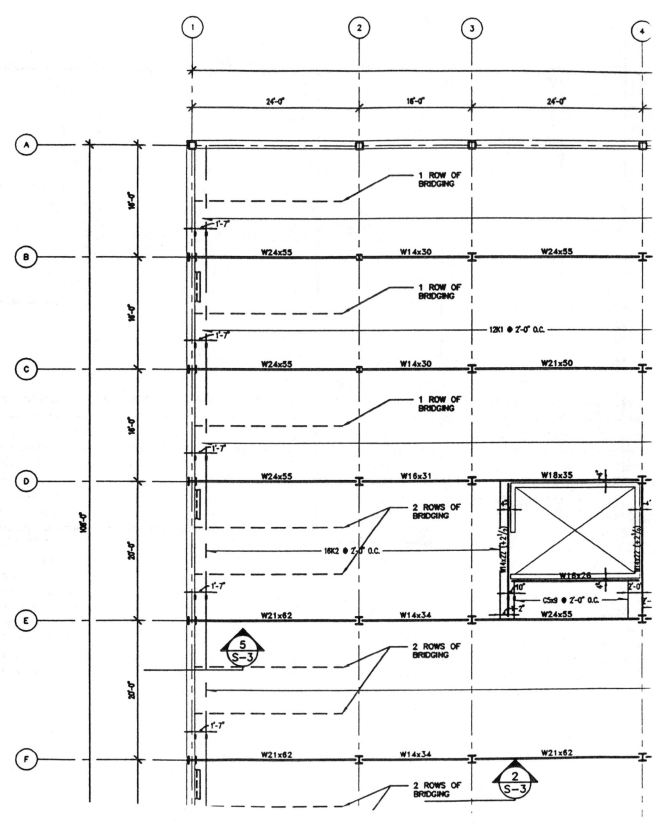

Figure 6–22 Structural steel floor framing plan

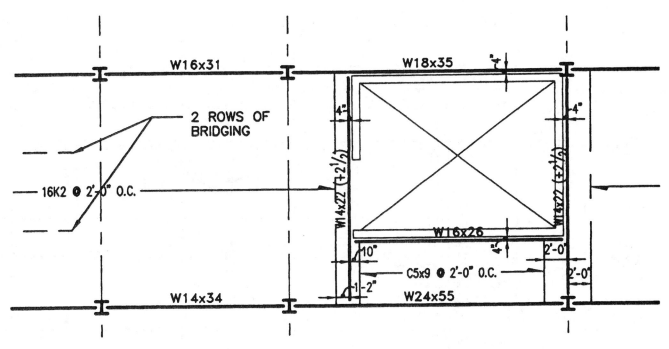

Figure 6–23   Partial structural steel floor framing plan

which means they will be listed on a column schedule. A *column schedule* is really a table showing such items as column size, baseplate size, floor-to-floor elevations, and sometimes even the quantity of anchor bolts required for the column. Column schedules will be discussed in more detail in Chapter 7.

Notice on this floor framing plan that all columns are located on the structural grid. The open-web steel joists are to be spaced at 2'–0" on center. The number of rows of bridging required for the open-web joists is indicated, and shelf angles are shown on the end walls to support the floor deck. Notice between grid lines Ⓓ and Ⓔ and ③ and ④ how the steel framing around a stairwell is shown. The enlarged view of this framing in Figure 6–23 shows that the stairwell is surrounded by wide-flange beams set very close to the walls. These beams, the webs of which are only 4" from the walls of the stairwell, support the steel decking and the end of several C5 × 9 joist substitutes along one wall.

Second Floor Framing Plan. Structural steel framing systems can support various types of floors. Figure 6–24 shows a framing plan for the second floor of a building in which the structural steel beams support a floor composed of 8" hollowcore precast concrete slabs with 2" of cast-in-place concrete topping. Like steel joist and deck systems, precast concrete slabs or plank make very efficient floor systems for apartment buildings, office buildings, and schools. Notice on this framing plan that the hollowcore precast plank are given an identification number such as PC1, PC2, PC3, etc., and a double arrow indicates the direction in which the precast plank will span between support beams. The W-shape beams supporting the hollowcore precast plank, such as the W24 × 55 and W18 × 35, are the largest beams. The W14 × 22 and W12 × 16 beams are basically non-loadbearing and are used primarily to help stabilize the columns. Figure 6–25, showing Detail ③⁄ₛ₋₂ on the floor plan, illustrates how weld plates embedded in the bottom of the

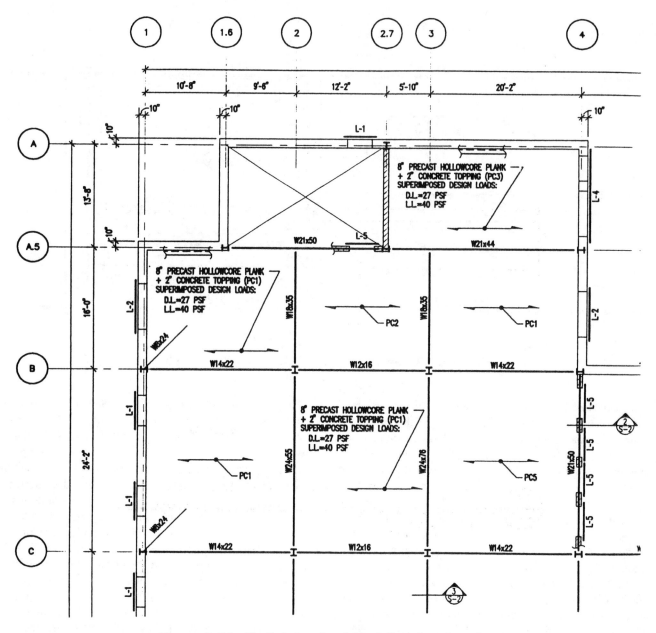

Figure 6–24   Partial structural steel floor framing plan

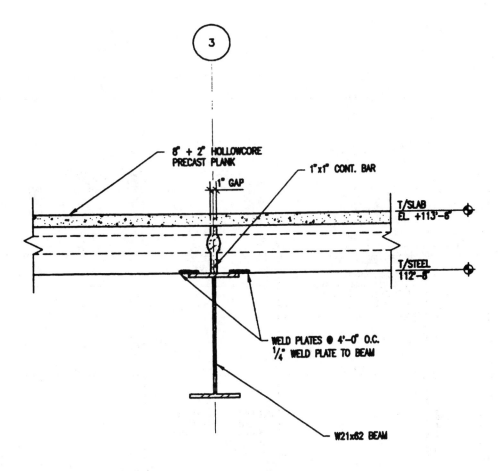

TYP. PLANKS ON BEAM DETAIL

$\frac{3}{S-2}$ SCALE: 1" = 1'-0"

Figure 6–25   Typical plank-on-beam detail

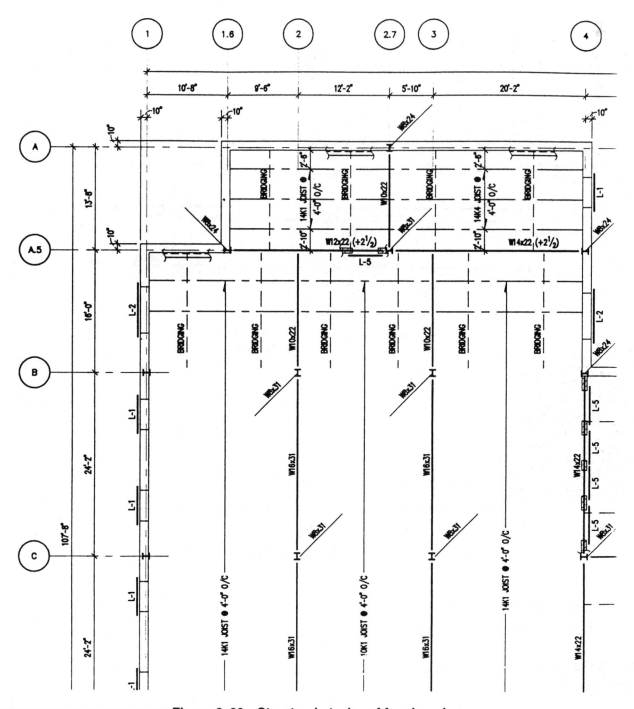

Figure 6–26   Structural steel roof framing plan

hollowcore plank can be used to secure the plank to their supporting beams.

Roof Framing Plans. Figure 6–26 shows a roof framing system for a small commercial building. Notice that the open-web roof joists are spaced at 4'–0" centers. The required bridging is again shown by dashed lines. The build-

ing walls below the roof are indicated, and lintels are identified over the windows.

Figure 6–27 illustrates a roof framing plan for a commercial building. Notice that this plan shows no columns and thus has no grid system. That is because the roof joists are supported by bond beams on loadbearing walls. Window openings and lintels are indicated and the 22K5 steel joists

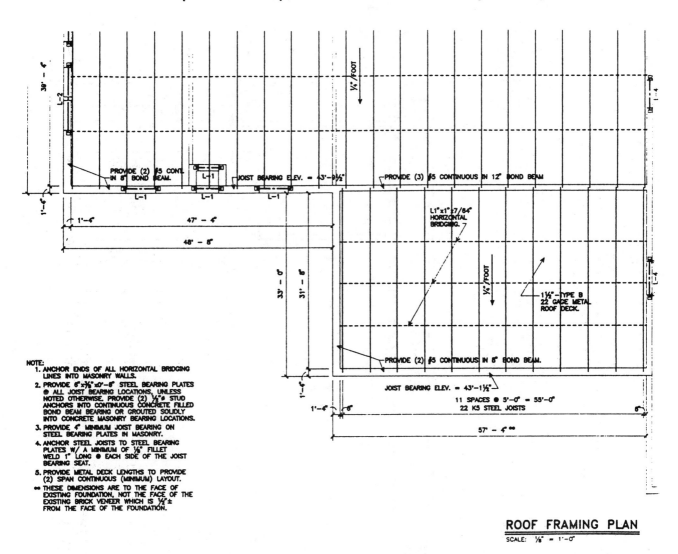

NOTE:
1. ANCHOR ENDS OF ALL HORIZONTAL BRIDGING LINES INTO MASONRY WALLS.
2. PROVIDE 6"x⅜"x0'-8" STEEL BEARING PLATES @ ALL JOIST BEARING LOCATIONS, UNLESS NOTED OTHERWISE. PROVIDE (2) ½"ø STUD ANCHORS INTO CONTINUOUS CONCRETE FILLED BOND BEAM BEARING OR GROUTED SOLIDLY INTO CONCRETE MASONRY BEARING LOCATIONS.
3. PROVIDE 4" MINIMUM JOIST BEARING ON STEEL BEARING PLATES IN MASONRY.
4. ANCHOR STEEL JOISTS TO STEEL BEARING PLATES W/ A MINIMUM OF ⅛" FILLET WELD 1" LONG @ EACH SIDE OF THE JOIST BEARING SEAT.
5. PROVIDE METAL DECK LENGTHS TO PROVIDE (2) SPAN CONTINUOUS (MINIMUM) LAYOUT.
∞ THESE DIMENSIONS ARE TO THE FACE OF EXISTING FOUNDATION, NOT THE FACE OF THE EXISTING BRICK VENEER WHICH IS ½"± FROM THE FACE OF THE FOUNDATION.

ROOF FRAMING PLAN
SCALE: ⅛" = 1'-0"

Figure 6–27   Structural steel roof framing plan

on the lower part of the building are shown starting right at the end walls, eliminating the need for deck support angles.

6.5 STRUCTURAL STEEL SECTIONS

On large projects, it is often necessary to make *section drawings* to clarify how the components of a structure fit together and to establish important height information such as top-of-steel elevations for floor and roof framing systems. Just as a framing plan cuts through the building in an imaginary horizontal cutting plane, a section slices through the structure at a vertical plane. Framing plans are parallel to the floor, while sections are perpendicular to the floor. Structural steel sections can be drawn either by single-line symbolic diagrams as in a framing plan or as larger-scale dou-

ble-line sections that clearly illustrate such features as beam and column sizes, top-of-steel and bottom-of-baseplate elevations, anchor bolt information, and the location of columns on foundation walls. Figure 6–28 shows a larger-scale double-line section. Such sections are usually drawn at scales of ½" = 1'–0" or ¾" = 1'–0".

One cardinal rule in structural drafting and design, in addition to the fact that the support system must safely carry all the loads, is that the structural components must fit together properly. Thus, it is much more desirable to find potential problems and interferences on the design drawings than to have them show up during the construction phase of the project. To that end, important to a set of structural design drawings are the *structural building sections*. The structural building sections serve as a way to verify that the various floor and roof framing systems will be compatible.

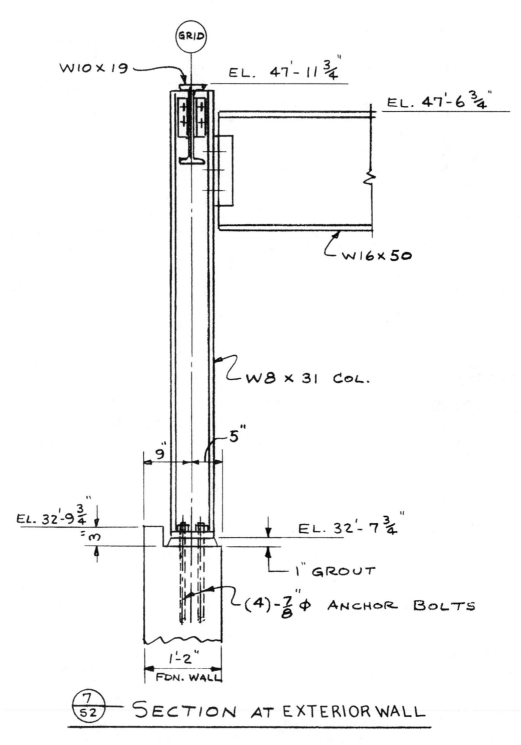

GRID

W10 × 19

EL. 47'-11¾"

EL. 47'-6¾"

W16 × 50

W8 × 31 COL.

5"

9"

EL. 32'-9¾"

3

EL. 32'-7¾"

1" GROUT

(4)-⅞"⌀ ANCHOR BOLTS

1'-2"

FDN. WALL

7/52 — SECTION AT EXTERIOR WALL

Figure 6–28   Structural steel section

These sections, also called full sections or cross-sections, are usually drawn to a small scale such as ⅛″ = 1′–0″. They are drawn in single-line symbolic representation and are taken at right angles to the major axes of the building. Typically, structural building sections cut through the entire length or width of a building and show only the main structural components. Very few dimensions are given other than the column grid dimensions and the top-of-steel elevation at each floor above the bench mark. Figure 6–29 shows a building section through a commercial or industrial steel-framed building structure. Note that, while most sections cut through the building in a straight line, the building cut does

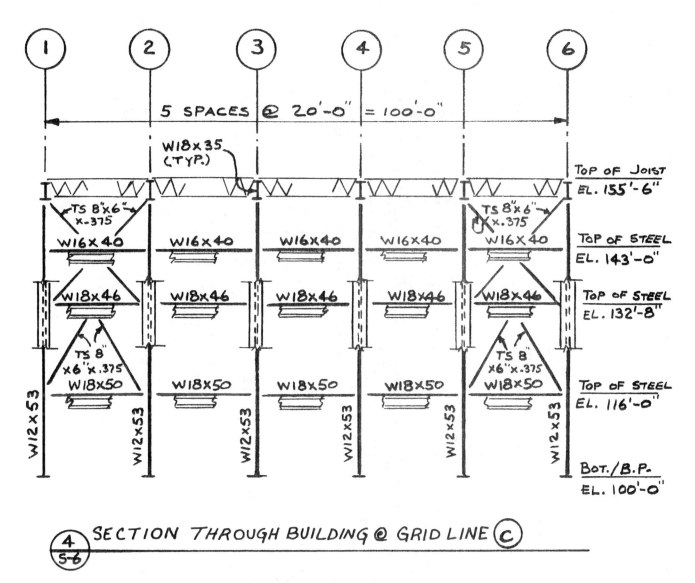

Figure 6–29   Structural steel building section

not necessarily have to run straight through the structure. It can jog back and forth as required to show interior space conditions that require clarification.

6.6 STRUCTURAL STEEL DETAILS

Structural details are a very important part of a set of structural design drawings because it is the all important detail that shows the contractor exactly how various structural members are to be connected together on the job site. Details are usually drawn at scales of ½″ = 1′–0″ to 1 ½″ = 1′–0″. Some examples of what might be shown on a structural detail are how a steel column is to be connected to the top of a footing by anchor bolts through the column base-

plate, how steel joists are to be located and field welded to the top of a wide-flange beam, or how a structural steel tube column is to be set in a column pocket in a foundation wall. Figure 6–25 is a detail that illustrates how hollowcore precast concrete plank were to be fastened to their supporting steel beams by weld plates, which would be cast into the bottom of the plank and then welded to the top flange of the beam. Figure 6–30 and 6–31 are two examples of structural details. Figure 6–30 illustrates that each of two anchor bolts supporting a W-shape steel column is to be furnished with two hexagon head nuts. The nut placed on the underside of the column baseplate is to be used as a "setting nut" or "leveling nut" to level up the column. After the column is level, the nut on top of the baseplate can be tightened to secure the column in place. Figure 6–31 is a detail showing how a K-series open-web steel joist is to be fastened to a bearing

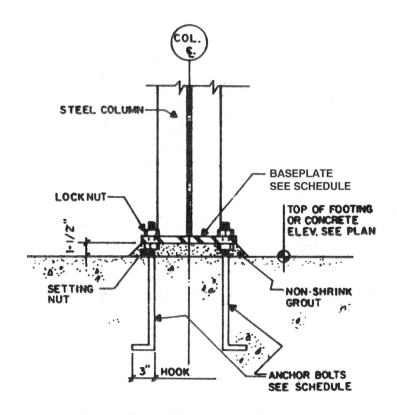

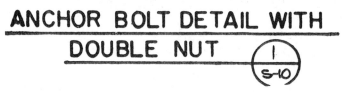

Figure 6–30 Structural steel column-to-footing detail

plate at the top of a 12″ concrete block wall with two ⅛″ × 1½ inch ling fillet welds made at the job site. These two excellent examples illustrate the importance of structural details. Figures 6–12, 6–13, 6–14, 6–15, 6–16, 6–17, and 6–25 are also very good examples of structural steel details.

6.7 INFORMATION REQUIRED ON STRUCTURAL STEEL DESIGN DRAWINGS

When preparing structural steel design drawings, the structural drafter must always be sure to show certain key information. This information is needed so that the general contractors can develop a competitive bid and the structural fabricators can prepare both their bids and the required shop or fabrication drawings. Following is a checklist of the specific information that should be shown on structural steel design drawings. Where the information might be found in a set of drawings is indicated in parentheses.

1. The center-to-center dimension of all columns. (Framing plans grid system)
2. The distance from the center line of the exterior columns to the outside of the exterior walls. (Framing plans)
3. Elevation at the tops of beams in relation to the finished floor line. (Framing plans, sections, and details)
4. When exterior walls are loadbearing, the distance from the outside of the wall to the end of the beam (Section or details)
5. The center-of-web location for all beams that may not line up with the beams located on the column grid lines. Examples would include beams around an elevator shaft or stair opening or beams that need to be

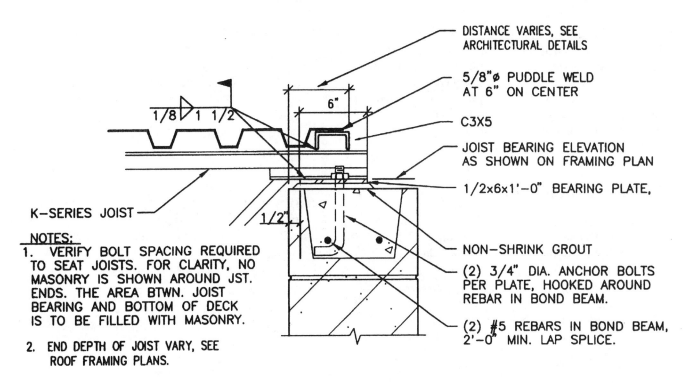

JOIST BEARING DETAIL
1 1/2" = 1'-0"

Figure 6–31   Structural steel joist bearing detail

off the center of columns due to exterior wall construction requirements. (Framing plans)

6. Steel joist sizes, spacings, and the required rows of bridging. (Framing plans)

7. When identifying beams, the nominal size (depths and weight per lineal foot), such as W14 × 31, W16 × 36, or W18 × 50. (Framing plans, sections, and details)

8. For composite construction, the size and quantity of shear studs. (Framing plans and details)

9. The required elevation at the bottom of column baseplates. Also, if grout is to be used between the bottom of the baseplate and the top of a wall or footing, the thickness of the grout must be shown. (Column section or detail)

10. Top or bottom of footing elevation and/or top of foundation wall elevation. (Foundation plan and/or sections and details)

11. Sizes of beam bearing plates and column baseplates. (Sections, details, or column schedules)

12. The quantity and size of anchor bolts. (Sections and details)

13. End reactions for beams when anything other than standard connection angles are required by loading or special conditions. (Framing plans)

14. Connection details. (Details)

15. Elevations at the bottom of lintels. (Sections or details)

16. The type of roof and floor construction. (Framing plans, sections, and details)

6.8   SUMMARY

The purpose of this chapter has been to acquaint the structural drafting student with the accepted methods and procedures used in structural design offices to prepare a set of structural steel design drawings for a commercial or industrial building. Specific topics have included a review of the basic objectives of structural design, the structural grid system, the symbols used to represent structural components and connections, structural steel sections, structural steel details, and information required on structural steel design drawings.

At the conclusion of this chapter, the student should have a clear understanding of what constitutes a set of structural steel design drawings. He or she should also know what should be drawn on the structural plans and how the drawings should be prepared and cross-referenced to convey the information required for the structural fabricator and general contractor to fabricate and erect the structural steel for a building project.

STUDY QUESTIONS

1. One of the first steps in preparing the structural steel design drawings for a commercial building is to establish the location of the columns, which is referred to as the _____.

2. The spacings of columns for most commercial buildings fall into the _____ to _____ range, with _____ being about the maximum unless the designer intends to use high-strength steel or joist girders.

3. Why is it good structural economy when as many column spacings as possible are the same?

4. When dimensioning framing plans for most commercial or industrial buildings, virtually everything on both the architectural and structural drawings is ultimately located off the _____.

5. Define a structural steel framing plan.

6. Structural steel framing plans for commercial or industrial buildings are usually drawn at a scale of _____.

7. How are structural steel beams indicated on framing plans?

8. Why are the structural columns usually drawn oversized on framing plans?

9. What are gage lines for steel members?

10. What is a beam pocket?

11. Define a lintel and a lintel schedule.

12. Sketch the symbol for a beam cantilevered over a pipe column.

Chapter 7

Structural Steel Sections and Details: Some Practical Examples

OBJECTIVES

Upon completion of this unit, you should be able to:

- read and comprehend fundamental structural steel design connection details.
- prepare structural steel section and detail drawings for a variety of construction conditions.
- prepare a column schedule for a steel-frame building.

7.1 INTRODUCTION

The late Chicago architect Mies van der Rohe, one of the giants of twentieth-century architecture, has often been quoted as saying: "God is in the Details." The longer one works in architectural or structural design, the truer that statement becomes. A prime example occurred at the Hyatt Regency Hotel in Kansas City, Missouri, on July 17, 1981, when 114 people died and more than 200 were injured in the collapse of two suspended walkways onto a crowded dance floor. An investigation determined that the cause of the collapse was an incorrect design detail. Specifically, a hanger rod had pulled through a box beam on a fourth-floor walkway, causing it to crash down onto the second-floor walkway below, which in turn caused both walkways to fall onto the floor of the hotel's atrium, proving once again that in a set of structural design drawings the importance of the details cannot be overstressed.

It is the structural drafter, designer, or engineer with a clear vision of the sections and details required to show exactly how the architectural or structural parts of a building fit together, who will be a key player on any drafting or design team.

Chapter 6 focused on the proper method of creating framing plans for a structural system. Various types of framing plans were shown in standard, single-line symbolic representation. Because of their simplicity, single-line drawings cannot show the building contractor or ironworker exactly how the various components are to be connected. That information is better illustrated on structural sections and details.

The experienced structural drafter or designer usually has a fairly good idea of how the structural connections will be made as he or she prepares the initial framing plans. However, the beginning structural drafter tends to be more unsure. As a structural drafting student, the more knowledge you can gain about sections and details, the better prepared you will be for the world of work. With that in mind, this chapter provides complete explanations of the fundamental types of structural steel connections. It also contains numerous illustrations of actual design sections and details carefully selected to show the currently acceptable standards and practices applied in structural engineering offices and departments.

7.2 COMMON FEATURES OF SECTIONS AND DETAILS

The dividing line between sections and details is sometimes ambiguous. In fact, except for full sections cut through a building as illustrated in Figure 6–29, the terms *section* and *detail* are often interchangeable. . . . For example, what one structural drafter might call a column base section, another might identify as a detail. Structural steel connection details are usually drawn to a larger scale than sections because more can be shown. However, the key purpose of both is to show how structural steel members fit together. In addition, either a section or a detail should give height information such as the top-of-steel elevation for beams, girders, or joists, or the bottom-of-baseplate and top-of-footing elevation for columns.

The remainder of this chapter consists of selected sections and details designed to help the student visualize how the components of a steel-framed structure are assembled.

7.3 COLUMN-TO-BASEPLATE CONNECTIONS

Generally, structural steel columns are connected to their supporting foundation walls, footings, pedestals, or piers at the bottom by being welded to a baseplate. (Calculations to determine the required size and thickness of a column baseplate were reviewed in Chapter 5.) The column is then firmly fastened, through the baseplate, to the top of the footing or wall with heavy bolts called anchor bolts. Column baseplates must have four anchor bolts to meet OSHA requirements.

Use of Anchor Bolts

Anchor bolts for commercial and industrial structures are primarily designed to resist tensile stresses, which would occur if, for example, strong lateral wind forces against the side wall of a building were to cause the columns along the exterior wall to overturn. Thus, anchor bolts must not only be large enough in area to resist the anticipated tensile stresses, but they must also be embedded into the concrete deep enough to prevent the anchor bolt from being pulled out. Anchor bolts in some heavy industrial applications may require special considerations for large traverse shearing stresses. However, in most ordinary commercial buildings, anchor bolt shear is not a problem because the vertical loads are large enough that frictional resistance under pressure will be sufficient to withstand any probable direct lateral force.

Realistically, construction workers cannot be expected to finish the entire top of a foundation wall or several interior pad footings to the exact desired elevation and with an absolutely level surface. For that reason, the required bottom-of-column baseplate elevation is usually calculated to be from ¾" to 1½" or more above the top of its supporting footing, pedestal, or foundation wall. The column is then set in place over the anchor bolts and leveled with metal shims or leveling nuts, after which the anchor bolts are tightened

to secure the column in place. The space between the bottom of the column baseplate and the top of the footing, pedestal, or foundation wall is then filled with non-shrinking grout. *Grout* is a special epoxy or mortar-type compound that can be made into a dry pack, plastic, trowelable, or flowable consistency. Properly prepared and installed grout will ensure full contact between the bottom of the column baseplate and the top of its supporting footing, pedestal, or foundation wall.

Figure 7–1 illustrates a typical column baseplate-to-footing connection detail for a W8 × 31 structural steel column. The W8 × 31 column is welded to a 14″ × 1¼″ × 1′–2″-long baseplate, which is in turn fastened to the top of the pad footing with four ¾″-diameter × 1′–4″-long anchor bolts. Full contact between the bottom of the column baseplate and the top of the footing is ensured by a 1½″ thickness of grout. The diamond-shaped, 2′–0″-square concrete encasement around the column base is isolated from the 4″-thick concrete floor by a ½″ thickness of expansion material. Notice that, in addition to showing the column size, baseplate size, anchor bolt size and quantity, and grout thickness, this detail gives the top-of-footing elevation, top-of-floor elevation, and bottom-of-baseplate elevation. All of this is important information that should be shown on engineering design details.

Figure 7–2 is another example of a concrete spread or pad footing supporting a W8 × 31 column. In this case, the column load is transferred to the footing by a 12″ × 12″-square concrete pedestal reinforced with four #6 (¾″-diameter) vertical reinforcing steel bars that are tied together with #3 (⅜″-diameter) reinforcing steel ties spaced at 12″ on center. The pedestal, which is always larger than the column baseplate, helps spread the column load over a larger area of the footing, which reduces the unit stress on the concrete footing. The height of the pedestal is determined by the difference between the top-of-footing and the top-of-floor elevation. Again, column size, baseplate size, anchor bolt size and quantity, and grout thickness are all shown on the detail.

Figure 7–3 illustrates a baseplate detail for a structural steel tube column. The weld symbol on the plan view is the AWS (American Welding Society) symbol. It indicates that the 4″ × 4″ tube column is fastened to the top of the baseplate by a ¼″ fillet weld all the way around the bottom. Chapter 9 will discuss weld symbols in more detail as connection design fundamentals are reviewed. Notice in Figure 7–3 that the column baseplate is set in a 1″-thick layer of grout and fastened to the top of the footing with four ¾″-diameter anchor bolts. Also notice that the AISC minimum recommended distance from the outside edge of the baseplate to the center of each anchor bolt has been shown on the plan view.

For small, one-story buildings, structural steel tube or pipe columns are often set on the top of a foundation wall to

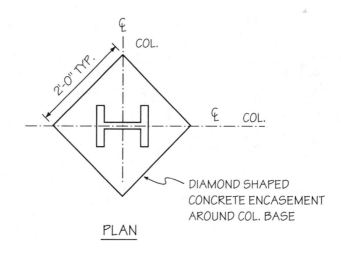

PLAN

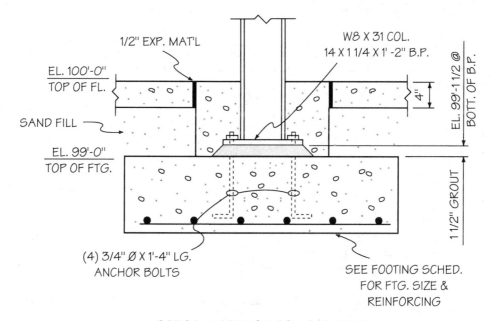

TYPICAL INTERIOR COL/FTG. DETAIL

Figure 7–1   Steel column-to-baseplate footing detail

help support the roof framing system. In these types of buildings, the effective length of the pipe or tube column is between 12′ and 14′, and the loads are relatively light. Such situations do not normally require heavy column baseplates, and for foundation walls between 12″ and 15″ wide, narrow-width baseplates are often preferred. Figure 7–4 shows the type of column baseplate the drafter or designer might select for a 3″ lightly loaded standard pipe column setting on top of a 12″-wide foundation wall. Notice that the longitudinal dimension of the plate runs parallel to the wall and that the column is held in place by four ¼″-diameter anchor bolts.

Columns are not always centered on their baseplates. Figure 7–5 illustrates a column baseplate for a 4″ × 4″ structural tube column located at the outside corner of a foundation wall. Notice that, while the column is centered on the grid lines, which in this case are 9″ in from the outside face of the wall, it is not centered on its baseplate.

Also, while most column baseplates are either square or rectangular, they are not always those shapes. Notice in Figure 7–5 that the AISC minimum recommended edge distance of 1½″ for the anchor bolts through the baseplate has been adhered to even though the baseplate has been notched out 2″ on the insider corner to fit properly on top of the foundation wall.

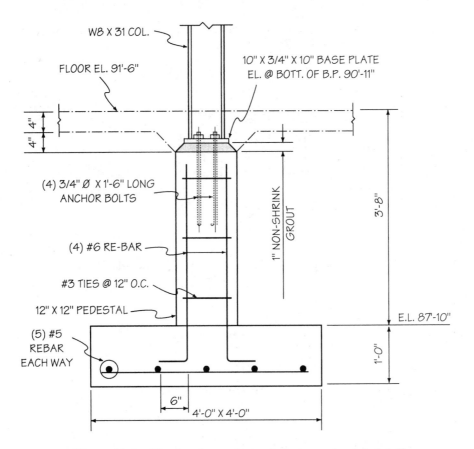

Figure 7–2   Steel column baseplate-to-pedestal detail

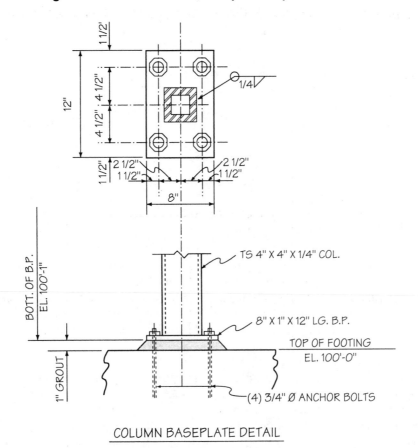

COLUMN BASEPLATE DETAIL

Figure 7–3   Steel tube column baseplate-to-footing detail

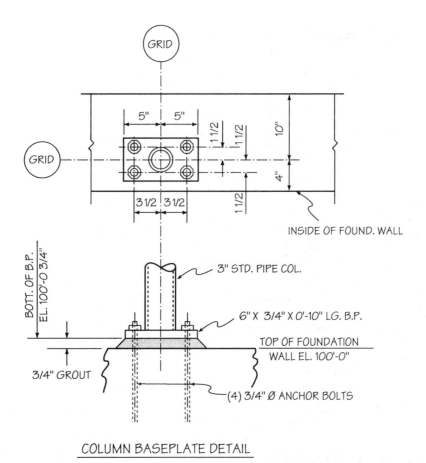

COLUMN BASEPLATE DETAIL

Figure 7–4   Steel pipe column baseplate-to-foundation wall detail

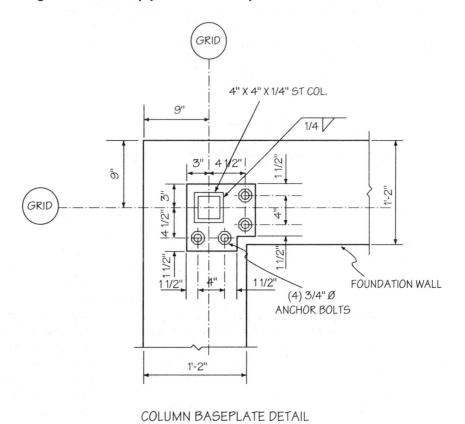

COLUMN BASEPLATE DETAIL

Figure 7–5   Corner column baseplate detail

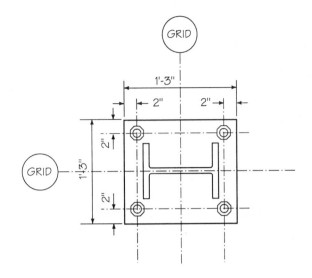

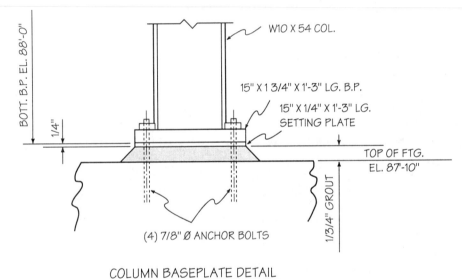

COLUMN BASEPLATE DETAIL

Figure 7–6   Column baseplate with setting plate detail

Use of Setting Plates

Figure 7–6 shows a column baseplate connection in which the designer has called for a setting plate. A *setting plate,* sometimes also called a leveling plate, is a steel plate set down over the anchor bolts, then shimmed, leveled, and grouted into place before the column is installed.

Setting plates are exactly the same dimension as the column baseplate except that they are usually about ¼″ thick. They are very useful when erecting larger, longer, and heavier structural steel columns, although some designers call for them with smaller columns as well. The W8 × 31 columns in figures 7–1 and 7–2 and the 4″ × 4″ structural tube and 3″ standard steel pipe columns in Figures 7–3, 7–4, and 7–5 are usually used in one- and two-story buildings, with or without a basement. Thus, they might be 14′ to 35′ or 40′ in

length for the W8s and 14′ to 25′ or 30′ for the small pipe and tube columns.

These smaller columns are commonly delivered to the job site with their baseplates already welded on. They are usually light enough that the ironworkers can install and level them to the proper elevation in the field. However, as buildings get taller and loads become heavier, the weights of the columns and their baseplates increase. Then it becomes much more economical for the steel erecting crews to have setting plates already set and leveled at the proper elevation and secured in place before the structural steel columns are delivered to the job site. Figure 7–6 illustrates how a setting plate is called for as part of the column baseplate connection detail for a W10 × 54 structural steel wide-flange column with a 1¾″-thick baseplate.

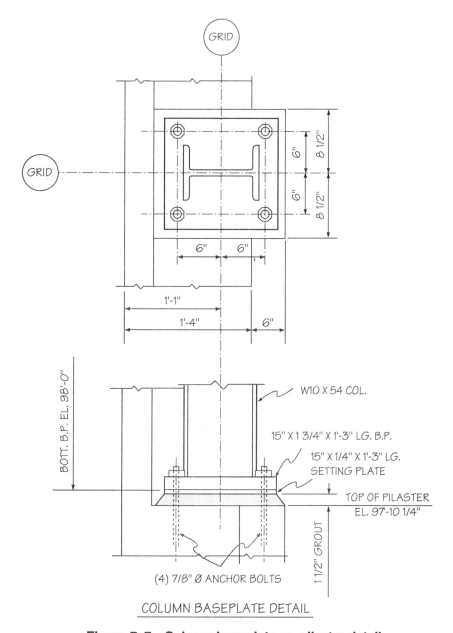

Figure 7–7   Column baseplate on pilaster detail

Use of Pilasters

Sometimes in larger commercial and industrial building projects, the exterior structural steel columns are carrying such heavy loads that the columns and their baseplates will not fit on the foundation wall as the pipe and tube columns did in Figures 7–4 and 7–5. When that happens, the usual solution is to jog the wall out of the column location for additional support. This thickening of the foundation wall under the steel column and its baseplate is called a *pilaster*. A pilaster can be thought of as a reinforced concrete column that happens to be located at the foundation wall.

Figure 7–7 is a detail showing a W10 × 54 column setting on a pilaster. The pilaster jogs 6″ in from the inside of the poured concrete foundation wall and is 17″ wide to accommodate the column and its baseplate. In this example, the column has a setting plate and is sitting in a *column pocket,* which allows exterior columns to be mounted below the top of a foundation wall. When the columns are mounted low enough, beams supporting the first floor can be connected to the columns. Examples of column pockets will be shown more clearly on beam-to-column connection details later in this chapter.

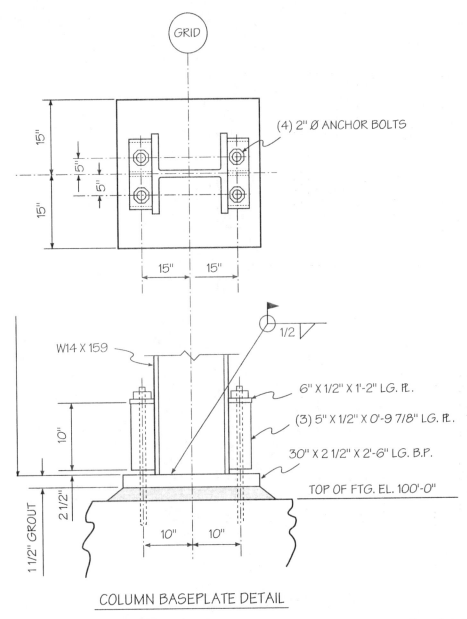

Figure 7–8   Moment-resisting column baseplate connection detail

Heavy Columns and Baseplates

Tall commercial buildings or large industrial buildings often require very heavy structural steel columns and baseplates. These members must be able to transfer gravity loads of hundreds or even thousands of kips and also to provide restraint against moments caused by wind loads, which can greatly increase the compression load on one side of the column while at the same time decreasing it on the other. The design procedures for such column baseplate connections are best left to very experienced engineers and designers.

Many types of details are possible for column baseplate connections of this type. Some of these baseplates can become very large—up to 36″ square and up to 8″ in thickness. Thus, the columns and their baseplates become so heavy and cumbersome that they must be shipped to the job site separately. Then, after the heavy baseplates are secured in position over the footings, the columns can be lowered into place and field-welded to the top of the baseplates. Two examples of moment-resisting column baseplate connection details are illustrated in Figures 7–8 and 7–9.

One last point should be made about column baseplates. The AISC now recommends that all column baseplates are to have four anchor bolts. For columns with heavy loads requiring large baseplates the anchor bolts are easily located

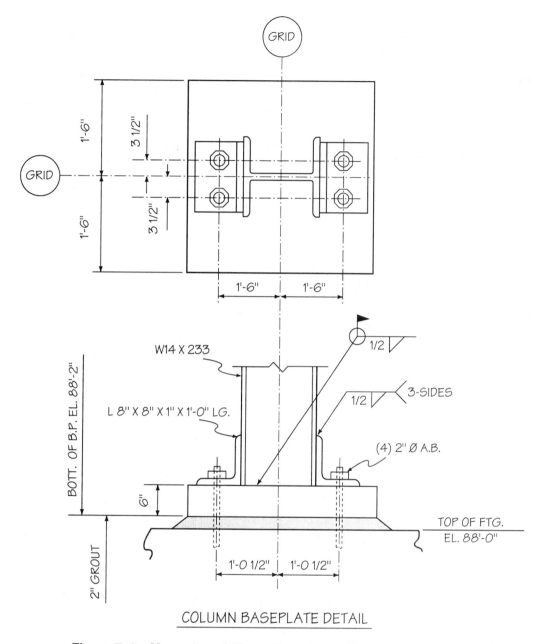

Figure 7–9   Moment-resisting column baseplate connection detail

on the outside of the column itself as shown on previous illustrations. However, when a smaller baseplate is sufficient because of lighter loads it is acceptable to locate the anchor bolt holes within the column area itself as shown in Figure 7–10.

7.4   COLUMN SCHEDULES

Columns are a very important part of any structural steel frame because they support the beams and girders and thus transfer the loads of floors and roofs down to the foot-ings and foundation walls. As discussed in Chapter 6, column shapes are shown on the structural framing plans (see Figures 6–20 through 6–26), but their sizes may not be identified. That information is usually found on the column schedule. *Column schedules* are tables that show such items as column sizes, column locations on the grid system, floor-to-floor elevations, column splice locations, column base-plate sizes, and bottom-of-baseplate elevations. Figure 7–11 illustrates what might be found on a column schedule for a small, two-story office building. Notice how column sizes and locations, baseplate sizes and elevations, and floor-to-floor elevations are shown.

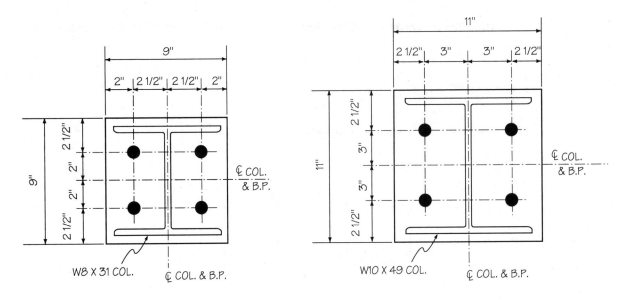

COLUMN BASEPLATES

Figure 7–10   Column baseplates

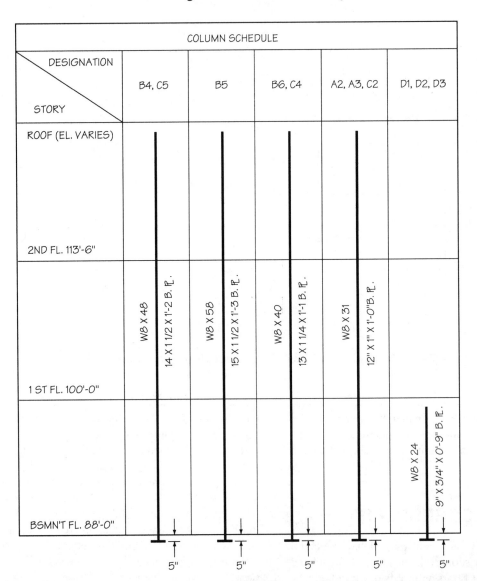

Figure 7–11   Column schedule for small commercial building

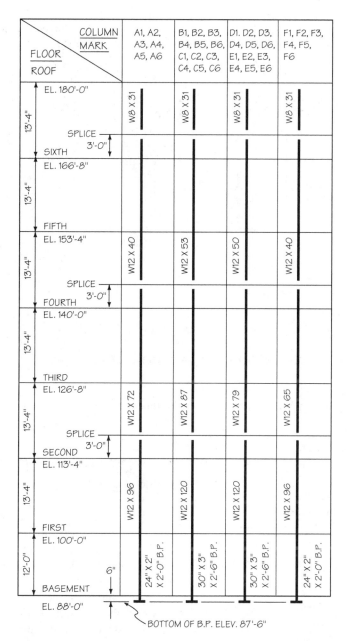

Figure 7–12   Column schedule for tall commercial building

Taller buildings have column schedules very much like the one shown in Figure 7–11, but with a few significant differences. For example, as buildings become taller, the columns at the upper levels are obviously not carrying as much load as those at lower levels. Thus, a column schedule might call for a W10 × 77 wide-flange column from the basement up through the first two floors, a W10 × 60 for the third, fourth, and fifth floors, and a W10 × 39 for the sixth floor, seventh floor, and roof.

In buildings fifteen to twenty stories high or more, steel column sizes may vary from heavy W14s at the lowest level,

to W12s and W10s at the intermediate levels, and then to W8s at the upper levels and roof. The reductions in column sizes would be shown on the column schedule as in Figure 7–12. Also notice in Figure 7–12 that the locations of column splices are indicated on the column schedule. *Column splices* are connections where column sizes change—for example, from a W10 × 49 to a W10 × 33 or from a W12 × 72 to a W10 × 49. Splice connections are usually located about 3′ above a floor so they will not complicate connections at the floor level.

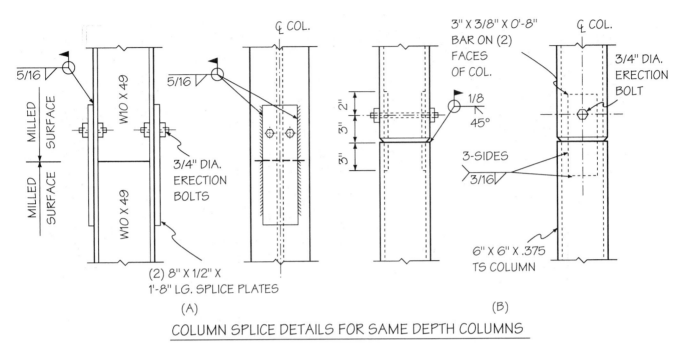

COLUMN SPLICE DETAILS FOR SAME DEPTH COLUMNS

Figure 7–13   Column splice details

Column Splice Connections

The column schedule for a six-story building shown in Figure 7–12 called for column splice connections at the second, fourth, and sixth floors. In taller commercial and industrial buildings, column splice connections are commonly used, both for columns of the same depth and in situations where column depth varies due to decreasing loads at the upper stories. Notice in Figure 7–13 that the W10 × 49 wide-flange columns in *A* and the 6 × 6 tube steel columns in *B* are spliced by first welding splice plates to the lower columns and then fastening the upper columns in place with ¾″-diameter erection bolts before finishing the column splice connection with field welds. Following usual practice, the butting surfaces of the W10 × 49 columns shown in Figure 7–13A are welded to ensure full column bearing at the connection.

Figure 7–14 shows column splice connection details for structural steel columns of varying depths. Notice in part *A* that a heavy bearing plate (also called a butt plate) is provided between the upper and lower columns. Notice also that the bearing plate and ends of the columns are milled to provide full contact between them.

Figure 7–14*B* is an example of a column splice plate connection in which the columns are spliced together by heavy flange plates bolted through the flanges of both the upper and lower columns. Filler plates are added to the flanges of the smaller upper column. Notice that a ⅛″ space between the filler plates and flange plates provides the required erection clearance. At erection, this space would be filled in with shim plates. With this type of connection, the

butting surfaces of the two columns must also be milled to provide full bearing between the columns.

7.5   BEAM-TO-COLUMN CONNECTIONS

As previously stated, the better the structural drafter understands how to component parts will fit together, the more competent and efficient he or she will be when drawing the plans. Nowhere is that visualization more important than in beam-to-column or beam-to-beam connections. Beam-to-column connection details are accepted methods by which the structural drafter or designer shows on the design drawings how structural steel floor and roof beams are to be fastened to structural pipe, structural tube, or W-shape columns.

Cap Plate Connections

One of the most common types of beam-to-column connections in commercial and industrial building construction is the beam bearing on column, or cap plate connection. A *cap plate connection* calls for a steel plate, commonly called a cap plate, to be welded to the top end of a wide-flange, pipe, or structural tube column at the fabricating shop. The connecting beam, which is usually a wide-flange beam, is then set in place on top of the cap plate by the erecting crew at the job site. Pre-drilled or punched holes on each side of the beam's web line up with matching holes through the cap plate, and the beam is bolted into place with high-strength

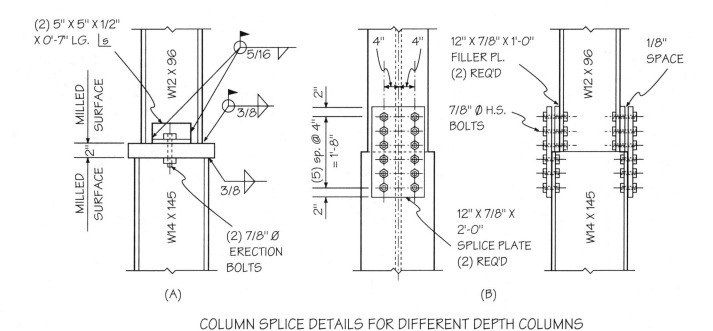

COLUMN SPLICE DETAILS FOR DIFFERENT DEPTH COLUMNS

Figure 7–14 Column splice details

bolts. Cap plates are usually as wide or slightly wider, and as thick or slightly thicker, than the bottom flange of the beam. Most drafters and designers try to keep the widths of column cap plates in full inches such as 6″, 7″, or 8″. Beam-to-column cap plate connections for most commercial and industrial buildings generally use ¾″-diameter high-strength bolts, and the minimum 1½″ distance from the center of the bolt holes to the edge of the cap plate as required by the AISC must be followed. Distances between holes through the cap plate and bottom flange of the connecting beam are governed by the width of the beam flange and required clearness from the column flanges or the surfaces of structural tube or pipe columns. Lengths of cap plates depend upon whether the column is an interior column or an end column and also whether one or two beams are bearing on the cap plate.

Figure 7–15 illustrates typical beam-to-column cap plate connections for small 4″ × 4″ structural tube columns suitable for a one-story office building or small shopping mall. Figure 7–15A shows an *end connection*, in other words a beam-to-cap-plate connection for columns located at the end of a building. For a cap plate end connection at the top of a structural tube or pipe column, the main consideration is that the cap plate extend on one side, beyond the column in the direction of beam span, far enough for two high-strength bolts to make the connection through the cap plate and the bottom flange of the beam. Notice, however, that at the end of the beam and cap plate, the extension beyond the face of the column is only ½″. This is required to provide space for the ¼″ fillet weld that connects the cap plate to the

top of the column. Fillet welds and their symbols will be discussed in much more detail in Chapter 9.

Figure 7–15B is an end view of the cap plate connection detail. This end view shows that the flange width of the W16 × 40 beam is 7″. In this instance, the drafter or designer has selected a cap plate the same width as the flange of the beam. He or she would find the flange width information in the AISC *Manual of Steel Construction,* Part 1, which lists dimensions and properties of structural shapes. See Table 2–1 in this text.

Figure 7–15C shows a beam-to-cap-plate connection in which two W16 × 40 beams are connected to the top of a 4″ × 4″ tube column. The 7″ × ½″ × 1′-0″-long cap plate is again welded to the top of the tube column, and each beam is fastened to the cap plate with two ¾″-diameter, high-strength bolts. Notice that a ½″ space is called for between the beams to allow for erection clearance.

Also notice in Figure 7–15A–C that the top-of-beam elevation of 113′–4½″ is shown. The top-of-beam elevation is very important for the general contractor, the structural steel detailers in the structural fabricator's office, and the steel erection crews, thus it should always be shown on the structural details prepared in the design office.

Figure 7–16 illustrates beam-to-cap-plate connection details at the top of a W8 × 35 structural steel column. Notice that the beams can be fastened to the cap plates with their webs either parallel to the column web as shown in part A or perpendicular to the column web as shown on part B. Splice plates, illustrated in part A, are usually optional. Some structural design firms use them as a standard part of

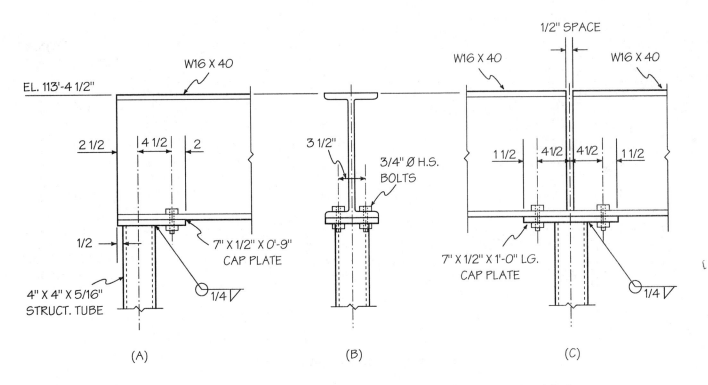

BEAM TO COLUMN CAP PLATE CONNECTIONS FOR THE TUBE COLS.

Figure 7–15 Beam-to-column cap plate connection details

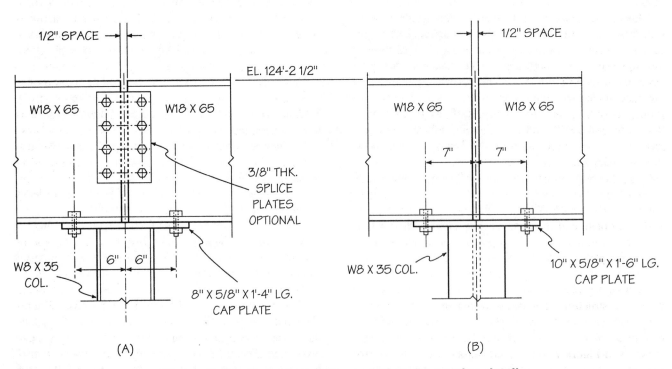

Figure 7–16 Beam-to-column cap plate connection details

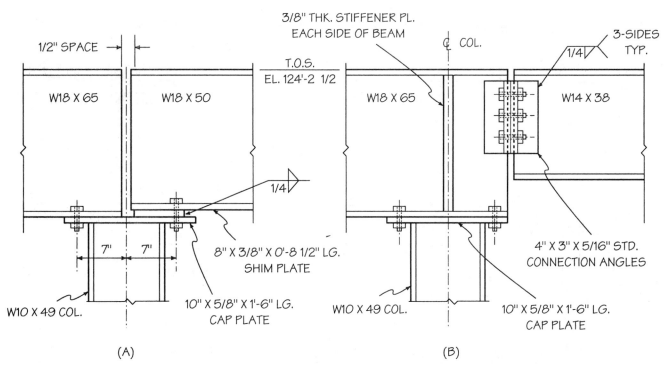

Figure 7–17   Beam-to-column cap plate connection details

the connection, while others do not. The 1½″ minimum edge distance from the edge of the cap plate to the center of the bolt holes must be adhered to, although it is not unusual for designers to use a 2″ edge distance as illustrated in Figure 7–16.

In preparing structural steel connection details, another very common situation occurs when beams of different depths connect at the same cap plate. Figure 7–17 illustrates this situation. Many times, if the beam depths are quite close, the solution is to weld a shim plate to the top of the cap plate in the fabrication shop, as shown in Figure 7–17A. In fact, as soon as the structural drafter or designer sees a W18 × 65 and a W18 × 50 setting on top of the same cap plate, he or she should immediately consult Part 1 of the AISC manual because, even though both beams are W18s, there is a very good chance they are not the same depth. When the depths are considerably different, as illustrated in Figure 7–17B, the usual solution is to fasten the larger beam to the cap plate and then connect the smaller beam to the larger beam either with connection angles as in part B or with splice plates as in Figure 7–16 part A.

Overhanging Beams. Figure 7–18 is a detail showing a W16 × 40 structural steel beam overhanging 4′–0″ beyond its supporting column, which in this case is a 4″ extra-strong steel pipe column. The cap plate shown is identical to the one used in Figure 7–15C. Overhanging beams are widely used in commercial and industrial buildings, particularly in roof framing construction where negative bending moments

produced by the overhang tend to offset the positive bending moment developed by load between the supporting columns. This makes the positive bending moment less than it would be if the beam were simply supported at each end, which in turn allows the drafter or designer to use a lighter, and thus more economical, beam. The common practice is to connect two overhanging beams between columns to an intermediate beam. The connection is usually made either with splice plates as illustrated in Figure 7–18 or with connecting angles as shown in Figure 7–17B. For this situation, the intermediate beam, even though it is simply supported at the ends, will often be smaller, and thus lighter, because its span is shorter than the distance between the columns. However, overhanging beams and intermediate beams may also be the same size, and this too is economical because all of the beams will be lighter than if they were simply supported. Refer to Figures 6–10 and 6–11 to see how this type of articulated framing would appear on a framing plan.

Splice Plate Connections. The most common type of connection between overhanging beams and intermediate beams are splice plate connections. *Splice plate connections* are simply a matching set of plates, usually ¼″ to ½″ thick, which connect the beams at the webs as shown in Figures 7–18 and 7–16A. The most common thickness for splice plates used in commercial buildings is ⅜″. The splice plate connection creates a double-shear flexible connection condition on the high-strength connecting bolts. The allowable double-shear stress on the bolts determines the required

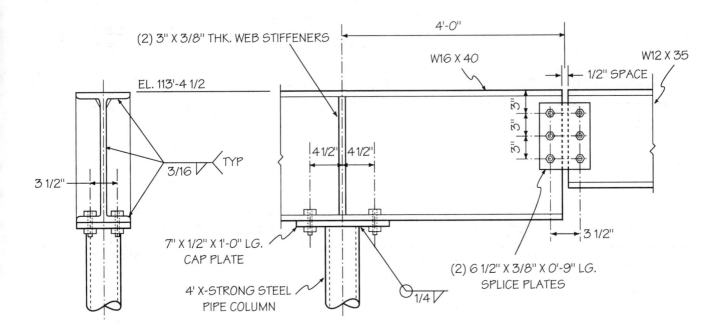

BEAM TO COLUMN CAP PLATE CONNECTION FOR OVERHANGING BEAM

Figure 7–18 Beam-to-column cap plate connection for overhanging beam

number of bolts to transfer the reaction of the intermediate beam into a concentrated load at the end of the overhanging beam. Sometimes, when the web of the overhanging beam is thicker than that of the intermediate beam, the splice plates are shop-welded to the end of the overhanging beam. Then, as the intermediate beam is bolted into place at the job site, shims are used to fill in the gap. Also, on lightly loaded beams, splice plates are sometimes required on only one side of the web, in which case the plate may be welded to the overhanging beam at the fabricating shop.

Use of Web Stiffeners. Notice in Figure 7–18 that ⅜″-thick web stiffeners have been provided on either side of the web of the W16 × 40, and that these stiffeners are located at the center line of the supporting column. *Web stiffeners* are steel bars, usually ¼″ to ½″ thick, which are used to stiffen or support the thin web of the beam at the supports. Web stiffeners are often required at the supports of overhanging beams because this is where the greatest stress occurs.

As we have already learned, loads at the top of the beam may be uniform along the entire length of the beam, or may occur at intervals of 2′-0″ or 4′-0″ or more, such as with the roof joists discussed in Chapter 5. But beam reactions usually occur at only two points. Thus, since the sum of the reactions always equals the sum of the loads, the greatest concentrated stress points on the beam will be at the reactions. Because steel mills typically make the webs of

W-shape beams as thin as possible for economy, the webs of W-shape beams tend to buckle at the supports.

Engineering design calculations can easily determine whether web stiffeners are required, and if so, they are welded into place at the fabricator's shop on both sides of the web as shown in Figure 7–18. If web stiffeners are required over W-shape columns, they are usually installed on both sides of the web as shown in Figure 7–19. Notice in Figure 7–19 that the web stiffener plates are in direct line with the flanges of the column.

Standard Framed Beam-to-Column Connections

In *framed beam connections,* beams are fastened to columns by means of framing angles. This usually consists of two structural steel angles, one on either side of the beam, which are either welded or bolted to the beam web in the fabricator's shop and then fastened to a column at the job site with high-strength bolts. For most commercial work, framing angles vary in thickness from ¼″ to ⅝″, and some typical angle sizes are 3″ × 3″, 4″ × 3″ or 4″ × 3½″. For heavier connections on industrial buildings, larger and thicker angles with double rows of bolts will often be used. For lighter connections, drafters and designers often call for *one-sided connections,* which means using one framing angle on one side of the beam web. Another common connection for lighter beam-to-column connections is the

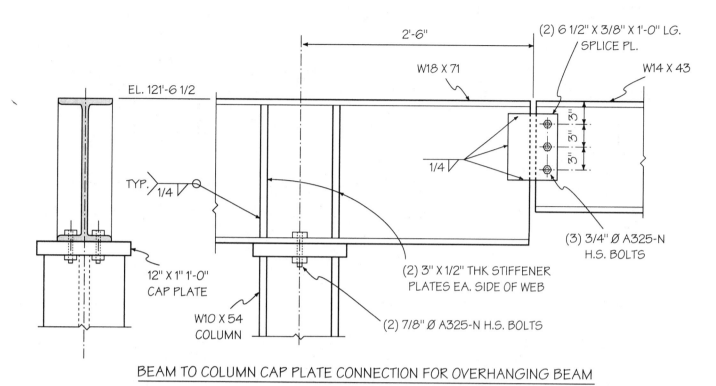

BEAM TO COLUMN CAP PLATE CONNECTION FOR OVERHANGING BEAM

Figure 7–19   Beam-to-column cap plate connection for overhanging beam

single-plate connection (sometimes called a shear tab connection) illustrated in Figure 7–28.

Shop-Bolted and Field-Bolted Connections. Figure 7–20 illustrates two structural steel wide-flange beams framing into the flanges of a W8 × 31 steel column. In this example, the W16 × 40 is connected to the column with a 4-row framing angle connection. A *4-row connection* has four rows of bolts, which for most commercial work are ¾″ in diameter and high strength. The framing angle connection for the W12 × 30 is a 3-row connection.

For framing angle connections two types of legs are identified: the *web legs,* which fasten to the web of the beam being connected, and the *outstanding legs,* which fasten to the supporting member. In Figure 7–20, the web leg would be the 3½″ leg, and the outstanding legs would be the 4″ leg of each angle. Notice also that the distance from the top of the beam to the first row of bolts is 3″, and the distance between bolts is 3″. This is the usual standard for commercial work, although the distance from the top of the beam to the first bolt can be more (or rarely, less) than 3″.

Note in Figure 7–20 that the framing angles, also called *clip angles,* are 2½″ longer than the sum of the distances between the rows of bolts. This conforms with the AISC-recommended minimum distance of 1¼″ from the centers of the top and bottom bolt holes to the end of the angle. As has already been pointed out, the majority of commercial work requires ¾″-diameter, high-strength bolts, although for

heavy connections, ⅞″ or even 1″-diameter bolts are not uncommon. The size and quantity of bolts are determined from engineering calculations, which will be discussed in more detail in Chapter 9.

One last point to notice about Figure 7–20 is that the connection is made with high-strength bolts through both the beam webs and the column flanges. This is called a *shop-bolted and field-bolted connection* because the bolts through the beam webs and web legs of the angles are installed in the fabricator's shop, while the bolts through the outstanding legs of the angles and the column flanges are installed by the erection crews at the job site or "in the field." This type of connection is also sometimes called a type 2, flexible *shear connection,* which is a connection assumed to transfer shear only and no moment. Connection types and bolt nomenclature will be discussed in greater detail in Chapter 9.

Shop-Welded and Field-Bolted Connections. Figure 7–21 is similar to Figure 7–20 except that the beams are connected to the web of the column, and the framing angles are welded to the beam and bolted to the column web with field bolts. Again, because the web legs of the framing angles are welded to the web of the beam in the fabricator's shop, this type of connection is called a shop-welded and field-bolted connection. According to AISC specifications, 4″ × 3″ angles can be used for this type of connection, with the 3″ leg being the web leg and the 4″ leg the outstanding

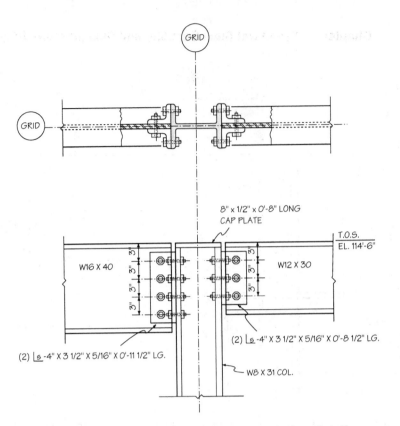

Figure 7–20 Beam-to-column connection detail

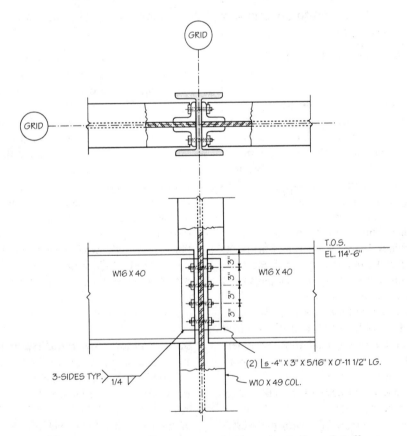

Figure 7–21 Beam-to-column connection detail

128

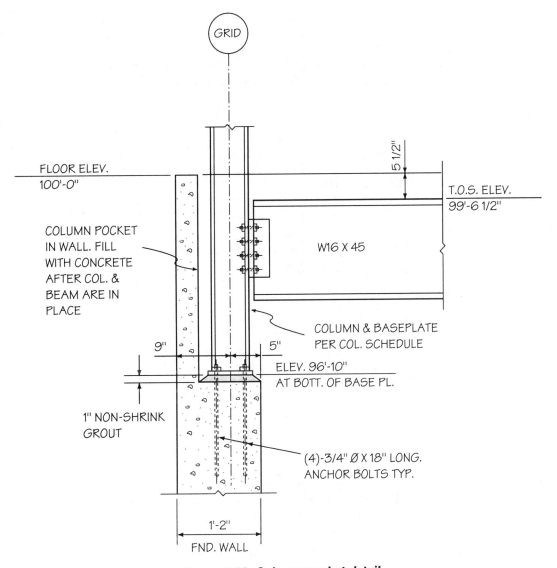

GRID

FLOOR ELEV.
100'-0"

5 1/2"

T.O.S. ELEV.
99'-6 1/2"

COLUMN POCKET
IN WALL. FILL
WITH CONCRETE
AFTER COL. &
BEAM ARE IN
PLACE

W16 X 45

COLUMN & BASEPLATE
PER COL. SCHEDULE

9"   5"

ELEV. 96'-10"
AT BOTT. OF BASE PL.

1" NON-SHRINK
GROUT

(4)-3/4" Ø X 18" LONG.
ANCHOR BOLTS TYP.

1'-2"

FND. WALL

Figure 7-22 Column pocket detail

leg. However, if the connection is framing into the web of a W8 column, the angles would have to be 3″ × 3″, and the distance between rows of bolts in the plan view through the column web would not be more than 4″. The structural drafter should also be aware that if, for instance, beams with flanges 8″ wide or wider were framing into the web of a W8 column, the flange width would have to be cut down to fit within the inside of the column flanges. How this is usually accomplished will be covered in Chapter 11 when beam detailing is discussed.

Figure 7–22 is a section view of a framed beam connection in which a W16 × 45 beam is connected to the flange of a W-shape structural steel column. In this example, the beam is supporting a floor, so both the top-of-beam and top-of-floor elevations are shown on the structural detail. Notice that the column is mounted in a column pocket provided in the 1′–2″-thick foundation wall. The column grid line is located 9″ in

from the outside of the wall. The column size and baseplate size are not shown on this detail because they are referred to on a column schedule. The elevation at the bottom of the baseplate, thickness of grout, and quantity and size of anchor bolts are shown for the benefit of both the general contractor and the structural detailer who will prepare the required anchor bolt setting plans, erection plans, and beam and column detail drawings in the fabricator's office.

Figure 7–23 illustrates a one-sided framing connection. The framing angle is intended to be shop-welded to the web of the W12 × 16 beam and field-bolted to the W8 × 31 column. Because the W12 × 16 is a relatively light beam, the loads imposed on the connection would also be quite small, thus three high-strength bolts should be sufficient for the connection. However, as with designing all structural connections, this must be verified by performing the required engineering calculations.

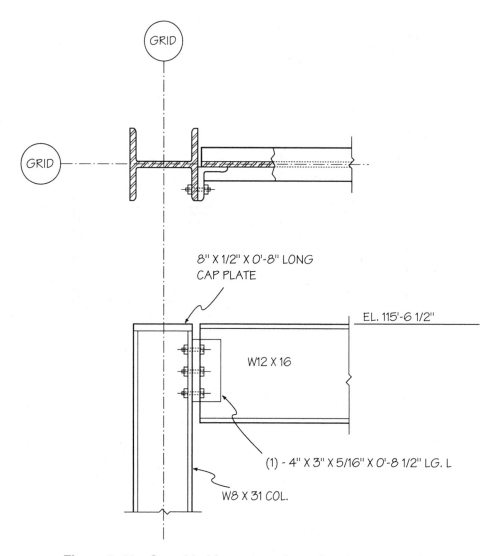

Figure 7–23   One-sided beam-to-column framing connection

Seated Beam-to-Column Connections. Figure 7–24 is an example of a beam connecting to the web of a W12 × 79 structural steel column with a *seated beam connection*. Seated beam connections may be either welded or bolted and either stiffened or unstiffened. Figure 7–24 is a bolted, unstiffened, seated connection in which the beam, a W21 × 73, is supported by an 8″ × 4″ × ⅞″ × 0′-8½″-long seat angle. The beam is stabilized by the 4″ × 4″ × 5/16″ top angle as required by AISC specifications.

Notice the nominal beam setback of ½″. The *setback* is the distance from the end of the beam to the column web. Notice also that the seat angle itself, being ⅞″ thick, is much thicker than the framing angles previously discussed. The greater thickness is usually required because the entire beam reaction is bearing on the 4″ horizontal leg of the seat angle. Part 4 of the AISC *Manual of Steel Construction* provides design data for seated beam connections.

Seated connections are widely used because they are economical in two ways. First, they are economical to fabricate in the shop because the beam is simply cut to length, and the required mounting holes are punched or drilled through the flanges. Second, they are economical to erect because the setback dimension makes them easier to install in the field than the tighter-fitting framed connections.

Figure 7–25 is a detail of a welded seated beam connection. With this type of connection, the seat angle is welded to the column, which in this example is the flange face of a W8 × 31. The required weld can be performed accurately and economically in the steel fabricator's shop before the column is delivered to the job site. The beam is then set in place and fastened to the column by the steel erecting crews. The two erection bolts for this type of connection may be standard A307 bolts rather than high-strength bolts because there is no shear stress on the bolts

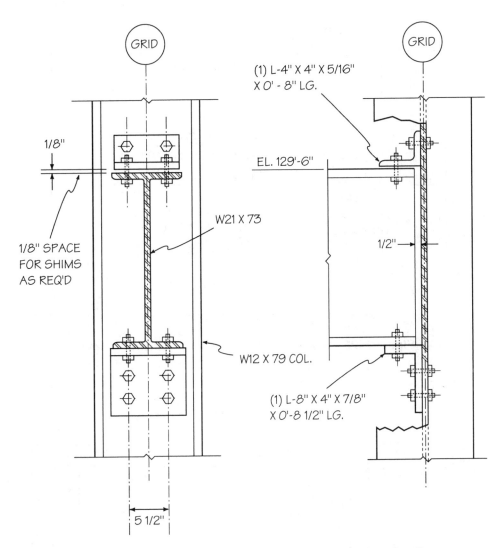

Figure 7–24 Seated beam-to-column connection detail

as was the case with the framed connections previously discussed.

After the beam has been secured in place by the erection bolts, it is ready to be field-welded to the seat angle. The supporting top angle is also field-welded to both the column and the top flange of the beam. When the structural drafter prepares a connection detail in which the seat angle is welded to the flange face of the structural column, he or she must know the length of the seat angle and the width of the column flange. The seat angle should usually be a minimum of 1″ longer or shorter than the flange width of the column so that the required fillet weld can be deposited along the vertical leg of the angle. If the flange width and angle length were the same, or very nearly the same, the welds (in this example the ⁵⁄₁₆″ fillet welds) could not be made. Also, seated connections are often not made to column flanges if the columns are to be enclosed because the horizontal legs of the angles would protrude through the architectural enclosure.

For heavy seated beam connections, when the beam reaction exceeds the capacity of a typical seat angle, *stiffened* seated beam connections are used. Stiffened connections, like unstiffened seated beam connections, may be either bolted or welded.

Figure 7–26 shows a stiffened seated beam-to-column web connection in which the welded connection consists of a horizontal seat plate and a vertical stiffener plate. The required sizes of seat plate, stiffener plate, and welds are determined by engineering calculation. The connection shown, assuming grade A36 steel for the beam, column, seat plate, and stiffener plate, would sustain a beam reaction value of approximately 100 kips.

Beam-to-Column End-Plate Shear Connections.
Figure 7–27 is a detail of an end-plate shear connection. *End-plate shear connections* consist of a single plate welded to the beam web with fillet welds on either side. They are

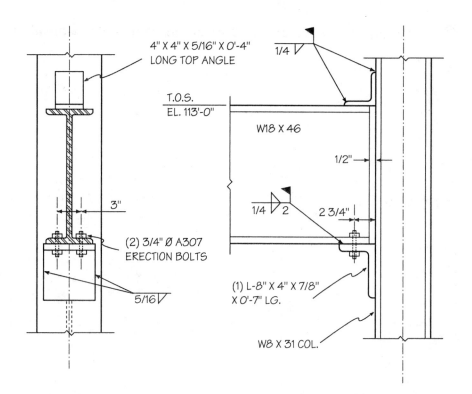

Figure 7–25   Seated beam-to-column connection detail

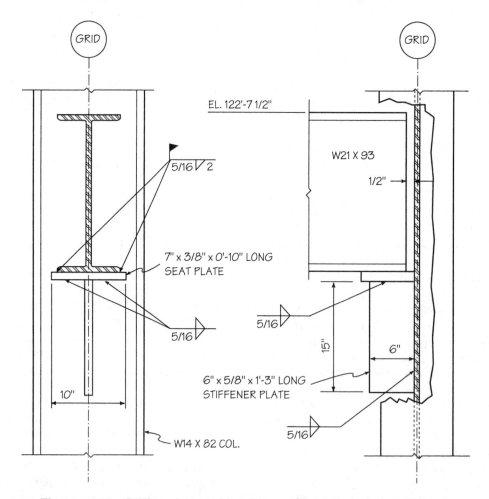

Figure 7–26   Stiffened seated beam-to-column connection detail.

132

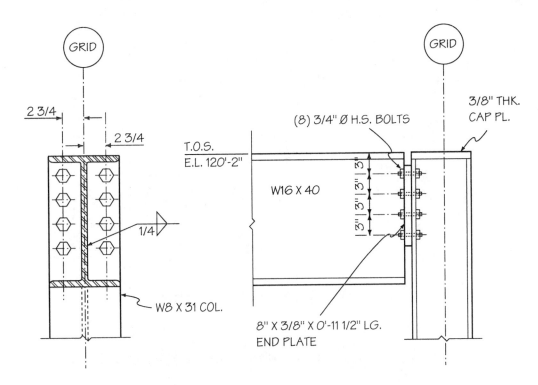

Figure 7–27 End-plate shear connection detail

very similar to the double-angle framing connections previously discussed. Part 4 of the AISC manual contains a very good explanation of end-plate shear connections.

The detail in Figure 7–27 illustrates a W16 × 40 beam connected to a W8 × 31 column with a 4-row end-plate shear connection. The 8″ × ⅜″ × 0′–11½″-long plate is welded to the end of the W16 × 40 beam with ¼″ fillet welds on both sides of the web. The beam is then fastened to the column flange with eight high-strength bolts. Not all fabricating shops are equipped to use this type of connection because it requires close control in cutting the beam to length and squaring the ends to ensure that the end plates at either end are truly parallel with each other.

Beam-to-Column Single-Plate Shear Connections.
On a nationwide assessment of various types of structural steel connections for commercial buildings, *single-plate shear connections* are possibly the most widely used of any structural steel beam-to-column connection. This type of connection is adaptable for numerous situations—from connecting the heaviest wide-flange beam to a wide-flange column to connecting a light beam to a 3″ standard steel pipe or 4″ structural steel tube column. By far the majority of connections between beams and round pipe, square, or rectangular steel tubes are single-plate shear connections.

The basic connection consists of a single steel bar or plate, usually from ¼″ to ½″ thick, welded to its support-

ing column and then connected to a beam with high-strength bolts. The most common thickness of plate used is probably ⅜″.

Single-plate shear connections are often used to connect beams to steel pipe or structural tube columns. The two most commonly used single-plate shear connections are the *shear tab connection,* in which the plate is welded to the face of the pipe or structural tube, and the *thru-plate connection,* in which the plate is inserted into a slot cut through the walls of the column. At present, both shear tab and thru-plate connections are being used by structural engineers. Although many structural engineers are more comfortable with thru-plate connections, recent research indicates shear tabs are often more economical because they avoid the labor costs involved in laying out and slotting the pipe or structural tube column to insert a thru-plate.

One feature of single-plate shear connections is that, if the center line of a beam is to be aligned with the center line of its supporting column (which is the usual case), the shear plates must be offset a distance of one-half of the beam web thickness on one side or the other of the center line of the column. Figures 7–28 through 7–31 illustrate some of the most common single-plate beam-to-column shear connections.

Figure 7–28 shows a W18 × 46 beam connected to the flange face of a W8 × 31 structural steel column by a single-plate shear connection. The connecting plate is welded to the face of the column flange with ¼″ fillet welds on each side of the plate. Notice that the plate is offset so that, after

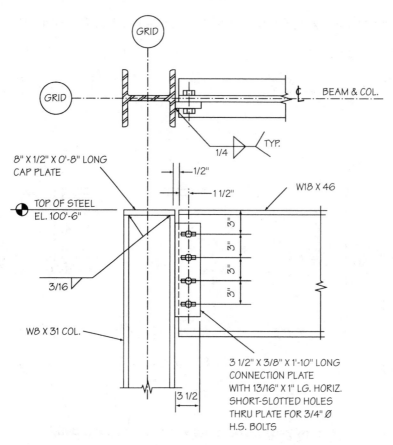

Figure 7–28   Single-plate shear tab connection detail

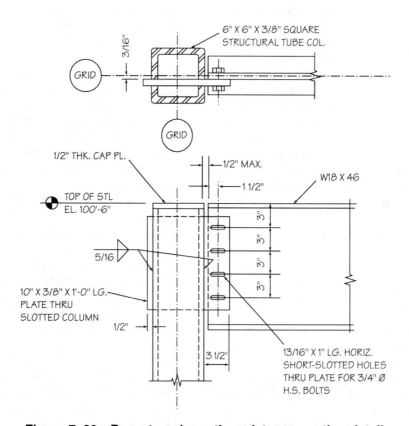

Figure 7–29   Beam-to-column thru-plate connection detail

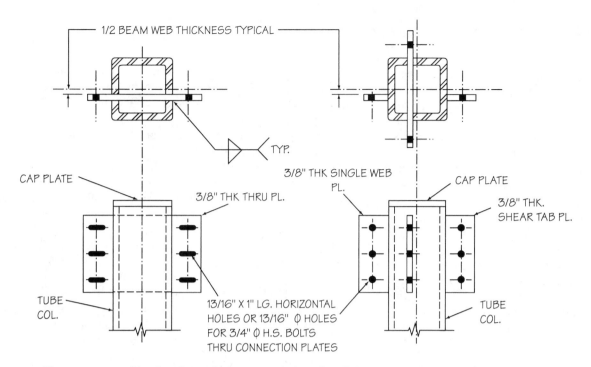

Figure 7–30   Single-plate shear connection detail for two to four beams

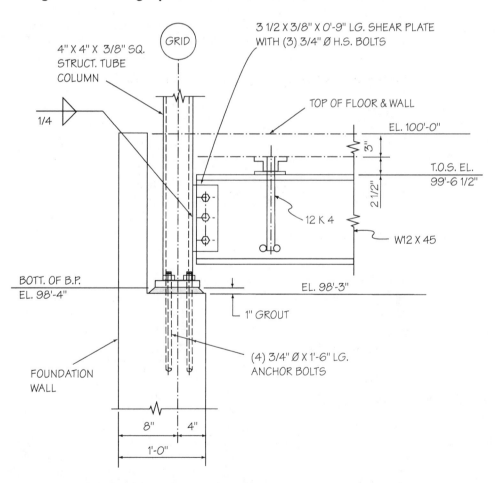

Figure 7–31   Structural tube column in column pocket

the connection is made, the center lines of the beam and column will align. Notice also that the holes through the plate are horizontal short-slotted holes rather than round holes. When calling for this type of connection, designers and fabricators commonly use horizontal slotted holes through either the beam or connecting plate to give the field erection crews more leeway in making the connection on the job site. Notice in this detail that the beam has a ½″ setback or clearance between the end of the beam and the face of the column.

Figure 7–29 is an example of a W18 × 46 beam connecting to a 6″ × 6″-square structural tube column by means of a single-plate shear connection. Again, horizontal short-slotted holes through the plate will help the erection crew if slight adjustments are required. In this detail, the connection plate is a thru-plate, which means it extends entirely through the tube column and is welded in place with 5/16″ fillet welds. Notice that the plate extends ½″ beyond the back face of the column to provide room for the fillet weld on that surface. The ½″ cap plate on top of the 6″ × 6″ tube column, and the W8 × 31 column in Figure 7–27, is installed to provide a surface at the top-of-beam and column elevation that could support a steel joist if one were required at that location.

Figure 7–30 illustrates how single-plate shear connections are often welded to tube or pipe columns to provide for the connection of two to four beams. Figure 7–30A is a typical thru-plate connection designed to fasten two beams to the column, one on either side. Figure 7–30B illustrates how one thru-plate and two surface-welded plates could be used in combination to connect four beams to one structural tube column. As previously mentioned, this type of knowledge is especially helpful to the structural drafter because the better understanding the drafter has of various types of framing connections, the more confidently and competently he or she can prepare practical framing plans that can be economically constructed and installed at the job site.

Figure 7–31 is a structural detail similar to Figure 7–22, except that in this case, a structural tube column is installed in a column pocket in a 12″ poured concrete foundation wall, and the W12 × 45 wide-flange beam supporting the floor joists is connected to the column with a single-plate shear connection.

Beam-to-Column Moment-Resisting Connections

While it is well beyond the scope of this book to illustrate all the possible arrangements of structural steel beam-to-column connections, the study of column connections should include at least a brief introduction to moment-resisting connections. In tall structural steel-frame buildings, the steel support frame, especially along the outside walls, must be designed to resist not only vertical gravity floor and roof loads, but also bending moments caused by the lateral forces of wind or earthquake loads. Although the design of moment connections is clearly the responsibility of the experienced engineer or designer, the entry-level structural drafter should also be aware of such connections.

Figure 7–32 shows one type of moment-resisting connection designed to resist both shear and bending. Like standard framing connections, moment-resisting connections may be welded, bolted, or both welded and bolted. In this example, the gravity loads are resisted by a single-plate shear connection welded to the face of the column flange, similar to the example in Figure 7–28. The plate is then fastened to the beam by four high-strength bolts going through the plate and the web of the W16 × 67 beam. The lateral bending forces are resisted by plates bolted to the top and bottom flanges of the beam and then field-welded with groove welds to the face of the column flanges. The 1″ square steel backing strips are used to ensure a full penetration weld to the column flanges.

In Figure 7–32, the plates are field-welded to the column, but it is also possible to have the flange plates and web plate welded to the column in the fabricator's shop and then field-bolted to the beam at erection. If this is desired, the distance between the flange plates must be greater than the depth of the beam, usually by about ⅜″. The gap between the top flange of the beam and the top plate is then filled in with thin strips of steel called shims.

Non-rectangular Beam-to-Column Connections

All beam-to-column connections discussed thus far have been *rectangular connections*—that is to say, they have been horizontal beams with vertical webs connecting at right angles (90 degrees) to the faces or webs of their supporting columns. However, the steel support systems of commercial and industrial buildings commonly require beam-to-column connections that are non-rectangular. In a *non-rectangular connection,* the beam and column come together at something other than a 90-degree angle.

Often roof beams, even on so-called "flat roofs," are actually pitched slightly, sloped downward to facilitate water drainage. Or many times an architect designs part of a building to angle gracefully away from another part, thus avoiding having all sharp 90-degree corners. These requirements demand that at least part of the structural framework be non-rectangular. While an in-depth discussion of non-rectangular framing would be impractical for a beginning textbook such as this one, a brief discussion of sloped and skewed beam-to-column connection details will introduce the subject of non-rectangular framing.

Sloped Beam Connections. Figure 7–33 is a detail of a beam-to-column connection in which the web of the W14 ×

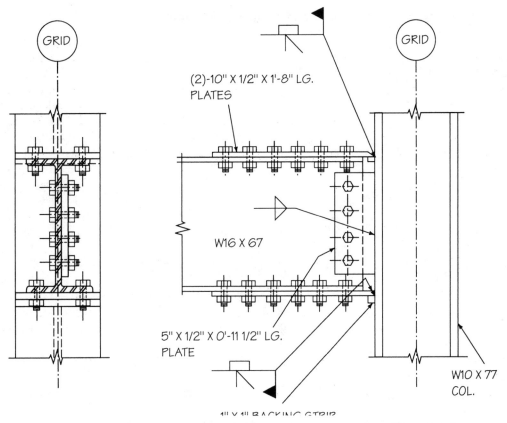

Figure 7–32   Beam-to-column moment-resisting connection detail

(2)-10" X 1/2" X 1'-8" LG. PLATES

W16 X 67

5" X 1/2" X 0'-11 1/2" LG. PLATE

1" X 1" BACKING STRIP

W10 X 77 COL.

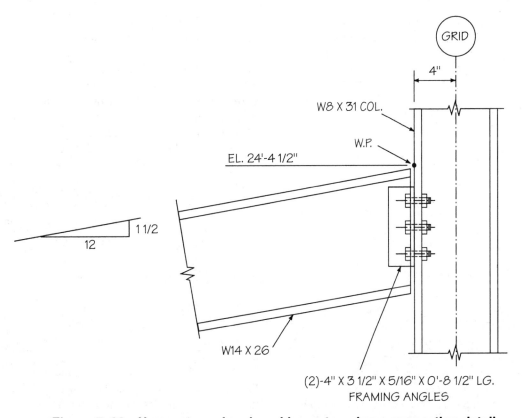

Figure 7–33   Non-rectangular sloped beam-to-column connection detail

4"

W8 X 31 COL.

W.P.

EL. 24'-4 1/2"

1 1/2

12

W14 X 26

(2)-4" X 3 1/2" X 5/16" X 0'-8 1/2" LG. FRAMING ANGLES

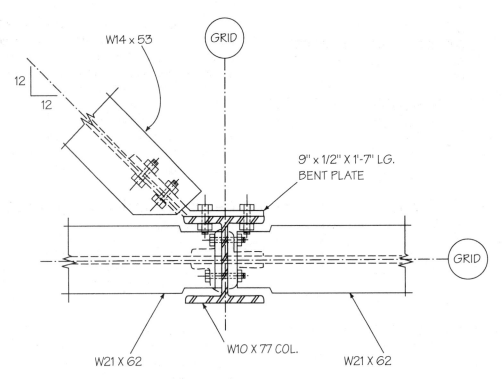

Figure 7–34   Beam-to-column skewed connection detail

26 beam is perpendicular or at right angles to the flange face of the W8 × 31 column. However, the flanges of the beam are not perpendicular to the column flange. This is called a *sloped beam connection* because the beam slopes downward away from the column. This type of connection might be called for in a situation where the W8 × 31 column continues up another story or more to a high roof while the W14 × 26 beam is part of the support system for a lower-level roof.

Notice first the small black dot on the face of the flange of the W8 × 31 column, the letters W.P. with the arrow pointing to the dot, and the elevation at the dot listed as 24'–4½". The black dot is a *work point*, and the structural drafter is indicating that the top of the W14 × 26 beam should fasten to the column at the 24'–4½" elevation. This is information the structural detailer in the fabricator's office must have later on in order to prepare the shop detail drawings of both the W14 × 26 beam and the W8 × 31 column.

On Figure 7–33, it can be seen that the beam does not come away from the column on the horizontal, but instead pitches downward at a slight angle. This angle is expressed on most structural steel working drawings in terms of slope, which is the ratio between the rise and run. The *run* (the horizontal distance) is shown on structural drawings as 12", while the *rise* varies depending upon how steep the designer wants the angle of slope. In this example, the symbol $\frac{|1½"}{12}$ indicates that, for every 12" of horizontal run, the beam is to drop or slope 1½", which is ample for water drainage. The rest of the detail indicates a standard double-angle framing connection in which the connecting angles

would be shop-welded to the W14 × 26 beam and field-bolted to the W8 × 31 column at the column flange.

Skewed Connections. Figure 7–34 is a plan view detail indicating two W21 × 62 beams framing into the web of a W10 × 77 column with a standard rectangular connection similar to that shown in Figure 7–21. However, the W14 × 53 beam connecting to the flange face of the column is coming in at an angle other than 90 degrees. This is an example of a skewed beam-to-column connection. In a *skewed connection,* the flanges of the beam and its supporting column are at right angles to each other, but the webs incline toward one another.

The angle of skew, like the angle of slope in Figure 7–33, is again given as a ratio. In this example, the ratio is $\frac{|12}{12}$, which means that, for every 12" of horizontal run, the beam skews off at 12" to the left in plan view. A 12/12 skew is an angle of 45 degrees. The detail also shows that the W14 × 53 beam is fastened to the face of the W10 × 77 column by a single bent plate, which is secured with double rows of high-strength bolts to both the column and the beam.

Notice also in Figure 7–34 that the flange widths of the W21 × 62 beams will need to be trimmed or narrowed to fit within the insides of the column flanges. When and how much they should be trimmed would be determined by the structural fabricator's detail drafter when preparing the shop drawings. The procedure for this will be discussed in Part 2.

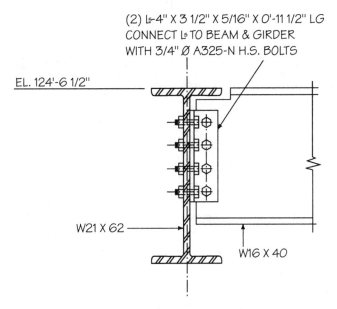

(2) L-4" X 3 1/2" X 5/16" X 0'-11 1/2" LG
CONNECT L TO BEAM & GIRDER
WITH 3/4" Ø A325-N H.S. BOLTS

EL. 124'-6 1/2"

W21 X 62

W16 X 40

**Figure 7–35   Beam-to-girder standard framed
connection detail**

The last item to notice about this detail is that the top flanges of the W14 × 53 and W21 × 62 beams are obviously at the same elevation, thus the top flange of the W14 × 53 has been trimmed off near the column so it will not interfere with the top flange of the W21 × 62 on the left side of the detail. The desired space between the two flanges after the top flange of the W14 × 53 has been cut back would usually be a minimum of about ½″.

7.6   BEAM-TO-GIRDER CONNECTIONS

Figure 7–35 shows a standard beam-to-girder framed connection in which the W16 × 40 structural steel beam shown in elevation view is connected to the W21 × 62 girder shown in section view. The connection is made with two 4″ × 3½″ × 5/16″ × 0'–11½″-long angles, one on either side of the web of the W16 × 40 beam. These angles are bolted through the webs of both the beam and its supporting girder with ¾″-diameter high-strength bolts. Notice that, because the top flanges of the beam and girder are at the same elevation, the top flange of the W16 × 40 beam has been cut out or *coped*. How to determine the proper dimensions of the cope or cut-out is usually a task for the detailer in the structural steel fabricator's office and will be discussed in greater detail in Part 2.

Coped Connections

Figure 7–36 illustrates two coped beam-to-girder connections. In Figure 7–36*A*, the W10 × 19 beam is fastened to

the top of the W18 × 35 girder with two 4″ × 3″ × ¼″ angles welded to the beam and bolted to the top flange of the girder. Because the top of the beam is to be 8″ above the top of the girder, the cope is made on the bottom of the beam. Notice also that the top-of-steel elevations for both the beam and girder are shown on the detail.

Figure 7–36*B* is similar to Figure 7–35 except that the top-of-steel elevation of the two W12 × 26 beams is required to be 2½″ above the top-of-steel elevation of the W16 × 40 girder. Thus, both top-of-steel elevations are shown on the detail.

Beam-to-Girder Single-Plate Connections

Figure 7–37 illustrates a beam-to-girder connection in which the W18 × 35 beam is fastened to the W21 × 68 girder with a single-plate connection. The ⅜″-thick plate would be welded to the girder web and between the girder flanges similar to the way the web stiffeners are welded to the W16 × 40 beam in Figure 7–18. The W18 × 35 is then fastened to the connecting plate with ¾″-diameter bolts similar to the single-plate connection detail shown in Figure 7–28. However, in Figure 7–36, notice that the top flange of the W18 × 35 must be coped so that it will not interfere with the top flange of the W21 × 68 girder, and the bottom flange of the W18 × 35 must be cut off so it does not interfere with the connection plate. Being able to visualize how components of the structural framework must fit together is an important skill for the structural drafter.

Seated Beam-to-Girder Connections

Figure 7–38 is an illustration of a seated beam connection in which a W14 × 34 beam is connected to the web of a W24 × 94 girder. Notice first that, because the top-of-steel elevation of the W14 × 34 beam is below the top-of-steel elevation of the girder, the beam does not have to be coped. It can simply be sawed off square on the end. The beam is connected to a 6″ × 4″ × ¾″ × 0'–8″-long seat angle that was welded in place on the girder web in the fabricator's shop with two 5/16″ fillet welds on either side of the vertical leg of the angle. With this type of beam-to-girder connection, the structural drafter must check very closely to make sure the girder is deep enough so that the vertical leg of the seat angle will not have an interference with the bottom flange of the girder.

After the W14 × 34 beam is connected to the seat angle with two ¾″-diameter erection bolts, the detail calls for two different field welds. As previously mentioned, field welds are performed at the job site during the erection of the steel. The beam is to be welded to the top of the seat angle with ¼″ fillet welds 2″ in length on both sides of the bottom flange of the beam. Also, the top stabilizing angle is to be field-welded where shown with ¼″ fillet welds.

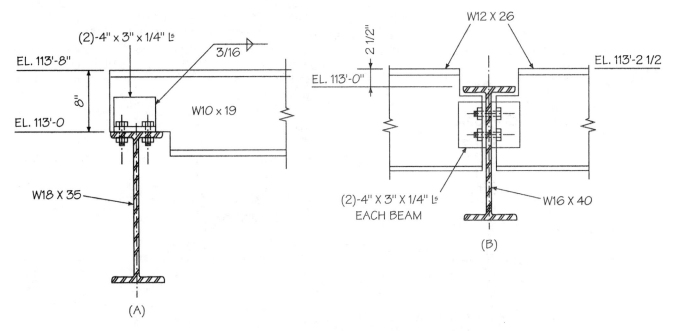

Figure 7–36 Coped beam-to-girder connections

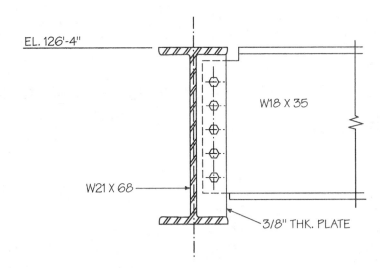

Figure 7–37 Beam-to-girder single-plate connection detail

Beam-to-Girder Flange Connection

Figure 7–39 is a detail in which a W14 × 34 beam is connected to the bottom flange of a W24 × 68 girder with four ¾″-diameter bolts. This is a simple, routine type of connection that the structural drafter can expect to encounter countless times during the course of his or her career.

Skewed Beam-to-Girder Connections

Figure 7–40 is an example of a skewed beam-to-girder connection. Notice on this detail that the work point is at the intersection of the center lines of the webs of both the W14

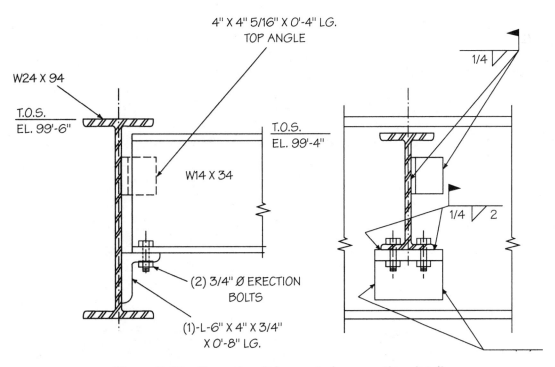

Figure 7–38   Beam-to-girder seated connection detail

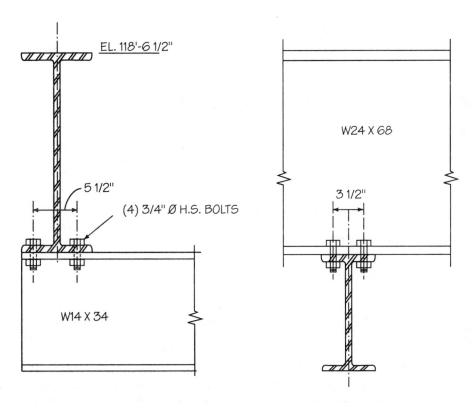

Figure 7–39   Beam-to-girder flange connection detail

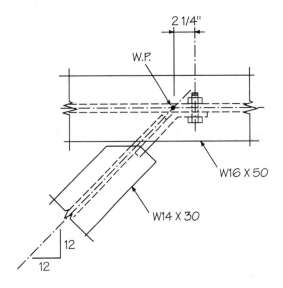

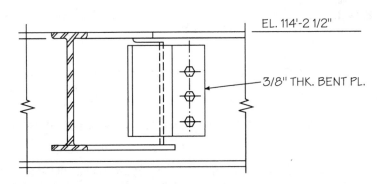

Figure 7–40 Skewed beam-to-girder connection detail

× 30 beam and the W16 × 50 girder. The connection is made with a single bent plate, and the single row of bolts is located 2¼″ to the right of the work point in the plan view. The angle of skew is shown as $\frac{|12}{12}$ and since the top flanges of both beam and girder are at the same elevation, the top flange of the W14 × 30 beam is shown cut off for clearance similar to the detail in Figure 7–34.

The bottom flange of the beam would have to be trimmed also, but it would be cut off close to the web of the girder. The structural design drafter in the structural engineering office would usually not show the bottom flange cut on his or her drawings; that would be the task of the structural detailer working in the fabricator's office.

7.7 STEEL JOIST-TO-COLUMN AND JOIST-TO-BEAM CONNECTIONS

In addition to being familiar with the types of open-web steel joists manufactured for commercial and industrial construction, the structural drafter must understand how steel joists are connected to beams and columns as part of the structural steel support frame. Figures 7–41 to 7–51 illustrate a variety of commonly used joist-to-column and joist-to-beam connection details.

Figure 7–41 shows a steel joist connection at the flange face of a W10 × 49 structural steel column. Notice that the joist is supported by a 4″ × 4″ × ⅜″ × 0′-5″-long seat angle that has been welded to the flange of the column in the fabricator's shop. This detail shows the joist secured to the angle with two ½″-diameter bolts. In most cases, the joist would subsequently be welded to the angle with a field weld. One reason these joists at column connections are initially bolted to the seat angle is to temporarily secure the joist to the angle, which is only 5″ long, thus ensuring that the steel joist is not accidentally knocked off the angle during the erection process. Notice also that the top-of-angle, or joist bearing, elevation is also shown on the detail. The structural detailer in the fabricator's shop must have this information in order to properly prepare the shop detail drawing required to fabricate the column. Incidentally, if the joists were being connected to the web of the column rather than to the flange, the same type of seat angle connection would be used.

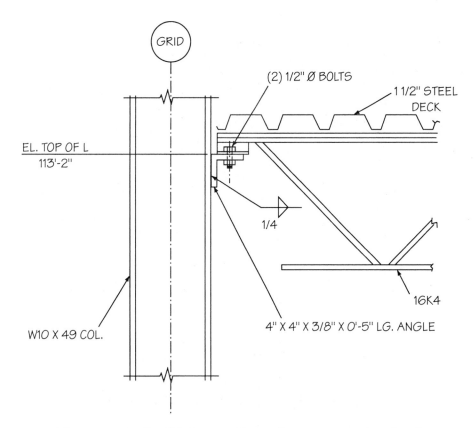

Figure 7–41 Steel joist-to-column flange connection detail

Figure 7–42 is an example of a single steel joist connected to the top of a W24 × 55 girder. The weld symbol indicates that a ³⁄₁₆″ × 2″-long field weld is to be made on both the near and far sides of the joist bearing plate. Notice the 3″ × 3″ × ¼″ continuous angle welded to the end of the joist. This angle acts as a stop, or form, for the concrete floor to be poured on the 1½″ steel deck. Notice also that both the joist bearing elevation and the top-of-floor elevation are shown on the detail.

Figure 7–43 is a detail showing two 18K4 open-web steel joists connecting to the top flange of a W16 × 40 girder with ⅛″ × 2″-long fillet welds performed at the job site by the ironworkers on the field erecting crew. Notice that the detail calls for a ¼″ space between the ends of the joists for erection clearance. Drafters usually allow ¼″ to ½″ for this purpose. When drawing this type of connection, the structural drafter must keep in mind that the top flange of the supporting girder should be at least 6″ wide. If for some reason this is impracticable, it would be permissible to stagger the joists so that they are not directly in line but sit one behind the other on the beam. As always, the joist bearing elevation is shown on the detail.

Figure 7–44 is, again, a detail showing two steel joists connecting to the top flange of a beam. In this example, the joists are of different depths, one being a 12K5 and the other a 16K4. Notice that the top-of-beam or joist bearing eleva-

tion is given. In this detail, the joists have extended bottom chords that would be field-welded to the short, 3″ × 3″ × ¼″ angles at erection time. The beam itself is shown connected with field bolts to the top of the 4″ extra-strong steel pipe column through a 6″ × ½″ × 1′–0″-long cap plate.

Figure 7–45 is a detail illustrating two 18K6 open-web steel joists connecting to the cap plate of a 5″ × 5″ × ⅜″ tube column with field welds. Notice that the top-of-column and joist bearing elevation is shown. This is information the structural detailer in the fabricator's shop must have in order to prepare the shop drawing required to fabricate the column.

Figure 7–46 is a detail showing one method of bracing steel joists at a column location. This is often done to help stabilize the column.

Figure 7–47 is a detail that shows a W14 × 38 beam supporting a 12K1 steel joist from one side and a joist substitute from the other side. A *joist substitute* is specifically fabricated to support a very short span of floor or roof load. Although the SJI steel joist tables do not list joists for spans of less than eight feet, joists for such short spans can be provided by the joist manufacturer. Still it is often more economical to make joist substitutes in the fabricator's shop.

The main point to keep in mind with joist substitutes is that the dimension from the bearing surface to the top of the joist substitute must be the same as that of the nearby joists,

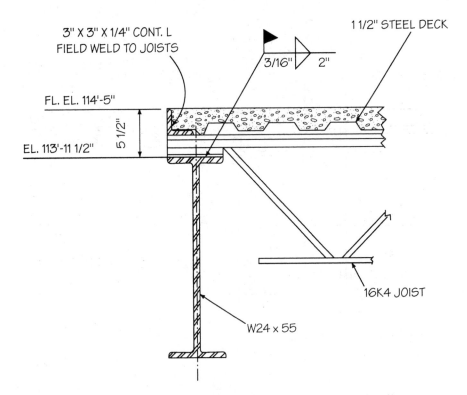

Figure 7–42 Steel joist-to-girder connection detail

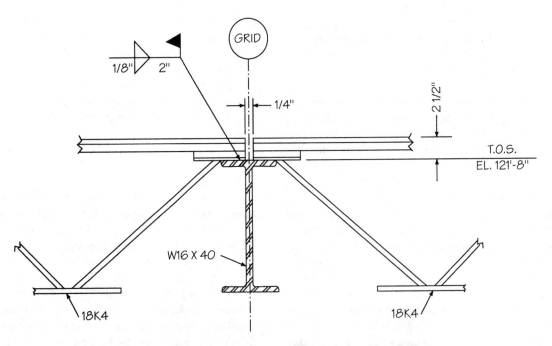

Figure 7–43 Steel joist-to-girder connection detail

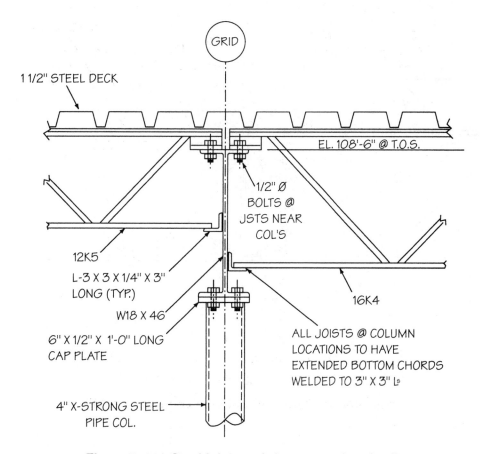

Figure 7–44 Steel joist-to-girder connection detail

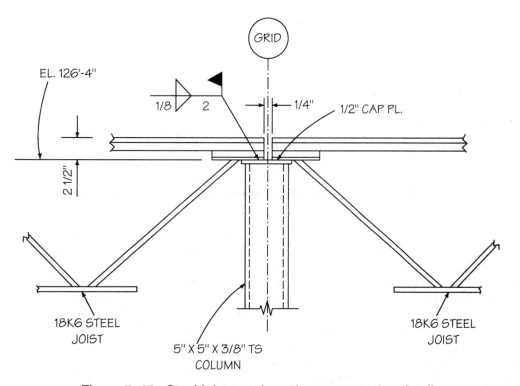

Figure 7–45 Steel joist-to-tube column connection detail

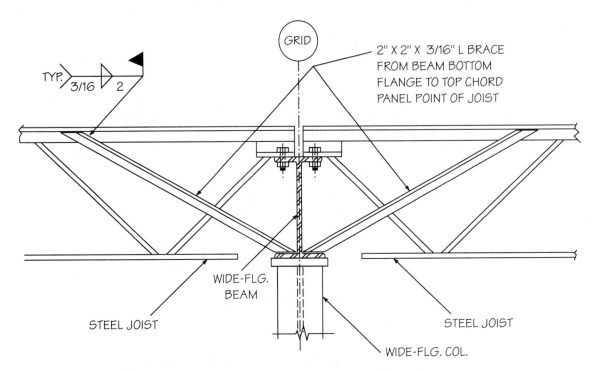

Figure 7–46 Steel joist bracing detail at column

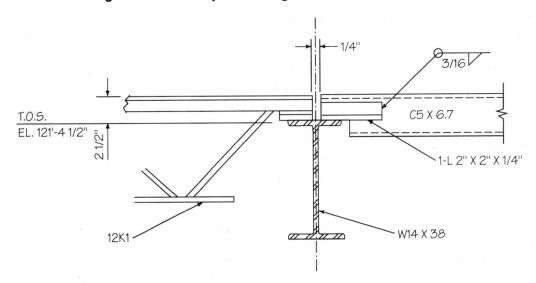

Figure 7–47 Joist substitute-to-girder connection detail

which in the case of a standard 12K1 would be 2½″. This detail calls for the joist substitute to be fabricated from one C5 × 6.7 structural channel that has been coped out on the bottom. A 2″ × 2″ × ¼″ angle is to be welded to the back side of the channel, and this angle will then be field-welded to the top of the W14 × 38 beam.

Figure 7–48 is a detail of a situation commonly found in structural steel drafting. A W16 × 45 girder supports a standard K-series steel joist on one side and a longspan joist on the other. Because the 12K3 has a 2½″ depth from the top chord to the bottom of its bearing plate and the longspan

joist has a 5″ depth from the top chord to the bottom of its bearing plate, a spacer of some type will be needed to keep the top-of-joist elevation the same across the connection. In this example, the structural drafter has elected to weld two 2 ½″ × 2½″ × ¼″ angles together and then to weld them to the top flange of the W16 × 45 as shown. Next, the 12K3 joist can be welded to the top of the spacer, and the 24LH10 welded to the top of the beam so that the tops of both joists will be at the same elevation. Rather than using two angles, a 2½ × 2½ × ¼ structural tube could also have been used.

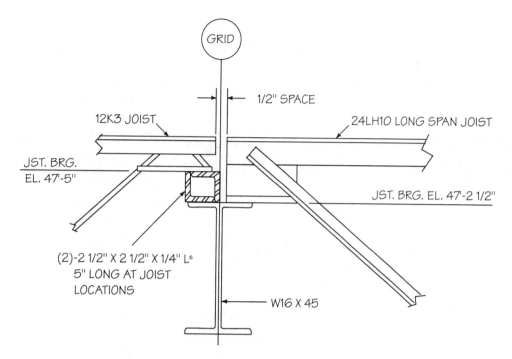

Figure 7–48 Joist-to-girder connection detail

When designing structural steel framing systems, it is not unusual for the joists in two adjacent bays to run at 90-degree angles to one another. In other words, the steel joists in one bay may be spanning in an east-to-west direction, while the joists in the next bay span north-to-south. When this happens, the steel deck sections setting on top of the joists also run in different directions. This requires a detail to show how the steel deck is to be supported at the point where the east-west and north-south sections of deck meet.

Figure 7–49 is an example of such a detail. The 20K6 joist welded to the top of the W18 × 50 beam supports its steel deck, while the end of the deck spanning at right angles to it is supported by a 2½″ × 2½″ × ¼″ continuous angle welded to the top of the beam flange. In this detail, ¼″-thick stiffener bars are welded to the bottom side of the angle and to the top of the beam at spaces of 4′–0″ on center.

Figure 7–50 is similar to Figure 7–49 except that a 2½″ × 2½″ × ¼″ continuous length of structural steel tube supports the end of the decking, which could either be the deck over the 16K3 joist or the end of steel deck at right angles to the joist as shown in Figure 7–49.

The discussion of open-web steel joists in Chapter 2 explained the need for the structural drafter or designer to specify the number of rows of bridging required to stabilize the joists against lateral buckling. The sample structural steel framing plans in Chapter 6 (Figures 6–22, 6–26, and 6–27) illustrate how joist bridging should be shown on a steel framing plan.

Figure 7–51 is a detail of one of the methods by which joist bridging is fastened to the building structure itself. In this example, a 3″ × 3″ × ¼″ structural steel angle is fastened to a cast-in-place concrete or concrete masonry wall with ½″-diameter expansion bolts. The rod bridging is then field-welded to the back of the angle, anchoring the bridging firmly to the building structure.

7.8 BEAM AND JOIST POCKET DETAILS

A discussion of structural steel framing connections would be incomplete without mentioning two of the most common types of structural steel connection details: beam bearing pockets and joist pockets. Both are methods by which steel beams and open-web steel joists are connected to poured concrete walls. Thus, they appear in virtually any steel-frame building structure that uses steel beams and/or joists.

Figure 7–52 is a detail of a beam bearing pocket. A *beam bearing pocket* is a pocket in a foundation wall constructed for the purpose of connecting a beam below the floor. The detail shown in Figure 7–52 illustrates a W18 × 46 beam set in a beam pocket. A 6″ × ¾″ × 1′–2″-long beam bearing plate has been mounted in the pocket over a ¾″ thickness of grout and fastened in place with two ¾″-diameter anchor bolts. While the length of the anchor bolts is usually not specified on the design detail, a ¾″-diameter anchor bolt of the type shown would usually extend at least 12″ down into the foundation wall. With a minimum 2″-long hook at the bottom end of the bolt and a required projection

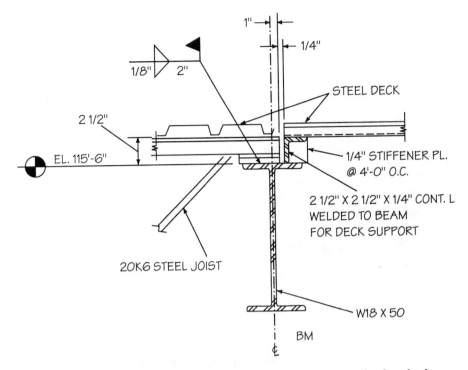

Figure 7–49   Deck support angle detail for perpendicular deck

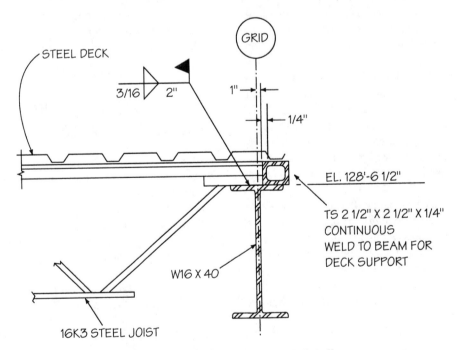

Figure 7–50   Deck support detail

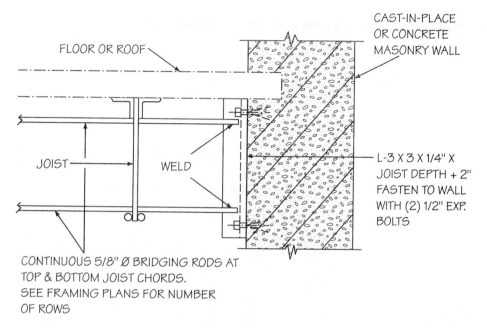

Figure 7–51 Joist bridging detail at foundation or masonry wall

FLOOR OR ROOF

CAST-IN-PLACE
OR CONCRETE
MASONRY WALL

JOIST

WELD

L-3 X 3 X 1/4" X
JOIST DEPTH + 2"
FASTEN TO WALL
WITH (2) 1/2" EXP.
BOLTS

CONTINUOUS 5/8" Ø BRIDGING RODS AT
TOP & BOTTOM JOIST CHORDS.
SEE FRAMING PLANS FOR NUMBER
OF ROWS

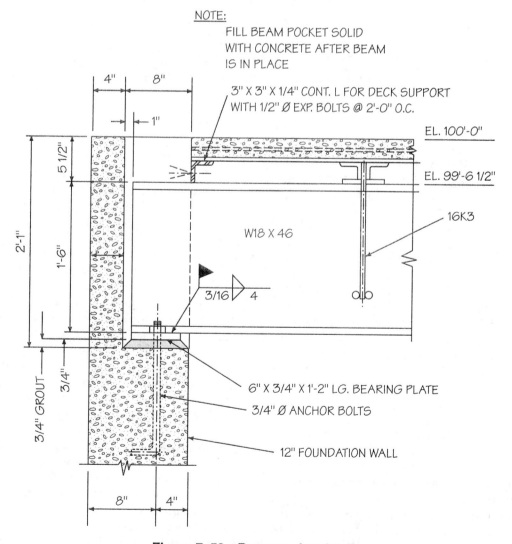

NOTE:
FILL BEAM POCKET SOLID
WITH CONCRETE AFTER BEAM
IS IN PLACE

3" X 3" X 1/4" CONT. L FOR DECK SUPPORT
WITH 1/2" Ø EXP. BOLTS @ 2'-0" O.C.

EL. 100'-0"

EL. 99'-6 1/2"

16K3

W18 X 46

3/16   4

6" X 3/4" X 1'-2" LG. BEARING PLATE

3/4" Ø ANCHOR BOLTS

12" FOUNDATION WALL

3/4" GROUT

Figure 7–52 Beam pocket detail

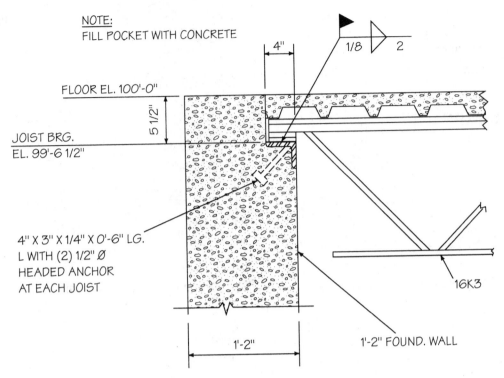

NOTE:
FILL POCKET WITH CONCRETE

FLOOR EL. 100'-0"

5 1/2"

JOIST BRG.
EL. 99'-6 1/2"

4"

1/8   2

4" X 3" X 1/4" X 0'-6" LG.
L WITH (2) 1/2" Ø
HEADED ANCHOR
AT EACH JOIST

16K3

1'-2"

1'-2" FOUND. WALL

Figure 7–53   Joist support angle detail

of approximately 4″ above the bottom of the beam pocket, the overall length of the anchor bolt before the hook is bent into it would be about 1′–6″. Anchor bolt details are the responsibility of the structural fabricator and will be discussed in more depth in Part 2.

After the beam bearing plate has been leveled and secured into position, the W18 × 46 beam would be set in place on the bearing plate and field-welded to the plate with $^3/_{16}$ × 4″-long fillet welds on the near and far side of the bottom flange. Notice that this detail also shows the anchor bolts located 4″ from the inside surface of the foundation wall. The structural detailer in the fabricator's drafting room would see from this detail that the bolts are in the center of the 6″-wide bearing plate. Thus, figuring a minimum edge distance of 1½″ from the center of the holes to the end of the plate in the longitudinal direction, he or she would locate the anchor bolts in that direction 11″ apart, or 5½″ on either side of the beam web.

Notice also that this detail calls for a continuous 3″ × 3″ × ¼″ shelf or deck support angle to run parallel to the joists along the inside of the foundation wall. Referring back to Figure 6–14 will show how this angle would be represented on the structural steel framing plan.

The top-of-floor and top-of-steel, or joist bearing, elevations are also shown on the detail. The dimensions of the beam pocket itself are often not shown on the design details but left up to the general contractor. However, the informa-

tion on this detail indicates that the drafter was visualizing a beam pocket 8″ long in the direction parallel to the beam, 2′–1″ deep from the top of the foundation wall, and 1′–4″ wide in longitudinal dimension parallel to the longitudinal dimension of the beam bearing plate. Notice also the note specifying that the beam pocket is to be filled in with concrete after the beam is mounted in place.

Figure 7–53 is a design detail showing a joist support angle embedded into a joist pocket at the top of a foundation wall. The support angle consists of a 4″ × 3″ × 0′–6″-long angle embedded into the concrete wall with two ½″-diameter headed anchors that would have been welded to the underside of the angle in the steel fabricator's shop. The detail specifies that the 4″ leg of the angle is the horizontal leg. Again, the top-of-support angle, or joist bearing, elevation is given along with the top-of-floor elevation. The joist is also to be welded to the top of the angle by the ironworkers at the job site with ¼″ × 2″-long fillet welds on the near and far side of the joist bearing plate.

The required length of the angle is determined by the reaction of the joist and the allowable strength of the concrete. The angle shown has a 4″-wide × 6″-long bearing surface, or 24 square inches, to spread the reaction of the joist onto the concrete wall. Thus, if the allowable stress of the grade of concrete being used for the wall was 800 pounds per square inch, this detail would be acceptable for a joist reaction of approximately 2,000 pounds, or 2 kips. It should be mentioned that sometimes structural drafters and designers, rather than specifying a joist support angle at each joist location, call for a continuous angle to run along the entire length of the wall.

7.9   SUMMARY

The ability to clearly visualize the details of how the various structural components of a steel framing system are connected for a commercial or industrial steel-frame building is the most important skill a structural drafter or designer can possess. The details included in this chapter, as well as those in Chapter 6, have barely scratched the surface of the hundreds of structural connection details the entry-level structural drafter, designer, or engineer can expect to encounter during the course of his or her career. The information covered in Chapters 6 and 7 of this textbook has been only a base upon which to build.

The end of this chapter is the end of Part 1 of *Structural Steel Drafting*. These chapters have attempted to take the student from the point of having no knowledge whatsoever of structural steel to the point of having a good fundamental perception of how to prepare the structural steel framing plans, sections, and details an entry-level drafter would be expected to make—using either manual or CAD techniques—in a typical structural engineering design office.

Along the way, we have discussed various other aspects of structural drafting and design. A brief overview of basic structural engineering calculations was intended to provide the beginning structural drafter with a background as to "where the numbers come from" for determining the required sizes of various structural steel members. Desirable personal characteristics for those training to enter this career field were discussed because it takes a thoughtful, patient individual with a genuine interest in structural drafting and design to perform this type of work on a day-to-day basis. What to look for and how to read the all-important architectural drawings was explained. Available strengths and types of structural steel shapes from which the steel framing systems are constructed were introduced. And finally, special emphasis was given to national organizations such as the American Institute of Steel Construction and the Steel Joist Institute as very important sources for structural drafters and designers.

We are now ready to move on, to learn the next step following the preparation of the structural steel design drawings for commercial and industrial buildings. That is when the structural detailer in the fabricator's office takes the information from the design drawings and produces the required anchor bolt plans, erection plans, and shop detail drawings. From the drawings prepared in the fabricator's office, structural members such as beams and columns are built or fabricated and the structural steel support frame is erected. This process will be the subject of Part 2 of *Structural Steel Drafting*.

STUDY QUESTIONS

1.   What is the main purpose of structural sections and details?
2.   The generally accepted method of connecting structural steel columns to their supporting footings or foundation walls is by welding the bottom of the column to a _____, which is in turn fastened to the footing or wall by _____.
3.   What is grout?
4.   Structural steel columns often fasten to a concrete pedestal, which in turn is tied into the footing. What is the purpose of a pedestal?
5.   What is a setting plate, or leveling plate, and what is its purpose?
6.   Structural steel columns are often mounted on foundation walls in column pockets. What is the purpose of a column pocket?
7.   What is the purpose of a column schedule?
8.   What is a column splice connection, and where are column splices usually located?
9.   What is a column cap plate, and what is its purpose?
10.   Why are overhanging beams commonly used in the design of structural steel support frames for building roofs?
11.   What are web stiffeners?
12.   What are framed beam connections?
13.   Define web legs and outstanding legs.
14.   What is another name for framing angles?
15.   What size and type of connecting bolts are used for the majority of structural steel connections for commercial buildings?
16.   What is a shop-welded and field-bolted connection?
17.   What is a one-sided framing connection?
18.   Why are seated beam connection angles thicker than framed connection angles?
19.   What is the purpose of the top angle when used with a seated beam-to-column or beam-to-beam connection?
20.   Not all fabricating shops are equipped to do end-plate shear connections. Why is this true?
21.   When drawing a detail of a beam-to-column single-plate shear connection, the plate is not located on the center line of the column. Why is this, and how far is it offset?
22.   What is a thru-plate connection? Draw a sketch if necessary to explain this.

23. Moment-resisting connections are designed to resist both _____ and _____ .

24. Name two types of nonrectangular beam-to-column connections.

25. When preparing details for nonrectangular framing, the drafter often shows a small black dot and refers to it with the letters "W.P." What do these letters signify?

26. What is a beam cope, and when is it required? When explaining this, draw a sketch if necessary.

27. When drawing a detail of two joists connecting to the top flange of a girder, the drafter usually shows a space of _____ to _____ between the joists for erection clearance.

28. The SJI steel joist tables do not list joists for spans of less than eight feet, thus _____ are often used to support very short spans of floor or roof load.

STUDENT ACTIVITY

Using either manual or CAD drafting technique, divide seven 12″ × 18″ sheets into two 12″ × 9″ parts by drawing a light line from top to bottom. At a scale of 1½″ = 1′–0″, draw the following structural details:

1. On the left side of sheet 1, draw a column baseplate connection detail. Select any of the column baseplate connections shown in Figures 7–1 through 7–9. On the right side of the sheet, draw a column cap plate connection consisting of either views *A* and *B* or *B* and *C* on Figure 7–15.

2. On sheet 2, draw the detail shown in Figure 7–20 on the left side and the detail shown in Figure 7–22 on the right side.

3. On sheet 3, draw the seated beam connection illustrated in Figure 7–25 on the left side and Figure 7–27 on the right side.

4. On sheet 4, draw the detail shown in Figure 7–27 on the left side and Figure 7–31 on the right side.

5. On sheet 5, draw the detail shown in Figure 7–33 on the left side and Figure 7–35 on the right side.

6. On sheet 6, draw the detail shown in Figure 7–42 on the left side and Figure 7–48 on the right side.

7. On sheet 7, draw the detail shown in Figure 7–52 on the left side and Figure 7–53 on the right side.

Note

It has long been said that the best way to learn is by doing. Thus, while the number of details drawn by the student should be left up to the instructor, the author firmly feels that the more details each student draws, the better grasp he or she will have of how structural steel connections are visualized—and this visualization is truly at the heart of structural drafting.

Whether drawing the details by manual or CAD methods, the student should remember to keep all object lines heavy and dark, arrowheads dark, extension and dimension lines thin but clear, and all lettering large enough and dark enough to be easily read, which means lettering should be about ⅛″ high. The student should also remember never to crowd lettering or dimensioning.

Part
2

Structural Steel Fabrication Drawings for Steel Construction

Chapter 8

An Introduction to Structural Steel Shop Drawings

OBJECTIVES

Upon completion of this chapter, you should be able to:

- understand the structural designer/fabricator relationship.
- distinguish between structural steel design drawings and structural steel shop, or detail, drawings.
- explain the general concepts of structural steel detailing as they relate to the preparation of structural steel shop drawings.

8.1 INTRODUCTION

The first seven chapters of this textbook provided the technical information required to prepare structural steel design drawings for commercial and industrial buildings according to the standards of a typical structural engineering design office. This chapter and those that follow will acquaint the student with generally accepted methods of preparing the very important structural steel shop drawings. *Shop drawings,* also called detail drawings, are the actual drawings of each individual building component, such as a beam, column, or anchor bolt. These drawings are prepared in the structural steel fabricator's office and then taken the fabricator's shop where the individual structural parts are built. After being fabricated, all these components must fit together at the job site to become the structural steel support system for a commercial or industrial building.

Drafters who prepare shop drawings are commonly referred to in the industry as *steel detailers.* Steel detailers are structural drafters who work in the drafting offices of structural steel fabricators rather than in structural engineering design offices. It is their responsibility to interpret the engineering design drawings, and from that information, prepare the required shop drawings. The steel detailer also designs the standard connections, lists the material required to fabricate the necessary structural components and prepares the anchor bolt and erection plans from which the structural steel framework is constructed at the job site.

All the information required for the structural steel detailer to prepare the shop or fabrication drawings comes from the design drawings. Thus, the relationship between the structural designer and the structural fabricator is very important and will be a major topic of this chapter. Subsequent chapters will discuss how the preparation of structural shop drawings is a specialized craft, unique in many ways from all other types of drafting. We will also learn that certain industry-wide standards must be followed when preparing shop detail drawings for structural beams, girders, and columns. Other material to be covered includes how to make anchor bolt plans and erection plans as well as how to design and draw various standard framed and seated connections. In other words, much new and absolutely essential material for structural drafters remains to be discussed in Part 2 of *Structural Steel Drafting.*

8.2 THE STRUCTURAL DESIGNER/FABRICATOR RELATIONSHIP

Preparing structural steel shop or detail drawings is very specialized work, and the construction industry always has a tremendous need for trained structural steel detailers. However, it is extremely important that all entry-level structural drafters, whether employed in a structural design office

155

or in a structural fabricator's office, have a complete understanding of how shop drawings are prepared. The reason is obvious when one considers that literally everything a structural drafter does on the drawing board or at the CAD workstation revolves to some degree around the shop drawing. The following paragraphs will explain specifically why it is so important for all structural drafters to be able to draw and understand shop drawings. Students should read this information carefully.

Previous chapters of this book have emphasized the production of structural steel drawings in the structural design office. The text explained how to develop a grid system, prepare a framing plan, size the structural members to sustain the loads, and prepare the necessary sections and details for a set of structural steel design drawings. Now some questions:

1. What is the purpose of the design drawings?
2. What happens to the design drawings after they leave the design office?
3. Who are the design drawings made for?
4. Are the individual steel beams, girders, and columns built from the design drawings?
5. Is the structural steel building frame assembled at the job site from the design drawings?

In answer to questions 1 and 2, the structural design drawings, along with the structural specifications, show how the steel framework for a building must be constructed. After the structural design drawings and specifications are completed, they are first used by the general contractor and the structural fabricator to estimate the structural steel requirements and costs for the project.

To answer question 3, it is very important to understand that, once the project is underway, the structural fabricator's office refers to the design drawings in its own drafting department while preparing the shop drawings. The shop drawings are then taken out into the shop, and from them, the required structural steel beams, girders, and columns are built. Thus, it is essential that *all* the information necessary to prepare the shop drawings is shown clearly on the design drawings. And the more every structural drafter knows about making shop drawings, the better are his or her chances of including the necessary information on the design drawings. Thus, the design drawings are in a very real sense made for the structural detailer working in the office of the structural fabricator.

The preceding paragraph has also answered question 4. That is, the individual structural steel components (beams, columns, etc.) are not built directly from the structural design drawings. They are built from shop drawings prepared in the fabricator's office.

Question 5 asked if the structural steel building frame is assembled at the job site from the design drawings. Again, the answer is no. As the structural detailer prepares the shop drawings, he or she assigns an identification mark to each part, such as a beam, column, lintel, or leveling plate, and then draws an *erection plan,* which is very similar to the framing plan within the set of structural design drawings. The erection plan gives the ironworkers the information required to install the structural steel support frame at the job site. The ironworkers also use the erection plan to set each beam or column in the right place and in the correct orientation. For example, the detailer might make a shop detail drawing of a beam identified as 1B1 and designed to fit between two columns, which the detailer has called 1C1 and 1C2. The steel detailer then makes an erection plan that clearly shows a beam with 1B1 painted on it and indicates it is to be installed between columns 1C1 and 1C2. As we shall see, it gets a little more complex than that, but the important point for now is to realize that the structural steel support frame is erected from the erection plan prepared in the fabricator's office, *not* from the design plans.

8.3   IMPORTANT CONSIDERATIONS IN STRUCTURAL STEEL SHOP DRAFTING

Learning to properly prepare shop drawings is somewhat like learning to read music. It may be difficult at first because so much is totally new. Also, when preparing shop drawings, there is a right way and a wrong way to do things, and they must be done right. Accepted structural detailing standards and practices must be followed. For example, whether working manually or at a CAD workstation, beams and columns must be drawn in accordance with industry standards, welding symbols must be understood and drawn correctly, all lettering must be dark and easy to read.

While some of the requirements may seem strict and difficult to grasp at first, students often find that the regimentation of the drafting process is precisely what makes structural detailing quite easy once the basic concepts have been mastered. The steel detailer does not have to guess how to dimension a beam or column, how to represent a desired weld, or what to show on an anchor bolt or framing plan. You know how to do it. You know what to show on your drawing.

From the beginning, every structural drafter should be aware of five important considerations in preparing structural steel shop drawings: accuracy, legibility, clarity, neatness, and speed. Although all five are essential, accuracy is unquestionably the most important. The numbers must add up; the pieces must fit. If dimensions are not accurately calculated, serious and costly errors can result.

A good general rule is that structural steel shop drawings should contain all of the required information in a form that can be quickly and correctly interpreted without loss of time in the fabricating shop or on the job site. Whether the shop drawing is prepared manually or at a CAD workstation, all linework and lettering must be legible—that is, easy to read and understand. Shop drawings should be simple, clear, and complete. They should never contain unnecessary lines, marks, symbols, or dimensions.

Neatness, a very important part of any type of drafting, means using the proper line thicknesses for object lines, extension lines, dimension lines, cutting plane lines, etc. Neatness also means having all lettering dark, clear, and easy to read whether done manually or with CAD software. But neatness involves even more than that. Neatness requires proper planning before the drafter begins to draw. Are you leaving enough room so that you will not have to crowd your lettering and/or dimensions? Do you have an idea of how many beams or columns you should show on one detail sheet? The neatness of the final drawing often depends on how much thought the drafter puts into a project before drawing a single line.

Speed is important in preparing structural steel shop drawings for two basic reasons. First, shop drawings should be produced as economically as possible. Suppose that a drafter working at firm A makes a drawing in one day (eight hours), while a drafter at firm B takes two days to produce essentially the same drawing. If the drafters at both firms are being paid $15 per hour, the same drawing will cost firm A $120 and firm B $240. Before long, firm B will be out of business.

A second reason for speed is that, once the drawing is complete, the shop needs time to fabricate the steel and deliver it to the job site on schedule. Steel fabrication offices are interested in production. Their profits depend upon how many tons of steel they can ship to various jobs each month. Production is simply a fact of life in the world of work and one that every structural drafter should be aware of.

The Importance of Clarity

When preparing shop detail drawings and erection plans, the steel detailer should keep in mind who will be reading them. Architectural drawings are usually prepared for businessmen, building committees, engineering consultants, or building contractors, but structural shop drawings are prepared for ironworkers, either in the fabricating shop or at the construction site. Thus, while the client or businessman looking at the architectural drawings might be favorably impressed by fine lines, shading, and sometimes

even fancy lettering, "pretty pictures" do not impress the ironworker at all.

What does the ironworker want? Remember, he or she is sometimes working under less than perfect conditions. The prints may become dirty, torn, or smudged in the fabrication shop or at the job site, so the ironworker needs heavy, dark object lines. Beams and columns should stand out so they can be easily seen. Lettering should be uppercase, clear, $\frac{1}{8}''$ high, and dark, with a straightforward technique.

Because clear communication is so important, structural shop drafting, like most types of drafting, have definite requirements for linework and lettering. Whether the drawings are produced manually or at a CAD workstation, correct line weights, lettering style, and heights must be used.

Another important practice when preparing shop drawings is that on structural steel shop drawings the individual beams and columns are rarely drawn to scale. For example, the depth of a beam or column might be drawn to a scale of $\frac{3}{4}'' = 1'-0''$, or $1'' = 1'-0''$, or even $1\frac{1}{2}'' = 1'-0''$ for small beams and columns, but the depth of a beam or column will never be drawn to a scale of $\frac{1}{2}'' = 1'-0''$ or $3'' = 1'-0''$ on a shop detail drawing. Also, the length of a beam or column is almost never drawn to scale. While the depth of a pipe, structural tube, or wide-flange column may be drawn to a scale such as $1'' = 1'-0''$ or $1\frac{1}{2}'' = 1'-0''$, it is extremely rare for the length of a column to be drawn to scale. The reasons for this seemingly strange practice will be explained as we learn to prepare structural steel shop drawings.

The Importance of Accuracy

As previously mentioned, accuracy is the most important consideration for the structural steel detailer. Even though structural steel shop drawings are rarely drawn to scale, the dimensions must be accurate. The numbers must add up correctly so that when the steel components of a structure are sent to the job site, they fit together. The importance of accuracy can best be understood by considering what could happen if a small error was not caught until the steel for a project was delivered to the job site.

Imagine that you are preparing the shop drawings for the steel support frame for a ten-story commercial office building. Each floor might need 50 floor beams exactly alike in size and length. So a ten-floor building might require 500 identical floor beams.

Following normal steel fabricating practice, the structural detailer would draw the beam one time, showing how it should be fabricated in the shop to fit in place out on the job. Then he or she would assign that beam a number (an identification mark) and instruct the shop to fabricate and ship 500 beams as shown. Now, suppose only one little dimension was wrong on the shop drawing, and no one caught the error. Then suppose that all 500 floor beams were

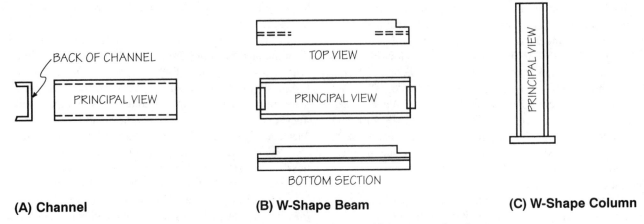

(A) Channel   (B) W-Shape Beam   (C) W-Shape Column

Figure 8–1   Typical principal views of beams and columns

shipped to the job site, and when the ironworkers began to put the first one in place, it was three inches short.

Obviously, there is no room in either the structural design or fabrication office for sloppy work. The hurry-up, slam-bang, throw-things-together individual has no place in this field. Structural drafting is a career for the conscientious, methodical person who is quality-oriented and does the best work he or she can on a day-to-day basis. Accuracy, neatness, and quality of work are characteristics employers of structural drafters and steel detailers look at very closely.

8.4 GENERAL RULES FOR PREPARING STRUCTURAL STEEL SHOP DRAWINGS

At this point, we will discuss some of the fundamental rules followed by steel detailers when preparing structural steel shop drawings.

In general, the arrangement of views on structural shop drawings is done in accordance with the principles of third-angle orthographic projection. According to convention, a drawing called the *principal view* is prepared to show the characteristic shape of each structural member. Once the principal view, which is usually an elevation, has been selected, the top view is shown above it, the right side or end to the right of it, and the left side or end to the left of it. Thus, the shop workers should be able to tell at first glance if they are looking at a wide-flange shape, a channel shape, or a structural pipe or tube.

One exception to the rules of orthographic projection for shop fabrication drawings is that the bottoms of members such as beams and girders are usually shown as a section looking down rather than a bottom view looking up, thereby reflecting their orientation on the design framing plans. These sectional views are understood by the fabricating shop, so no section cutting planes or labeling of bottom sectional views is necessary. Figure 8–1 shows typical principal views of a channel, a beam, and a column.

Note in Figure 8–1 that the principal view is usually shown in the same position on the shop detail drawing as it has in the structure. Thus, beams are drawn horizontally and columns vertically with the top up. Exceptions are very long columns, which are sometimes detailed horizontally because of space limitations on the sheet.

Most structural steel components including beams, girders, columns, and bearing plates, can be sufficiently detailed with only one view, the principal view. However, if required, additional views can be drawn to detail the member clearly and completely. For example, the flange cuts on the top and bottom flanges of the beam in Figure 8–1B could not be shown on the principal view alone.

The structural detailer should also be aware that, it is just as important to draw enough views for the fabricating shop as to avoid drawing unnecessary views. Creating unnecessary drawings not only adds time, and thus added expense, to the steel detailer's employer, but also increases the possibility of costly errors.

When detailing shop fabrication drawings, the depth of beams and columns is usually drawn to a scale. However, as previously mentioned, the longitudinal dimension of beams and columns is usually foreshortened. The most common minimum scale for detailing structural steel beams and columns 14″ deep or more is ¾″ = 1′–0″. For beams and columns with depths of 8″ to 12″, the most commonly used scale is 1″ = 1′–0″. For beams and columns less than 8″ deep, the most common scale is 1 ½″ = 1′–0″. Other scales

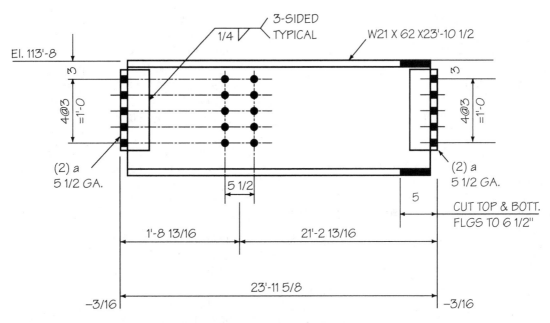

Figure 8–2   A typical fabrication shop beam detail

may be used as the structural detailer feels necessary to clearly show how the structural piece is to be built, always keeping in mind that, ultimately, the drawing must be easily read and understood by the fabrication shop.

Notice on the structural steel shop detail in Figure 8–2 that the object lines on the principal view of the W21 × 62 are firm and black, showing a definite contrast with the center lines and dimension lines. Notice also that two longitudinal dimensions are obviously not drawn to scale: the 1′–8¹³⁄₁₆″ from the left end of the beam to the center of the set of holes through the web of the beam, and the 21′–2 ¹³⁄₁₆″ dimension from the connection point to the right end of the beam. Other characteristics of structural steel shop illustrated in Figure 8–2 include:

- The overall dimension of 23′–11 ⅝″ on the detail is not the length of the W21 × 62, but the end-to-end dimension between the connecting angles.
- The length of the W21 × 62 itself is listed as 23′–10 ½″, or about 1+″ shorter than the overall end-to-end dimension between the connecting angles.
- Inch marks (″) are not used. The overall dimension is shown as 23′–11 ⅝, *not* 23′–11 ⅝″. This feature is unique to shop fabrication drawings.
- The holes through the web of the beam are drawn as black dots, but the hole size is not given. This practice, also unique to structural steel detailing, will be explained in Chapters 9 and 10.

- Dimensioning and welding symbols are read from the bottom of the sheet. It is also permissible for dimensions and welding symbols to be read from the right-hand side of the sheet.
- All dimensions are placed above the dimension line.
- Detail dimensions such as hole spacing are located closest to the principal view. Summations of space, including overall dimensions, are located further away from the principal view of the beam.
- All letters and numbers are written in a straightforward, easy-to-read style designed to clearly relay the desired information. Like object lines, all lettering on shop drawings must be firm and dark.
- Beam flange cuts, which will be discussed further in Chapter 12, are shown in elevation by being blackened. Cross-hatching of the flange cuts is also permissible.

The beam detail in Figure 8–2 is intended to introduce some of the fundamental rules and unique concepts of structural steel shop or fabrication drafting. Much of the information in this and subsequent chapters can be supplemented by referring to either *Detailing for Steel Construction* or *Engineering for Steel Construction,* two AISC publications that are the best references available on structural steel shop drafting. Purchasing these two books is strongly recommended for anyone contemplating a career in structural drafting, and certainly for those intending to become structural detailers.

8.5 SUMMARY

The main objective of this chapter has been to introduce the structural drafting student to the world of the structural steel detailer who, from information found on the design drawings, prepares the shop or fabrication drawings. Structural steel detailing is a highly specialized area of drafting with unique rules and concepts. Yet, it is a discipline that must be thoroughly understood by every structural drafter who produces drawings for commercial and industrial building construction, whether in a structural design office or a structural fabricator's office.

To that end, we have reviewed the structural designer/fabricator relationship, with emphasis on both the differences and the interdependence between engineering design drawings and structural steel shop drawings. We have also discussed some basic structural steel detailing concepts, citing a typical shop detail drawing for a steel beam to illustrate the preparation of structural steel fabrication drawings.

STUDY QUESTIONS

1. Structural drafters who prepare shop fabrication drawings are commonly referred to in the industry as _____.

2. Certain information is required before structural steel shop fabrication drawings can be prepared. What is this information, and where can it be found?

3. What drawings do the ironworkers use at the job site to set each beam or column in the proper location as they construct the structural steel support frame?

4. What five important considerations must the drafter keep in mind when preparing structural steel shop drawings?

5. In structural steel shop drawings, the view usually drawn as an elevation to show the characteristic shape of a member is called the _____ view.

6. When preparing shop fabrication drawings, the depths of beams and columns 14″ deep or more are usually drawn to a scale of _____.

7. For beams and columns 8″ to 12″ deep, the most commonly used scale for showing section depth is _____.

8. For beams and columns less than 8″ deep, the usual scale is _____.

9. What is the usual scale for showing the longitudinal dimension (length) of structural steel beams and columns on shop fabrication drawings?

10. *The Manual of Steel Construction* published by the American Institute of Steel Construction is a very important reference for drafters and designers working in either the structural design or structural fabrication office. Two additional AISC publications that are also excellent sources for structural drafters and designers are _____ and _____.

Chapter 9

Structural Connections

OBJECTIVES

Upon completion of this chapter, you should be able to:

- name the common types of structural connections.
- identify the various types of structural bolts.
- name some common types of structural welds and recognize basic weld symbols.
- describe the design procedures typically followed for standard framed and seated connections.

9.1 INTRODUCTION

An axiom often quoted by structural engineers and designers states that, if a structural steel-frame building is going to fail, it will fail at the connections. The history of recorded failures of structural systems in the United States, including the 1981 tragedy at the Hyatt Regency Hotel in Kansas City, Missouri, has proven the axiom about connection failure to be all too true.

AISC specifications define three types of steel-frame construction based upon the type of framing connection used. They are:

Type 1: commonly designated as rigid frame or continuous frame, which assumes that the connections have sufficient rigidity to hold unchanged the original angles between intersecting members.

Type 2: commonly referred to as simple, unrestrained, free-ended framing, which assumes that, insofar as gravity loading is concerned, the ends of the beams and girders are connected for shear only are free to rotate under gravity loads.

Type 3: commonly designated as semi-rigid or partially restrained framing, which assumes that the connections of beams and girders possess a dependable and known moment capacity intermediate in degree between the rigidity of Type 1 and the flexibility of Type 2.

Type 2 connections are generally known as flexible connections. It is common practice throughout the construction industry for the structural steel fabricator to always provide flexible Type 2 connections unless the structural design plans and specifications clearly indicate that rigid or semi-rigid connections are to be used and give the proper design information for such connections. With this in mind, we will review the basic materials and procedures for selecting some of the more common types of standard structural steel connections. However, before the structural drafting student can successfully learn the fundamentals of designing standard structural steel connections, he or she must have a basic knowledge of the types of bolts used to erect structural steel support frames.

9.2 STRUCTURAL BOLTS

Currently, the primary methods of connecting the major components of a steel-frame building are bolting and welding. Essentially, two types of bolts are used for connections: common bolts and high-strength bolts.

Common bolts, sometimes called unfinished, machine, plain, or rough bolts, are the least expensive, but their applications are restricted. Because of their low carbon content, common bolts, usually referred to as A307 bolts, are not recommended for high-stress situations and thus are not used for major connections. Instead, they are used for noncritical connections such as temporary erection bolts for

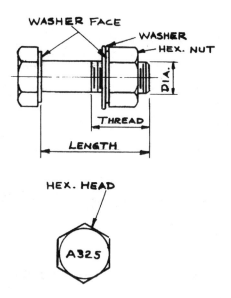

Figure 9–1   High-strength bolt nomenclature

Figure 9–2   Tension control bolt. (Courtesy of Le Jeune Bolt Company)

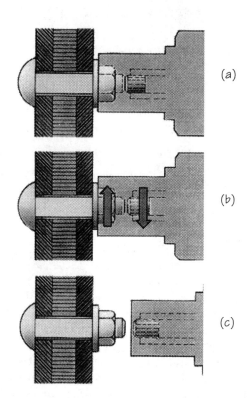

Figure 9–3   Tension control bolt installation process. (Courtesy of Le Jeune Bolt Company)

welded seat angle connections or for fastening secondary structural members such as purlins, girts, stairs, or platforms, which are not part of the main structural support frame for a building.

High-strength bolts, used for the vast majority of field connections, are considerably stronger than common bolts. They are furnished in two grades: ASTM A325 high-strength carbon steel and A490 high-strength alloy steel. Of the two types, the A325 is clearly the one more commonly used in commercial and industrial construction. For most types of bolted connections, a round, plain, hardened washer is furnished with the bolt. Figure 9–1 illustrates typical nomenclature for high-strength bolts. The structural drafter should know this nomenclature (i.e., length, thread, hex, head, washer, etc.) because he or she will work with it very routinely when preparing the *field bolt list,* an important part of any structural steel framing project. The field bolt list enumerates all the different lengths and sizes of high-strength bolts required for the ironworkers to fasten together the beams, girders, columns, and other components of the structural steel support frame at the job site.

In recent years, a new and very efficient type of high-strength bolt, called a *tension control bolt,* has been introduced. The tension control bolt combines the following advantages: low-cost, one-man/one-side installation, built-in positive tension control, and visual inspection, which eliminates the need for continuous torque testing. Figure 9–2 is an illustration of a tension control bolt.

Tension control bolts are high-strength bolts available in either the ASTM A-325 or A-490 specification. They are very economical because they can be installed by one iron-

worker with the aid of a double-socket electric wrench, which engages the bolt tip and the nut simultaneously as shown in Figure 9–3A.

When the electric wrench is activated, the bolt remains stationary while the outer socket rotates the nut, causing it to tighten the connection. When the nut is tightened to the proper bolt tension, the outer socket stops, and the inner socket of the electric wrench, rotating counterclockwise, shears off the calibrated tip of the bolt as illustrated in Figure 9–3B. The wrench is then removed from the bolt as shown in Figure 9–3C, the sheared tip is ejected, and the ironworker is ready to install another bolt.

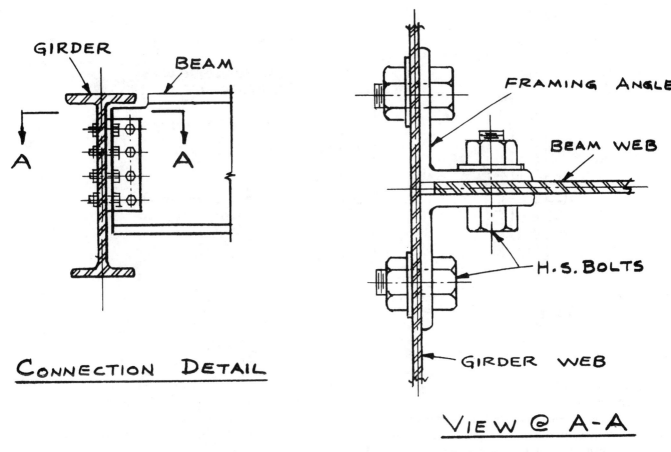

Figure 9–4 Beam-to-girder bolted connection

9.3 BOLTED CONNECTIONS

Figure 9–4 illustrates a beam-to-girder connection. In this illustration, the connection is made with framing angles bolted to the beam through the beam web with high-strength bolts as shown in plan view in Section View A–A. The legs of the angles that fasten to the beam are called web legs. The legs of the angles that complete the connection by fastening to the girder through the girder web are called the outstanding legs.

A load on the beam, such as the dead and live loads of a floor, would cause a reaction at the end of the beam, which itself becomes a concentrated load on the web of the girder. The eight bolts connecting the beam to the girder web through the framing angles and the four bolts connecting the angles to the beam web would be required to withstand the stress created in them by this load. That stress would be a shearing stress on the bolts because, if they were to fail, they would fail by shearing off between the face of the angles and the faces of either the beam web or girder web.

Bearing-Type and Slip-Critical Connections

Two basic types of bolted connections are used to transmit loads. They are *bearing-type connections,* in which the high-strength bolts are assumed to bear against the sides of the holes in the connected material (i.e., the angle legs and the beam and girder webs in Figure 9–4), and *slip-critical connections,* in which the high-strength bolts are assumed to clamp the connected parts together with such pressure that the shearing force is resisted by the friction between the connected parts, and not by shear stress on the fasteners. In slip-critical connections, the connecting bolts are not considered to come in contact with the sides of the holes through which they pass. Slip-critical connections must, however, meet all requirements for bearing connections as well as provide slip-resistance, because it must be assumed that they could eventually slip into bearing.

When a fastener transmits a shear load in a bearing-type connection, the designer assumes a bearing stress is present

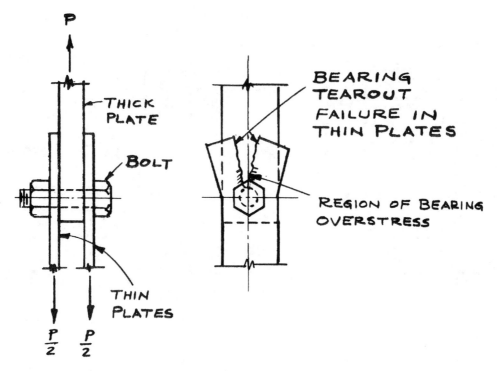

Figure 9–5   Bearing tear-out failure

in both the high-strength bolt and the connected material. Since any structural connection can only be as strong as its weakest element, the material joined together must be strong enough to withstand the bearing stress caused by the bolt. For example, if a ¾″-diameter bolt were connecting two pieces of cardboard together and a load or stress were placed on the joint, the bolt would undoubtedly hold very well, but the joint would surely fail because the cardboard would tear. By the same reasoning, the material being connected in any bearing-type structural joint must be strong enough to withstand the stresses it will develop under load, or else the joint will fail by a bearing or tear-out failure as illustrated in Figure 9–5. Thus, in bearing-type connections, the bearing value of the connected material must be checked.

Common Nomenclature

Before explaining the usual design procedure for standard bolted connections, a few more basic considerations should be discussed. First, the student should be aware of the nomenclature commonly used for computations of standard connections. The following symbols recommended by the American Institute of Steel Construction (AISC) are widely used throughout the structural steel industry and will be used for all calculations in this textbook. The symbols are:

A_b = Nominal body area of a fastener (cross-sectional area based on nominal diameter), sq. in.

F_t = Allowable axial tensile stress, ksi

F_v = Allowable shear stress, ksi

F_p = Allowable bearing stress of the type of steel being used, ksi

F_y = Specified minimum yield stress of the type of steel being used, ksi

F_u = Specified minimum tensile strength of the type of steel or fastener being used, ksi

f_t = Computed tensile stress, ksi

f_v = Computed shear stress, ksi

f_p = Computed bearing stress on support, ksi

f_R = Computed shear or bearing value of one fastener, kips

r_v = Allowable shear or bearing value of one fastener, kips

r_t = Allowable tension value of one fastener, kips

Single Shear and Double Shear

Another consideration in the design of standard bolted connections is whether the bolts themselves are in a single-shear or double-shear condition. Whenever two materials are joined together in a shear connection as illustrated in the beam-to-girder connection in Figure 9–4, these materials (i.e., the faces of the angles and the faces of the beam and girder webs) have a tendency to slide past each other along their contact surfaces. This tendency is resisted by the bolts. When there is one traverse section where this sliding might occur, the bolts are said to be in *single shear*. When there are two traverse sections, or shear planes, the bolt is said to be

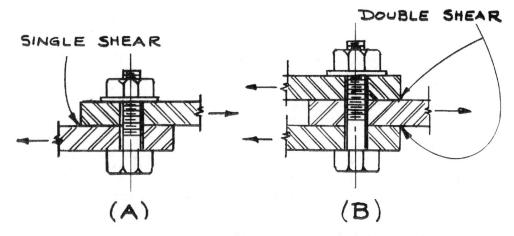

Figure 9–6 Single and double shear

in double shear. Figure 9–6*A* is an example of single shear. Figure 9–6*B* is an example of a double-shear condition. Examination of the bolted connection illustrated in Figure 9–4 shows that the bolts connecting the framing angles to the girder web are in a single-shear condition, while those connecting the framing angles to the beam web are in a double-shear condition.

Determining whether a connection is in single or double shear is a frequent task of the experienced structural drafter or designer in his or her day-to-day work. Another consideration when designing standard bolted connections is that every bolt has a specific load capacity for a specific loading condition. If this capacity were exceeded by the applied load, the fastener would obviously fail. For example, if the high-strength bolts shown in Figure 9–4 were to be overstressed, they would shear off at the shear plane and the connection would fail. This, of course, is exactly what the structural drafter or designer *does not* want to happen.

The ability of a high-strength bolt to resist shear stress at any traverse plane depends upon its diameter or nominal body area (A_b) and its allowable shear stress (F_v). The product of these two quantities, $A_b \times F_v$, is the allowable single-shear value of the bolt. In a double-shear condition having two shear planes, such as the angle-to-beam web connection in View A–A of Figure 9–4, the allowable double-shear value of the bolt doubles simply because there are now two planes of shear. If a connection had a triple-shear condition, the allowable triple-shear value would be three times the value of $A_b \times F_v$, and so on.

In the design of structural steel bolted connections, the single- and double-shear conditions illustrated in Figure 9–4 are by far the most widely used for most day-to-day structural steel connections. Also, the most common practice is to install the bolts that fasten the framing angles to the beam web in the fabricating shop and then connect those that join the framing angles to the web of the girder at the job site or "in the field" by the ironworkers. Thus, the terms shop bolts

and field bolts are very familiar to structural steel drafters and designers.

The allowable shear value or total shear load a single bolt can support in a specific condition of shear is labeled r_v. Thus, shear relationships may be expressed by the equations:

$$r_v \text{ (single shear)} = (F_v \times A_b)$$
$$r_v \text{ (double shear)} = 2 (F_v \times A_b)$$
$$r_v \text{ (triple shear)} = 3 (F_v \times A_b)$$

To assist the structural drafter or designer in designing standard structural steel bolted connections, Table 9–1 from the AISC *Manual of Steel Construction* lists the allowable shear stresses (F_v) for various sizes and types of bolts in kips-per-square-inch (ksi) for both single- and double-shear conditions.

Allowable Shear Stresses

Notice first on Table 9–1 that most of the bolts have various allowable stress ratings for different connection types and loading conditions. For example, A325 high-strength bolts are listed for three connection types. They are A325-SC (slip-critical), A325-N (bearing-type connections with threads *included* in the shear plane), and A325-X (bearing-type connections where the length of the bolt is such that the bolt threads are *excluded* from the shear plane.

Figure 9–7 illustrates what is meant by the terms "threads included" and "threads excluded" from the shear plane. Figure 9–7*A* shows that the bolt threads pass through the shear plane of the connection. This would be designated as an A325-N bolt because the threaded part of the bolt is included in the plane of shear. Because the grooves of the threads decrease the diameter (and thus the area) of the bolt at that point, the allowable shear stress (F_v) will be less than if the threads were excluded from the shear plane and the full diameter of the bolt could be used to resist shear as

TABLE 9–1
Allowable Bolt Shear Loads

BOLTS, THREADED PARTS AND RIVETS
Shear
Allowable load in kips

TABLE I-D. SHEAR

	ASTM Designation	Connection Type[a]	Hole Type[b]	F_v ksi	Loading[c]	5/8	3/4	7/8	1	1 1/8	1 1/4	1 3/8	1 1/2
						.3068	.4418	.6013	.7854	.9940	1.227	1.485	1.767
Bolts	A307	—	STD	10.0	S	3.1	4.4	6.0	7.9	9.9	12.3	14.8	17.7
			NSL		D	6.1	8.8	12.0	15.7	19.9	24.5	29.7	35.3
	A325	SC[a] Class A	STD	17.0	S	5.22	7.51	10.2	13.4	16.9	20.9	25.2	30.0
					D	10.4	15.0	20.4	26.7	33.8	41.7	50.5	60.1
			OVS, SSL	15.0	S	4.60	6.63	9.02	11.8	14.9	18.4	22.3	26.5
					D	9.20	13.3	18.0	23.6	29.8	36.8	44.6	53.0
			LSL	12.0	S	3.68	5.30	7.22	9.42	11.9	14.7	17.8	21.2
					D	7.36	10.6	14.4	18.8	23.9	29.4	35.6	42.4
		N	STD, NSL	21.0	S	6.4	9.3	12.6	16.5	20.9	25.8	31.2	37.1
					D	12.9	18.6	25.3	33.0	41.7	51.5	62.4	74.2
		X	STD, NSL	30.0	S	9.2	13.3	18.0	23.6	29.8	36.8	44.5	53.0
					D	18.4	26.5	36.1	47.1	59.6	73.6	89.1	106.0
	A490	SC[a] Class A	STD	21.0	S	6.44	9.28	12.6	16.5	20.9	25.8	31.2	37.1
					D	12.9	18.6	25.3	33.0	41.7	51.5	62.4	74.2
			OVS, SSL	18.0	S	5.52	7.95	10.8	14.1	17.9	22.1	26.7	31.8
					D	11.0	15.9	21.6	28.3	35.8	44.2	53.5	63.6
			LSL	15.0	S	4.60	6.63	9.02	11.8	14.9	18.4	22.3	26.5
					D	9.20	13.3	18.0	23.6	29.8	36.8	44.6	53.0
		N	STD, NSL	28.0	S	8.6	12.4	16.8	22.0	27.8	34.4	41.6	49.5
					D	17.2	24.7	33.7	44.0	55.7	68.7	83.2	99.0
		X	STD, NSL	40.0	S	12.3	17.7	24.1	31.4	39.8	49.1	59.4	70.7
					D	24.5	35.3	48.1	62.8	79.5	98.2	119.0	141.0
Rivets	A502-1	—	STD	17.5	S	5.4	7.7	10.5	13.7	17.4	21.5	26.0	30.9
					D	10.7	15.5	21.0	27.5	34.8	42.9	52.0	61.8
	A502-2 A502-3	—	STD	22.0	S	6.7	9.7	13.2	17.3	21.9	27.0	32.7	38.9
					D	13.5	19.4	26.5	34.6	43.7	54.0	65.3	77.7
Threaded Parts	A36 (F_u=58 ksi)	N	STD	9.9	S	3.0	4.4	6.0	7.8	9.8	12.1	14.7	17.5
					D	6.1	8.7	11.9	15.6	19.7	24.3	29.4	35.0
		X	STD	12.8	S	3.9	5.7	7.7	10.1	12.7	15.7	19.0	22.6
					D	7.9	11.3	15.4	20.1	25.4	31.4	38.0	45.2
	A572, Gr. 50 (F_u=65 ksi)	N	STD	11.1	S	3.4	4.9	6.7	8.7	11.0	13.6	16.5	19.6
					D	6.8	9.8	13.3	17.4	22.1	27.2	33.0	39.2
		X	STD	14.3	S	4.4	6.3	8.6	11.2	14.2	17.5	21.2	25.3
					D	8.8	12.6	17.2	22.5	28.4	35.1	42.5	50.5
	A588 (F_u=70 ksi)	N	STD	11.9	S	3.7	5.3	7.2	9.3	11.8	14.6	17.7	21.0
					D	7.3	10.5	14.3	18.7	23.7	29.2	35.3	42.1
		X	STD	15.4	S	4.7	6.8	9.3	12.1	15.3	18.9	22.9	27.2
					D	9.4	13.6	18.5	24.2	30.6	37.8	45.7	54.4

[a] SC = Slip critical connection.
N: Bearing-type connection with threads *included* in shear plane.
X: Bearing-type connection with threads *excluded* from shear plane.
[b] STD: Standard round holes (d + 1/16 in.)   OVS: Oversize round holes
LSL: Long-slotted holes normal to load direction   SSL: Short-slotted holes
NSL: Long-or short-slotted hole normal to load direction
(required in bearing-type connection).
[c] S: Single shear   D: Double shear.
For threaded parts of materials not listed, use $F_v = 0.17F_u$ when threads are included in a shear plane, and $F_v = 0.22F_u$ when threads are excluded from a shear plane.
To fully pretension bolts 1 1/8-in. dia. and greater, special impact wrenches may be required.
When bearing-type connections used to splice tension members have a fastener pattern whose length, measured parallel to the line of force, exceeds 50 in., tabulated values shall be reduced by 20%. See AISC ASD Commentary Sect. J3.4.

(Courtesy of the American Institute of Steel Construction Inc.)

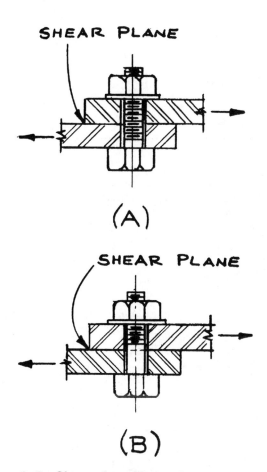

SHEAR PLANE

(A)

SHEAR PLANE

(B)

Figure 9–7   Shear plane illustrations

illustrated in Figure 9–7B. That is why Table 9–1 lists the F_v of the A325-X bolts as 30 ksi and the F_v of the A325-N bolts as 21 ksi. This might lead one to believe that the A325-X bolt would be the type most widely used in structural steel connections, but actually, it is not. In design practice, the A325-N type of connection is most widely used because the designer does not have to worry about whether or not the threads might fall in the shear plane of the connection.

As previously discussed, the A325-SC, or slip-critical, connections assume that the connecting bolts do not touch the sides of the holes through which they pass but hold the connected parts together so tightly that the friction between the connected parts keeps them from slipping past each other. In addition, with slip-critical connections, the structural drafter must always add a note that paint should not be applied between mating surfaces unless the paint has been qualified according to the RCSC (Research Council on Structural Connections). Paint can act as a lubricant and thus reduce the friction of steel against steel. Notice in Table 9–1 that the F_v value in ksi varies from 17 ksi if standard round holes are used, to 15 ksi if oversized or short-slotted holes are used, and down to only 12 ksi if long-slotted holes are used. In other words, the allowable load in kips-per-bolt is

always less in slip-critical connections than for either A325-N or A325-X applications.

To the right of the columns in the table that list the ASTM Designation, Connection Type, Hole Type, and F_v values is a column marked Loading. In this column, for each ASTM designation, are the letters *S* and *D,* which simply identify if the connection is in *single* shear or *double* shear. The rest of the table lists various diameters of bolts from $\frac{5}{8}$″-diameter us to 1½″-diameter in $\frac{1}{8}$″ increments, the areas of the various bolts, and the allowable load in kips-per-bolt depending upon bolt size and whether the connection is in single or double shear. For example, a ¾″-diameter A325-N high-strength bolt is rated for 9.3 kips in single shear and 18.6 kips in double shear. Examination of the connection in Figure 9–4 shows four bolts in double shear connecting the angles to the beam web and eight bolts in single shear connecting the framing angles to the girder web. Thus, the connection would be good for:

(8) bolts × 9.3 kips per bolt = 74.4 kips at girder web
(4) bolts × 18.6 kips per bolt = 74.4 kips at beam web

Thus, it would be safe to say the connection shown in Figure 9–4, insofar as the ¾″-diameter high-strength bolts are concerned, is good for 74 kips using the A325-N type of connection.

Since we are considering the connection shown in Figure 9–4 to be a bearing-type A325-N connection, the bearing value of the material being connected must be checked to ensure that the connection will not fail because of the web of the beam or girder "tearing out" as illustrated in Figure 9–5. This simple procedure requires referring to Table 1–E in the AISC manual, reproduced here as Table 9–2.

Bearing Capacities

Notice in Table 9–2 that the column on the far left lists material thicknesses ranging from $\frac{1}{8}$″ to 1″. The next four columns list various grades of structural steel. Each of those columns is divided into three columns for bolt diameters of ¾″, $\frac{7}{8}$″, and 1″. Assuming that the material being used is A36 steel for which the ultimate stress (F_u) is 58 ksi, the allowable bearing load in kips per bolt on 1″-thick material is 52.2 kips. If the material were ½″ thick, the allowable bearing load per bolts would be 26.1 kips per bolt, and so on.

Bearing capacities can be easily calculated using the 52.2 kip value listed for 1″-thick material as a basis. For instance, if the beam shown in Figure 9–4 were a W16 × 50 with a web thickness of .380″, the allowable bearing value of four bolts through the web of the beam would be:

(4 bolts × 52.2) × .380″ = 79.3 kips

If the girder in Figure 9–4 were a W21 × 44, a check of W-Shapes Dimensions in Part 1 of the AISC manual would

TABLE 9–2
Allowable Bolt Bearing Loads

BOLTS AND THREADED PARTS
Bearing
Allowable loads in kips

TABLE I-E. BEARING
Slip-critical and Bearing-type Connections

Material Thickness	F_u = 58 ksi Bolt dia.			F_u = 65 ksi Bolt dia.			F_u = 70 ksi Bolt dia.			F_u = 100 ksi Bolt dia.		
	3/4	7/8	1	3/4	7/8	1	3/4	7/8	1	3/4	7/8	1
1/8	6.5	7.6	8.7	7.3	8.5	9.8	7.9	9.2	10.5	11.3	13.1	15.0
3/16	9.8	11.4	13.1	11.0	12.8	14.6	11.8	13.8	15.8	16.9	19.7	22.5
1/4	13.1	15.2	17.4	14.6	17.1	19.5	15.8	18.4	21.0	22.5	26.3	30.0
5/16	16.3	19.0	21.8	18.3	21.3	24.4	19.7	23.0	26.3	28.1	32.8	37.5
3/8	19.6	22.8	26.1	21.9	25.6	29.3	23.6	27.6	31.5	33.8	39.4	45.0
7/16	22.8	26.6	30.5	25.6	29.9	34.1	27.6	32.2	36.8		45.9	52.5
1/2	26.1	30.5	34.8	29.3	34.1	39.0	31.5	36.8	42.0			60.0
9/16	29.4	34.3	39.2	32.9	38.4	43.9		41.3	47.3			
5/8	32.6	38.1	43.5		42.7	48.8		45.9	52.5			
11/16		41.9	47.9		46.9	53.6			57.8			
3/4		45.7	52.2			58.5						
13/16			56.6									
7/8			60.9									
15/16												
1	52.2	60.9	69.6	58.5	68.3	78.0	63.0	73.5	84.0	90.0	105.0	120.0

Notes:

This table is applicable to all mechanical fasteners in both slip-critical and bearing-type connections utilizing standard holes. Standard holes shall have a diameter nominally 1/16-in. larger than the nominal bolt diameter (d + 1/16 in.).

Tabulated bearing values are based on $F_p = 1.2\ F_u$.

F_u = specified minimum tensile strength of the connected part.

In connections transmitting axial force whose length between extreme fasteners measured parallel to the line of force exceeds 50 in., tabulated values shall be reduced 20%.

Connections using high-strength bolts in slotted holes with the load applied in a direction other than approximately normal (between 80 and 100 degrees) to the axis of the hole and connections with bolts in oversize holes shall be designed for resistance against slip at working load in accordance with AISC ASD Specification Sect. J3.8.

Tabulated values apply when the distance l parallel to the line of force from the center of the bolt to the edge of the connected part is not less than 1½ d and the distance from the center of a bolt to the center of an adjacent bolt is not less than 3d. See AISC ASD Commentary J3.8.

Under certain conditions, values greater than the tabulated values may be justified under Specification Sect. J3.7.

Values are limited to the double-shear bearing capacity of A490-X bolts.

Values for decimal thicknesses may be obtained by multiplying the decimal value of the unlisted thickness by the value given for a 1-in. thickness.

(Courtesy of the American Institute of Steel Construction Inc.)

TABLE 9–3
Edge Distance Requirements for Bolts

BOLTS AND RIVETS
Bearing
Allowable loads in kips

TABLE I-F. EDGE DISTANCE

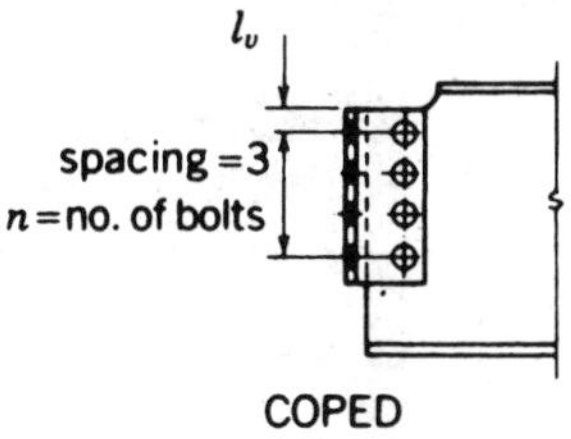

spacing = 3
n = no. of bolts

l_v

COPED

Edge Distance[b] l_v In.	Allowable Loads, Kips[a] (for one fastener, 1-in. thick material)			
	$F_u = 58$	$F_u = 65$	$F_u = 70$	$F_u = 100$
1	29.0	32.5	35.0	50.0
1⅛	32.6	36.6	39.4	56.3
1¼	36.3	40.6	43.8	62.5
<1½	43.5	48.8	52.5	75.0

Bolt Dia.	1½ d In.	Values when edge distance is 1½ d or greater [c]			
1	1½	69.6	78.0	84.0	120
⅞	1⁵⁄₁₆	60.9	68.3	73.5	105
¾	1⅛	52.2	58.5	63.0	90.0

[a] Total allowable load $= \Sigma\left[(\text{tabular value}) \times n\right] t$
where
   t = thickness of critical connected part, in.
   n = number of fasteners.

[b] $l_v \geq 2P/F_u t$ (AISC ASD Spec. J3.9) distance center of hole to free edge of connected part in direction of force, in.

where
   F_u = specified minimum tensile strength of material, ksi
   P = force transmitted by one fastener to the critical connected part, kips
[c] $P = 1.2 F_u d$ (AISC ASD Spec. Sect. J3.7).

(Courtesy of the American Institute of Steel Construction Inc.)

show that the web thickness of a W21 × 44 is .350″. Thus, the allowable bearing value for the eight bolts through the web of the girder would be:

$$(8 \text{ bolts} \times 52.2) \times .350″ = 146 \text{ kips}$$

Since the bearing values of both the beam and girder webs exceed the 74-kip values of the bolts, the connection is adequate for all bearing stresses.

Notice in Figure 9–4 that the top-of-steel elevation of both the beam and the girder are the same. This is a very common occurrence in structural steel design, which requires that the beam be cut out or *coped* at the top to avoid interference between the two top flanges. Therefore, an edge distance check should be made to see if the distance from the bottom of the beam web cope to the first fastener is

adequate for the 74-kip load value of the four bolts in double shear.

Edge Distance

Table 9–3 is a reproduction of Table 1–F in the AISC manual. This table is used to determine if the edge distance from the first bolt hold to the bottom of the beam web cope is adequate. To practice using the table, assume the bolts are ¾″ in diameter, with the standard 3″ space between them, and assume the AISC recommended minimum edge distance from the center of the top bolt hole to the bottom of the beam web cope is 1¼″. Enter the table in the $F_u = 58$ ksi column, which is the column for A36 steel. Since an edge distance of 1¼″ is greater than 1½ × the ¾″ bolt diameter (1.5

× .75 = 1.125″), the proper coefficient to use is 52.2 kips per inch of web thickness for all four bolts.

$$(52.2 \times 4 \text{ bolts}) \times .380'' = 79.3 \text{ kips}$$

Because the 79.3-kip value exceeds the 74-kip value of the high-strength bolts, the edge distance of 1¼″ is adequate for the load.

When relatively thin beam webs are coped and high fastener values are used, high bearing stresses may cause a *block shear,* or web tear-out condition in which a portion of the beam web tears out along the perimeter of the holes as shown in Figure 9–8. Although block shear is usually not a problem with ¾″ or ⅞″-diameter bolts and F_y = 36 ksi or F_y = 50 ksi steel using standard framing connections, the entry-level structural drafter or designer should be aware of it.

Figure 9–8A illustrates the beam-to-girder bolted connection shown in Figure 9–4 with emphasis on the framing angle to beam web part of the connection. Notice that the end of the W16 × 50 beam is shown set back ½″ from the face of the girder web. The holes through the web of the beam are located 2″ back from the end of the beam. We will assume that the center of the top connecting hole is 1¼″ below the bottom of the beam web cope and check for block shear. Block shear is caused by a combination of shear in the vertical plane through the connecting holes and tension along the horizontal in the area resisting web tear-out. In other words, block shear is failure by web tear-out of the shaded area of the beam web shown in Figure 9–8B.

Table 9–4 is an illustration of Table 1–G in the AISC *Manual of Steel Construction.* This table of coefficients is used to calculate a beam end's resistance to block shear. The table considers several different framing conditions of 1_v (distance from the bottom of the beam cope to the first connecting hole), 1_n (distance from the end of the beam to the row of bolt holes), and bolt diameters of ¾″, ⅞″, and 1 ″ in standard-sized holes at the standard 3″ spacing. The allowable block shear reaction is found by adding coefficient C_1 for whatever 1_v and 1_n is to be used to coefficient C_2 for the quantity and size of bolts, and then multiplying this sum by the web thickness of the beam and the given value of F_u (58 ksi for A36 steel).

To practice checking block shear, let us use the example shown in Figure 9–8A:

To check for C_1, enter the table at the 1_v = 1¼″ in the far left column and read across to the 1_n = 2″. At this point, coefficient C_1 = 1.38.

The value of coefficient C_2 for n = 4 is 1.64
The allowable reaction for block shear (R_{BS}): R_{BS} = (1.38 + 1.64) × 58 ksi × .380″ = 66.6 kips

In this case, block shear would limit the connection to 66 kips. However, a reaction of 66 kips is very large for a W16 × 50 beam and would be found only in a situation

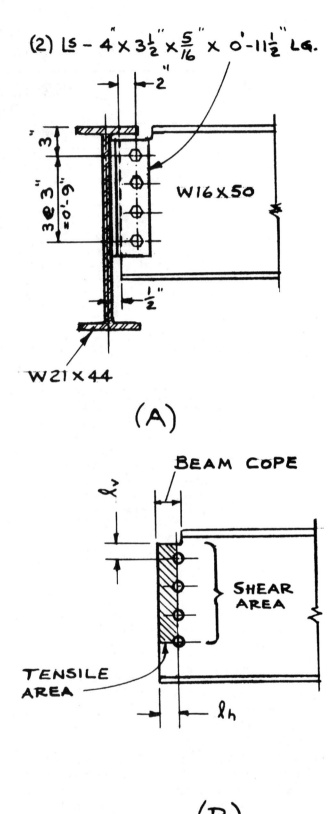

(A)

(B)

Figure 9–8   Block shear in beams

TABLE 9–4
Coefficients for Block Shear

BOLTS AND RIVETS
Bearing

TABLE I-G. COEFFICIENTS FOR WEB TEAR-OUT (BLOCK SHEAR)
Based on standard holes and 3-in. fastener spacing

Coefficient C_1

l_v In.	l_h, In.												
	1	1⅛	1¼	1⅜	1½	1⅝	1¾	1⅞	2	2¼	2½	2¾	3
1¼	.88	.94	1.00	1.06	1.13	1.19	1.25	1.31	1.38	1.50	1.63	1.75	1.88
1⅜	.91	.98	1.04	1.10	1.16	1.23	1.29	1.35	1.41	1.54	1.66	1.79	1.91
1½	.95	1.01	1.08	1.14	1.20	1.26	1.33	1.39	1.45	1.58	1.70	1.83	1.95
1⅝	.99	1.05	1.11	1.18	1.24	1.30	1.36	1.43	1.49	1.61	1.74	1.86	1.99
1¾	1.03	1.09	1.15	1.21	1.28	1.34	1.40	1.46	1.53	1.65	1.78	1.90	2.03
1⅞	1.06	1.13	1.19	1.25	1.31	1.38	1.44	1.50	1.56	1.69	1.81	1.94	2.06
2	1.10	1.16	1.23	1.29	1.35	1.41	1.48	1.54	1.60	1.73	1.85	1.98	2.10
2¼	1.18	1.24	1.30	1.36	1.43	1.49	1.55	1.61	1.68	1.80	1.93	2.05	2.18
2½	1.25	1.31	1.38	1.44	1.50	1.56	1.63	1.69	1.75	1.88	2.00	2.13	2.25
2¾	1.33	1.39	1.45	1.51	1.58	1.64	1.70	1.76	1.83	1.95	2.08	2.20	2.33
3	1.40	1.46	1.53	1.59	1.65	1.71	1.78	1.84	1.90	2.03	2.15	2.28	2.40

Coefficient C_2

n	Bolt Dia., In.		
	¾	⅞	1
2	.33	.24	.16
3	.99	.86	.74
4	1.64	1.48	1.32
5	2.30	2.10	1.90
6	2.96	2.72	2.48
7	3.61	3.34	3.06
8	4.27	3.96	3.64
9	4.93	4.58	4.23
10	5.58	5.19	4.81

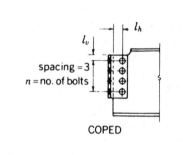

spacing = 3
n = no. of bolts

COPED

Notes:
R_{BS} = Resistance to block shear, kips
= $0.30 \, A_v F_u + 0.50 \, A_t F_u$ (from AISC ASD Sect. J4)
= $\{(0.30 \, l_v + 0.5 \, l_h) + 0.30 \, [(n-1)(s-d_h) - d_h/2] - d_h/4\} \times F_u t$
= $(C_1 + C_2) \, F_u t$

where
A_v = net shear area, in.²
A_t = net tension area, in.²
F_u = specified min. tensile strength, ksi
d_h = dia. of hole (dia. of fastener + ¹/₁₆), in.
l_h = distance from center of hole to beam end, in.
l_v = distance from center of hole to edge of web, in.
n = number of fasteners
s = fasteners spacing, in.
Tabular values are based on the following:
   AISC ASD Specification Sects. D1, J4, J3.9, J5.2.
   AISC ASD Commentary Sects. D1, J4, J3.9.

(Courtesy of the American Institute of Steel Construction Inc.)

where the beam had a very short span and a very heavy load—for instance, if the beam were supporting a very heavy machine, electrical transformer, or some other type of heavy equipment. In normal building construction, very few W16 beams would be used at spans of less than 10′–0″, and at a span of more than 10′–0″ for a W16 × 50 beam, it is unlikely the loads would be such that block shear would govern. The most logical step for the drafter or designer to take at this point, assuming the beam size is limited to a W16, would be to try a W16 × 57 beam with a web thickness of .430″.

$$R_{BS} = (1.38 + 1.64) \times 58 \text{ ksi} \times .430'' = 75.3 \text{ ksi}$$

Because the 75.3-ksi allowable block shear is greater than the 74-ksi load value of the bolts, the W16 × 57 would be adequate for block shear. If space permitted, another solution might be to try a W18 × 55 beam and go to a 5-row (n = 5) connection.

We have examined most of the AISC Table 1 data (1–D, 1–E, 1–F, and 1–G) used to help the structural drafter or designer investigate various areas of concern in the design of bolted framed beam connections. AISC Table 11–A is reproduced here as Table 9–5. This is a widely used and very helpful table for selecting bolted framed beam connections using A307, A325-SC, or A490-SC bolts in slip-critical connections and A325-N or -X and A490-N or -X bolts in bearing-type connections with standard or slotted holes.

To gain practice using this table, we will begin by examining data for the most common shop-bolted and field-bolted framed beam connection. Assume that we are using ¾″-diameter A325-N bolts, high-strength bolts for which the bolt threads are included in the shear plane of the connection.

Notice in Table 9–5 (AISC Table II–A for bolts in bearing-type connections) that the column for A325-N bolts lists first the allowable F_v per bolt as 21 ksi, and under that, bolt sizes ¾″, ⅞″, and 1″ in diameter. The recommended connecting angle thicknesses listed directly under the bolt sizes are 5/16″ for ¾″-diameter bolts, ⅜″ for ⅞″-diameter bolts, and ⅝″ for 1″-diameter bolts. Under the angle thickness and bolt size columns are listed the allowable loads in kips for various-sized connections.

Framed beam connections are listed according to the number of rows of bolts in the connection. For example, the connection detail in Figure 9–4 contained an example of a 4-row connection. Calculations previously discussed determined that a 4-row connection using ¾″-diameter, A325 high-strength bolts could support a load of approximately 74 kips. The column labeled *n* in Table 9–5 for a bearing-type connection using ¾″-diameter A325-N bolts shows that with *n* equal to 4 (in other words a 4-row connection), the allowable load is 74.2 kips. A 3-row connection is good for 55.7 kips, a 5-row connection is good for 92.8 kips, and so on.

Notice the two columns directly to the left of the *n* column in the table. These columns are listed under the headings *L* and *L′*. The *L* and *L′* columns give the recommended lengths of connection angles for various sizes of framed beam connections. For standard connections with standard or slotted holes, the L column tells us that the length of the angle should be 5½″ for a 2-row connection, 8½″ for a 3-row connection, 11½″ for a 4-row connection, etc. Figure 9–9 illustrates how the angle length for standard framed beam connections is determined.

Both Table 9–5 and Figure 9–9*A* show that the length of angles for standard framed beam connections is based upon the number of rows of bolts plus 2½″. Because the spacing between bolts, or *pitch,* in standard connections is assumed to be 3″ and the recommended edge distance from the center of both the first and last hole to the end of the angle is 1¼″, the required angle length is simply the sum of the 3″ spaces + 1¼″ at each end. Since there is always one more row of bolts than there are spaces, the length of framing angles becomes very easy to calculate. For example, a 4-row connection has (3) spaces at 3″ center-to-center + 2½″, or (3) spaces × 3″ = 9″ + 2½″ = 11½″ long. A 2-row connection has (1) space × 3″ + 2½″ = 5½″ long, etc. The AISC table for bolted framed beam connections lists recommended angle lengths ranging from 2½″ for a 1-row connection to 29½″ for a 10-row connection. Sometimes—for example, when beams frame into both sides of a girder web—it becomes necessary to stagger the shop (web) bolts and field (outstanding leg) bolts as shown in Figure 9–9*B* to provide enough clearance for the ironworkers to insert and tighten the field bolts. When bolt holes are staggered like this, the angle length must be increased by 1½″. This is illustrated in both the AISC table that shows the increased length of *L′* and in Figure 9–9*B,* which shows how the web leg bolt holes are staggered between the outstanding leg bolt holes. This topic will be covered in more depth in Chapter 11.

Common Bolt Symbols

Bolt hole and bolt symbols will be illustrated in greater detail in subsequent chapters as we discuss beam and column detailing, but it is worthwhile at this point to explain some of the symbols used in Table 9–5 and Figure 9–9. The symbol commonly used on structural steel shop drawings to designate bolts in bolted connections is the diamond symbol ◇, which indicates a *shop bolt* through the web leg of the connection angle to be installed in the fabricating shop. The solid round black symbol ● indicates an *open hole*—in other words, a hole through a framing angle, beam web, beam flange, column baseplate, through which a field bolt will be inserted by the ironworkers out on the job to make a connection. The symbol ⬬ indicates a standard horizontal slot for field bolts by the ironworkers at the job site. You

TABLE 9–5
Framed Beam Connections

FRAMED BEAM CONNECTIONS
Bolted
TABLE II Allowable loads in kips

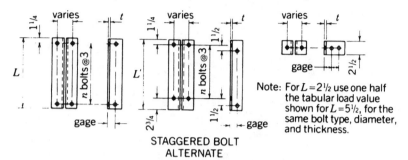

STAGGERED BOLT
ALTERNATE

Note: For $L = 2\frac{1}{2}$ use one half
the tabular load value
shown for $L = 5\frac{1}{2}$, for the
same bolt type, diameter,
and thickness.

TABLE II-A Bolt Shear[a]

For A307 bolts in standard or slotted holes and for A325 and A490 bolts in **slip-critical** connections with standard holes and Class A, clean mill scale surface condition.

Bolt Type			A307			A325-SC			A490-SC			Note:
F_v, Ksi			10.0			17.0			21.0			For slip-critical connections with oversize or slotted holes, see Table II-B.
Bolt Dia., d In.			¾	⅞	1	¾	⅞	1	¾	⅞	1	
Angle Thickness t, In.			¼	¼	¼	¼	⁵⁄₁₆	½	⁵⁄₁₆	½	⅝	
L In.	L' In.	n										
29½	31	10	88.4	120	157	150	204	267	186	253	330	
26½	28	9	79.5	108	141	135	184	240	167	227	297	
23½	25	8	70.7	96.2	126	120	164	214	148	202	264	
20½	22	7	61.9	84.2	110	105	143	187	130	177	231	
17½	19	6	53.0	72.2	94.2	90.1	123	160	111	152	198	
14½	16	5	44.2	60.1	78.5	75.1	102	134	92.8	126	165	
11½	13	4	35.3	48.1	62.8	60.1	81.8	107	74.2	101	132	
8½	10	3	26.5	36.1	47.1[b]	45.1	61.3	80.1	55.7	75.8	99.0	
5½	7	2	17.7	24.1	31.4[b]	30.0	40.9	53.4	37.1	50.5	66.0	

Notes:

[a]Tabulated load values are based on double shear of bolts unless noted. See RCSC Specification for other surface conditions.

[b]Capacity shown is based on double shear of the bolts; however, for length L, net shear on the angle thickness specified is critical. See Table II-C.

(Courtesy of the American Institute of Steel Construction Inc.)

TABLE 9–5 Continued
Framed Beam Connections

FRAMED BEAM CONNECTIONS
Bolted
TABLE II Allowable loads in kips

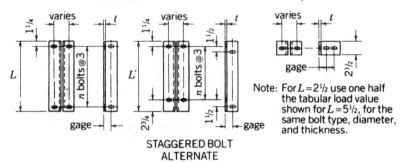

STAGGERED BOLT
ALTERNATE

Note: For $L = 2\frac{1}{2}$ use one half
the tabular load value
shown for $L = 5\frac{1}{2}$, for the
same bolt type, diameter,
and thickness.

TABLE II-A Bolt Shear[a]
For bolts in **bearing-type** connections with standard or slotted holes.

Bolt Type			A325-N			A490-N			A325-X			A490-X		
F_v, Ksi			21.0			28.0			30.0			40.0		
Bolt Dia., d In.			¾	⅞	1	¾	⅞	1	¾	⅞	1	¾	⅞	1
Angle Thickness t, In.			5/16	⅜	⅝	⅜	½	⅝	⅜	⅝	⅝	½	⅝	⅝
L In.	L' In.	n												
29½	31	10	186	253	330	247	337	440[b]	265	361	[c]	353	481	[c]
26½	28	9	167	227	297	223	303	396[b]	239	325	[c]	318	433	[c]
23½	25	8	148	202	264	198	269	352[b]	212	289	[c]	283	385	[c]
20½	22	7	130	177	231	173	236	308[b]	186	253	[c]	247	337	[c]
17½	19	6	111	152	198	148	202	264[b]	159	216	283	212	289	377
14½	16	5	92.8	126	165	124	168	220[b]	133	180	236	177	242	314
11½	13	4	74.2	101	132	99.0	135	176[b]	106	144	188	141	192	251
8½	10	3	55.7	75.8[b]	99.0	74.2	101[b]	132[b]	79.5[b]	108	141	106[b]	144	188
5½	7	2	37.1	50.5[b]	66.0	49.5	67.3[b]	88.0[b]	53.0[b]	72.2	94	70.7[b]	96	126

[a]Tabulated load values are based on double shear of bolts unless noted. See RCSC
Specification for other surface conditions.

[b]Capacity shown is based on double shear of the bolts; however, for length L, net shear on
the angle thickness specified is critical. See Table II-C.

[c]Capacity is governed by net shear on angles for lengths L and L'. See Table II-C.

(Courtesy of the American Institute of Steel Construction Inc.)

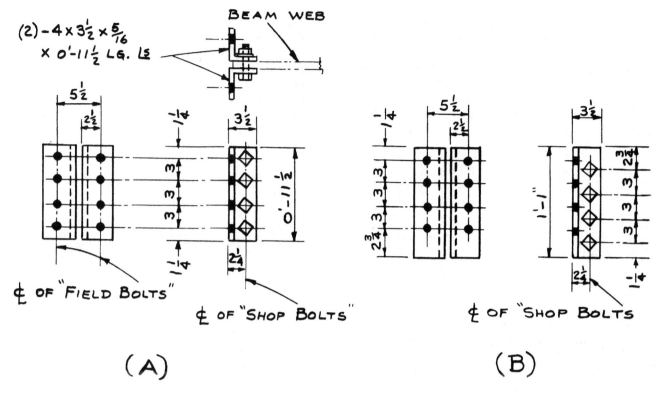

Figure 9–9 Connection angle lengths

may refer back to Figures 7–28, 7–29, and 7–30A to see more examples of standard horizontal slots used for beam-to-column connections.

Given a fundamental understanding of what has been discussed thus far regarding standard framed beam bolted connections, Table 9–5 (AISC Table II-A) is very easy to read. Just keep in mind that the tabulated values are based upon a double-shear condition of the bolts, and that the single-shear value of a high-strength bolt is half the double-shear value.

For example, notice that, for ¾″-diameter A325-N bolts, the load rating for a 4-row connection is listed as 74.2 kips. This tells us that the four shop bolts (those through the beam web as illustrated in Figure 9–4) are rated at 74.2 kips. Referring to Table 9–1, we see that, at a rating of 18.6 kips-per-bolt in double shear for ¾″-diameter A325-N bolts, the same answer could be found by multiplying 18.6 kips-per-bolt × 4 bolts. However, this table saves the structural drafter or designer from making the calculation. If the eight bolts through the web of the connecting girder are in single shear (again as illustrated in Figure 9–4), they are also adequate for the load because the single-shear value of 9.3 kips-per-bolt is multiplied by the eight bolts in the outstanding legs of the framing angles. Table 9–5 also tells us that the connecting angles must be ⁵⁄₁₆″ thick and 11½″ long for a 4-row connection.

When designing a structural steel bolted beam connection, once a certain connection capacity is surpassed, proper procedure dictates that the structural drafter or designer simply move up to the next higher strength connection. In other words, suppose that a W18 × 60 beam with an end reaction of 57 kips were fastening to the web of a girder or column or to the flange face of a column. The structural drafter or designer, seeing that the 55.7-kip capacity of a 3-row connection was exceeded for ¾″ A325N bolts, would immediately try a 4-row connection and work from there, even though the 74.2-kip allowable load for a 4-row connection greatly exceeds the required 57-kip reaction.

Another very important point concerning the selection of framed beam connections is that the *minimum length* of a connection angle, according to AISC recommendation, must be at least one-half the T-dimension of the beam it is supporting to provide stability during field erection. This means the minimum length of framing angle for a W16 × 36, which has a T-distance of 13⅝″ (see Table 2–1), is 6¹³⁄₁₆″. Thus, even if the beam were very lightly loaded with a 15-kip reaction for instance, a 3-row (8½″ angle length) would be required, although obviously a 2-row (5½″ angle length) connection rated at 37.1 kips would be strong enough to support the load. Every structural drafter and designer should be aware that the T-distance requirement is often the governing factor even before the first connection calculation is made. One example would be the design of roof systems where relatively long spans and relatively light loads are often encountered.

The excellent reference book *Detailing for Steel Construction* is published by the AISC to provide sufficient information for an entry-level structural steel detailer to learn how to detail the structural members and connections for a simple steel-frame building. The book states:

> The design drawings prepared by the designer convey the information necessary for the preparation of structural shop drawings by the structural detailer. *They should indicate the type or types of construction, and should provide all necessary data on loads, shears, moments and axial forces to be resisted by all members and their connections.*

Perhaps they should, but at the present time, unless the contract drawings are for composite construction or continuous framing, or unless a framed connection is in some way quite unique, very few structural design drawings list beam and girder reactions on their framing plans. Knowing this, the AISC goes on to say: "In detailing a beam, the detailer (structural drafter preparing the shop detail drawings) must first design the end connections to transmit the beam load to its supporting members."

The AISC then adds in both the *Manual of Steel Construction* and *Detailing for Steel Construction*: "If the beam reactions are not shown, connections *must* be selected to support one-half the total uniform load capacity shown in the Allowable Uniform Load Tables, Part 2 of this Manual, for the given beam, span, and grade of steel specified."

With the AISC statements in mind, and referring to Tables 2–1 and 2–3 from Chapter 2 of this textbook, it is apparent that these tables, as well as the AISC tables discussed in this chapter, are very important in the day-to-day work of the structural drafter or designer. We will now look at the proper procedure for selecting structural steel bolted framed beam connections.

Selecting Bolted Framed Beam Connections

Figure 9–10 represents part of a steel-frame floor system for a commercial or industrial building as it would appear on a floor framing plan. It shows a W18 × 65 beam framing into a W24 × 84 girder on grid line ① and to the flange face of a W10 × 60 column on grid line ② . The span from the center of the column to the center of the girder is shown as 16′–0″.

Checking Table 2–1, the drafter or designer sees that the T-distance for a W18 × 65 beam is 15½″, thus the framing angle length can vary from a maximum of the 15½″ T-

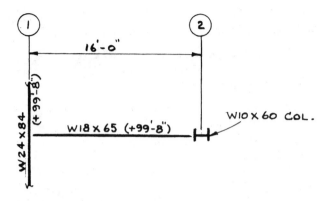

NOTE:
MATERIAL: A36 STEEL
FASTENERS: $\frac{3}{4}$″ φ A325-N HIGH-STRENGTH BOLTS.

Figure 9–10   Partial steel-frame floor plan

distance to a minimum of T/2 of the W18 × 65. T/2 of the W18 × 65 is 15.5″/2 = 7¾″, so the drafter or designer immediately knows that the minimum angle length permissible is a 3-row, 8½″-long framing angle.

Table 2–3 indicates that, at a span of 16′–0″, a W18 × 65 beam is rated for a total allowable uniform load of 116 kips. Because the reaction is one-half the total allowable load, each reaction must be assumed to be 116 kips/2 = 58 kips, since the reactions are not shown on the framing plan.

By checking Table 9–5, the structural drafter or designer will find that the allowable load for a ¾″-diameter A325-N bolted connection is listed as 55.7 kips. The 55.7-kip allowable load is less than (<) the 58-kip assumed load, so a 4-row connection rated at 74.2 kips with ⁵⁄₁₆″-thick and 11½″-long connecting angles will be used.

Checking the W18 × 65 for bearing on Table 9–2 shows that, for a 3″ pitch and using 1″-thick material as a basis:

(4 bolts × 52.2 kips) × .450″ (the web thickness of the W18 × 35) = 93.6 kips allowable bearing load. 93.6 kips > 58 kips   OK

A check of Part 1 of the AISC manual would show that the web thickness of the W24 × 84 girder (.470″) and the flange thickness of the W10 × 60 column (.680″) are both greater than the .450″ web thickness of the W18 × 65 beam. Also, with the bearing stress on the outstanding legs of the support angles distributed over eight bolts, a bearing check of that part of the connection is not required. Assuming the AISC recommended minimum edge distance of 1¼″ and following the procedures previously described to check for edge distance and block shear, a 4-row connection would be deemed more than adequate for the condition shown in Figure 9–10. Thus, the connection would require:

(2) 4 × 3½ × ⁵⁄₁₆ × 11½″-long framing angles
(8) ¾″-diameter A325-N shop bolts
(16) ¾″-diameter A325-N field bolts

This discussion of bolted framed beam connections has not attempted to teach all there is to know about the subject. Rather, it has tried to cover some basic bolted framed beam connection considerations while at the same time explaining why it is so important for every structural drafter to become as proficient as possible in using the information and tables available in the AISC *Manual of Steel Construction* in his or her day-to-day work.

9.4   STRUCTURAL WELDS AND WELD SYMBOLS

Besides bolting, another method of fastening structural components together is welding. Welding is a method of transferring load between connected parts without using mechanical fasteners. The welding process most commonly used in building construction is electric arc welding. In *electric arc welding,* one terminal of a power source is connected to the structural steel components to be joined, and the other power source is connected to an electrode through the welding rod, which the welder holds in an insulated holder. When the electrode is positioned close to the steel members to be welded, an electric arc is formed, and the intense heat at the end of the electrode melts a small part of each member along with the filler metal of the electrode. As these areas cool and solidify, they become fused into one homogenous piece.

Technological advances in automatic and semi-automatic welding processes have made the use of shop-welded and field-bolted beam framing connections very common because they enable the structural steel fabricators to economically weld together a wide variety of structural steel components in the controlled conditions of their shops. Welding is done out on the job site as well. Called field welding, it is usually kept to a minimum because field welds tend to be time-consuming and expensive.

So many different types of welds are used in structural steel construction that an in-depth discussion of all of them would be inappropriate in this book. However, every entry-level structural drafter should be familiar with the most common types of welds used in construction and know how to show them by symbolic representation on either design or shop fabrication drawings. He or she should also know how to determine the load-carrying capacity of the fillet weld, the weld specified for approximately 80 percent of all structural

steel welded connections, and how to use tables in the AISC manual to determine the load-carrying capacity of various sizes of fillet welds for the very widely specified shop-welded and field-bolted framing connection.

Three basic classifications, or types, of welded joints are used for the majority of welded connections in structural steel construction: lap joints, tee joints, and butt, or groove, joints. Lap and tee joints are the most common and are usually welded with fillet welds. Butt joints are usually made with square, single or double-vee groove, or bevel welds. Figure 9–11 illustrates several examples of lap, tee, and butt-welded joints.

Basic Weld Symbols

Welded joints are represented on structural steel drawings by standard basic and supplementary weld symbols. Through these symbols, the drafter or designer specifies the size and type of weld desired for various structural steel connections. Every structural drafter must understand how to work with weld symbols because, used properly, they can convey a great deal of information, yet take up a minimal amount of space on structural steel design or shop fabrication drawings. Figure 9–12, reproduced from the AISC manual, illustrates the basic and supplementary welding symbols used in structural steel construction.

Basically, weld symbols can be broken down into three parts: (1) a horizontal weld line or reference line containing such information as the size and length of weld required; (2) a basic weld symbol indicating the type of weld required, such as a fillet weld or bevel weld; and (3) a straight inclined leader line terminating with an arrow pointing to the joint. Sometimes a fourth part of the welding symbol, a tail, is shown at the end of the horizontal reference line to supply additional information when required. With various combinations of basic and supplementary weld symbols, the structural drafter can use the symbol to indicate exactly the type and size of weld required for a specific connection. Figure 9–13 gives two examples of how the standard parts of a weld symbol are put together by the structural drafter to illustrate the type of weld he or she desires for a certain type of connection.

Figure 9–13A shows the standard symbol for a simple fillet weld. Notice first that the leader line with the arrowhead at the end, which points to the joint to be welded on a drawing, is located at the end of the horizontal reference line. The leader and arrowhead may be at either end of the reference line and may slant upward or downward from it. Notice also that, in this case, the symbol for a fillet weld, a right triangle with the vertical leg *always* on the *left side,* is drawn below the reference line. Placing the symbol for any basic weld type below the reference line indicates that the

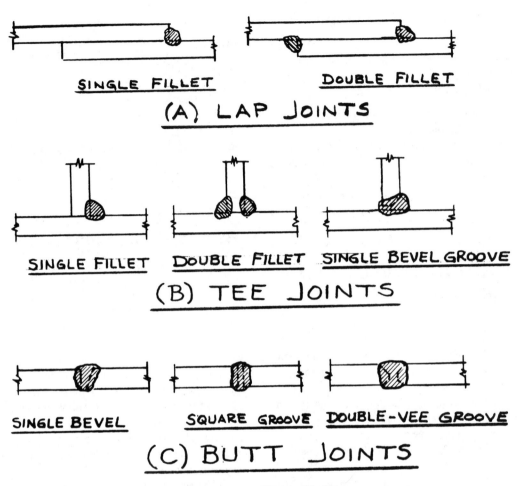

SINGLE FILLET   DOUBLE FILLET

(A) LAP JOINTS

SINGLE FILLET   DOUBLE FILLET   SINGLE BEVEL GROOVE

(B) TEE JOINTS

SINGLE BEVEL   SQUARE GROOVE   DOUBLE-VEE GROOVE

(C) BUTT JOINTS

Figure 9–11 Welded joints

weld is to be made on the arrow side or near side of the joint—in other words, the part of the joint the arrowhead is touching. If the weld symbol is shown above the weld line or reference line, the weld is to be made on the far side or back side of the joint. And if both the near and far sides of the joint are to be welded, all the required information should be shown on each side of the weld line.

It should also be mentioned that the near and far side welds do not have to be either the same size or the same type of weld. The structural steel drafter could show a symbol calling for a ⅜″ fillet weld on the near side (arrow side) of the joint and a ¼″ fillet weld on the far side (back side) of the joint. In addition, it is not unusual to call for a fillet weld on the near side and a butt weld on the far side of a joint.

The size of the weld is always shown on the left side of the basic weld symbol, which in the case of a fillet weld, is to the left side of the vertical leg of the triangle. However, every weld symbol does not require that a size be shown. For example, fillet welds and plug welds show a weld size, but bevel welds and butt welds do not.

The required length and/or spacing of welds is shown to the right of the weld symbol. The "2 @ 6" below the weld symbol in Figure 9–13*B* indicates that the drafter desires the butt weld to be an intermittent weld. *Intermittent welds* are those in which short lengths of weld in a series are separated by regular spaces. These welds are commonly called for in situations where even the smallest continuous weld would be much stronger than required and thus very uneconomical. An example might be a lintel over an eight-foot-wide door or window opening consisting of a light W-shape beam welded to the top of a 11½″ × ¼″-thick plate. In that case, a 2 @ 6 weld length would mean each short length of

WELDED JOINTS
Standard symbols

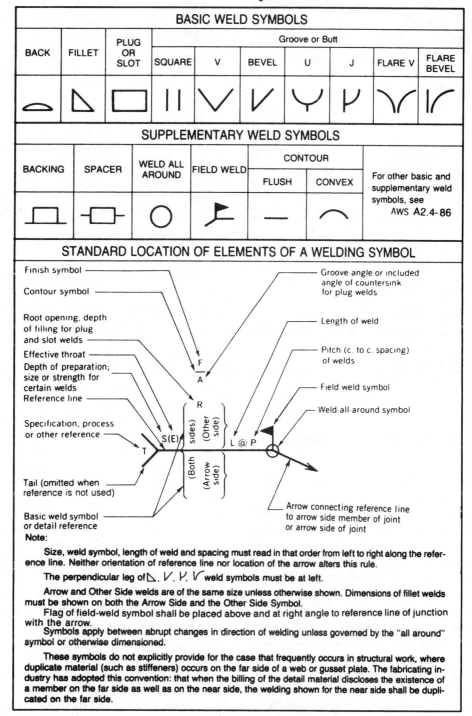

BASIC WELD SYMBOLS									
BACK	FILLET	PLUG OR SLOT	Groove or Butt						
			SQUARE	V	BEVEL	U	J	FLARE V	FLARE BEVEL

SUPPLEMENTARY WELD SYMBOLS						
BACKING	SPACER	WELD ALL AROUND	FIELD WELD	CONTOUR		For other basic and supplementary weld symbols, see AWS A2.4-86
				FLUSH	CONVEX	

STANDARD LOCATION OF ELEMENTS OF A WELDING SYMBOL

Finish symbol

Contour symbol

Root opening, depth of filling for plug and slot welds

Effective throat

Depth of preparation; size or strength for certain welds

Reference line

Specification, process or other reference

Tail (omitted when reference is not used)

Basic weld symbol or detail reference

Groove angle or included angle of countersink for plug welds

Length of weld

Pitch (c. to c. spacing) of welds

Field weld symbol

Weld-all-around symbol

Arrow connecting reference line to arrow side member of joint or arrow side of joint

Note:

Size, weld symbol, length of weld and spacing must read in that order from left to right along the reference line. Neither orientation of reference line nor location of the arrow alters this rule.

The perpendicular leg of ⊿, V, ⊬, ⟋ weld symbols must be at left.

Arrow and Other Side welds are of the same size unless otherwise shown. Dimensions of fillet welds must be shown on both the Arrow Side and the Other Side Symbol.

Flag of field-weld symbol shall be placed above and at right angle to reference line of junction with the arrow.

Symbols apply between abrupt changes in direction of welding unless governed by the "all around" symbol or otherwise dimensioned.

These symbols do not explicitly provide for the case that frequently occurs in structural work, where duplicate material (such as stiffeners) occurs on the far side of a web or gusset plate. The fabricating industry has adopted this convention: that when the billing of the detail material discloses the existence of a member on the far side as well as on the near side, the welding shown for the near side shall be duplicated on the far side.

Figure 9–12   Basic weld symbols (Courtesy of the American Institute of Steel Construction Inc.)

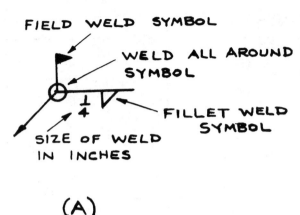

(A)

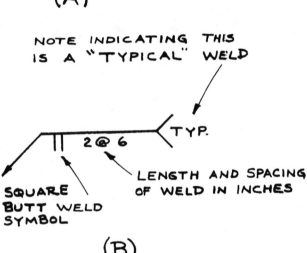

(B)

Figure 9–13   Standard weld symbols

9.5   THE DESIGN OF SIMPLE FILLET WELDS

The purpose and scope of this textbook do not permit an in-depth discussion of the design of welded connections. However, an explanation of the basics of welding connection design, with emphasis on the fillet weld, is warranted because fillet welds are the type most commonly used in structural steel construction.

A fillet weld is assumed to have a cross section similar to that of a 45-degree isosceles right triangle as illustrated in Figure 9–15. This type of weld is used primarily to hold a structural member in place and keep it from sliding over another member. Examples would include the lap joint formed between the web leg of a framing angle and the web of a beam or the tee joint created when a structural tube column sits on its baseplate. Fillet weld sizes are specified in $\frac{1}{16}$-inch increments such as $\frac{1}{8}''$, $\frac{3}{16}''$, $\frac{1}{4}''$, $\frac{5}{16}''$, and $\frac{3}{8}''$. The weld size relates to the length of one of the two equal legs of the isosceles triangle formed by the cross-section of the fillet.

A fillet weld can be loaded in tension, compression, or shear, but because the weld is weakest in shear, the allowable shear stress through the throat of the weld is the governing factor in design, regardless of the direction of the applied load. The weld throat, as shown in Figure 9–15, is the perpendicular distance from the bottom of the inside corner or root of the weld to the hypotenuse. For a 45-degree, equal-leg fillet weld, the throat dimension is equal to the size of the weld multiplied by 0.707, and the allowable shear stress on the effective area of the weld is 0.30 times the nominal strength of the weld metal.

The filler metal welding electrodes are rated as E60XX, E70XX, E80XX, etc. The letter E designates an electrode, and the numbers (60, 70, 80, etc.) indicate the nominal or minimum ultimate tensile strength of the weld in ksi. The XX digits indicate the type of coating used with a certain electrode. Strength is the most important consideration for the structural drafter or designer, so he or she will often refer to welding electrodes as E70 for an electrode with an ultimate tensile strength of 70 ksi. The two most commonly used electrodes for structural steel connections are E70XX for steels with F_y ranging from 36 to 60 ksi and E80XX for steels of $F_y = 65$ ksi. Thus, E70 electrodes are used when welding A36 structural steel.

The allowable working strength of fillet welds is usually given in kips-per-lineal-inch. As a practice exercise, let us determine the working strength or allowable load in kips-per-lineal-inch that may be applied to a $\frac{5}{16}''$ (0.3125") fillet weld. Assume we are using the shielded metal arc process and E70XX electrodes for which the allowable unit shear stress at the throat is $(0.30 \times 70) = 21$ ksi.

weld is to be 2″ long, and the welds are to be spaced 6″ center-to-center.

The TYP. at the end of weld 9–13B indicates that this weld is a "typical" weld. In other words, all similar welds made on the drawing sheet where weld 9–13B appears are to be made exactly like this one. This TYP. note eliminates the need for redrawing the same weld symbol over and over again on the same sheet.

The circle at the bend in the reference line on weld 9–13A means the weld should be made all the way around the joint. This is routinely done, for example, when steel pipe or structural tube columns are welded to their baseplates. The black flag on the weld symbol in Figure 9–13A indicates that the weld is to be made at the job site by the ironworkers rather than in the structural steel fabricator's shop. Figure 9–14 illustrates several examples of various fillet welds and how they apply to structural work. After studying the weld symbols in Figure 9–14, apply your new knowledge of weld symbols to see how many you can identify and interpret in the detail figures in chapter 7.

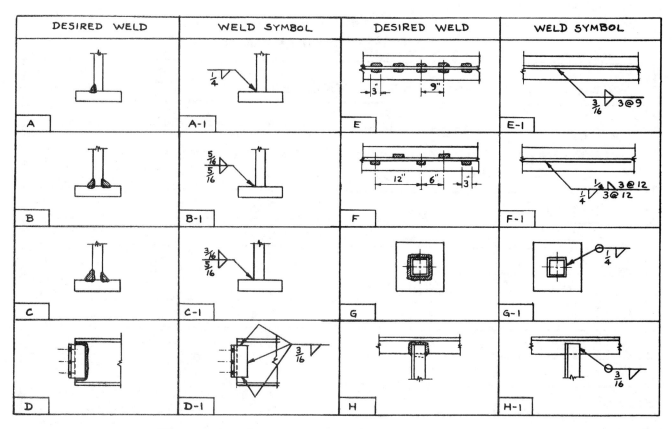

Figure 9–14 Common structural welds and their symbols

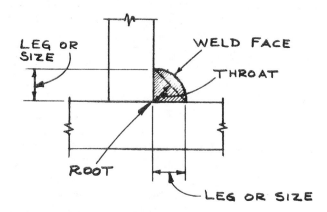

Figure 9–15 A standard fillet weld

Solving first for the throat thickness of the weld:

Throat thickness = 0.707 × 0.3125 = 0.221 in.

Solving next for the allowable weld capacity:

Weld capacity = 0.221 in. × 21 ksi = 4.64 kips
per lin. in.

Based on these calculations, the allowable working strengths of $\frac{3}{16}''$ to $1''$ fillet welds in kips-per-lineal-inch are listed in Table 9–6 for E60XX, E70XX, and E80XX electrodes. Values in this table are rounded to $\frac{1}{10}$ kip.

The majority of welded connections are designed either on the basis of strength of weld per lineal inch, as illustrated in Table 9–6, or on the strength of the base metal being connected, whichever is smaller. However, other Allowable Stress Design considerations apply in the selection of fillet welds, including minimum weld length, maximum and minimum weld sizes, and end returns.

The minimum recommended *effective length* of a fillet weld should not be less than four times the nominal size of the weld. This is rarely a problem because, for instance, a structural drafter or designer would rarely call for a $\frac{1}{4}''$ fillet weld to be less than one inch long. However, if necessary, a shorter length can be used if the considered effective weld size is reduced to one-fourth of the weld length.

The *maximum size* of fillet weld permitted along the square edge of a base metal less than $\frac{1}{4}''$ thick is a weld equal to the material thickness. For base metal $\frac{1}{4}''$ thick or thicker, the maximum weld size is limited to $\frac{1}{16}''$ less than the

TABLE 9–6
Working Strengths of Welds

ALLOWABLE WORKING STRENGTH OF FILLET WELD			
SIZE OF FILLET WELD (INCHES)	ALLOWABLE LOADS (KIPS PER LIN. INCH)		
	E60XX ELECT. F_{vw} = 18 KSI	E70XX ELECT. F_{vw} = 21 KSI	E80XX ELECT. F_{vw} = 24 KSI
$\frac{3}{16}$ (.1875)	2.4	2.8	3.2
$\frac{1}{4}$ (.250)	3.2	3.7	4.2
$\frac{5}{16}$ (.3125)	4.0	4.6	5.3
$\frac{3}{8}$ (.375)	4.8	5.6	6.4
$\frac{1}{2}$ (.5)	6.4	7.4	8.5
$\frac{5}{8}$ (.625)	8.0	9.3	10.6
$\frac{3}{4}$ (.75)	9.5	11.1	12.7
1 (1)	12.7	14.8	17

thickness of the base metal. Figure 9–16 from the AISC manual illustrates the maximum sizes of fillet welds along edges of material.

Minimum sizes of fillet welds recommended by the AISC are based, not so much on strength considerations, but on the fact that thick materials have a rapid cooling effect on small welds, which can result in a loss of weld ductility and even cause the weld to crack. With this in mind, minimum sizes of fillet welds recommended by the AISC for various thicknesses of base metal are shown in Table 9–7.

When fillet welds extend to the corner of a member, it is common practice to continue the weld around the corner for a short distance as shown in Figure 9–17. This continuation of the weld is known as the *end return,* and its length should be at least twice the nominal size of the weld. Thus, a ¼″ fillet weld would have a minimum ½″-long end return. The purpose of the end return, which may be included in effective length computations for the load capacity of the fillet weld, is to reduce high-stress concentrations at the end of the weld.

In the practical design of fillet welds, the structural drafter or designer should always be thinking about economical weld selection. Welds cost money, and the competent drafter or designer is aware that, on a large steel-frame building, it would be very expensive if all the welds were larger than required. In general, considerable economy can

Fillet welds

Base metal less
than 1/4 thick

(A)

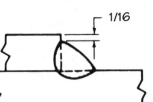

Base metal 1/4 or
more in thickness

(B)

Maximum detailed size of fillet weld along edges

**Figure 9–16   Fillet welds (Courtesy of the
American Institute of Steel Construction Inc.)**

TABLE 9–7
Minimum Size of Fillet Welds

Minimum Size of Fillet Welds

Material Thickness of Thicker Part Joined (in.)	Minimum Size of Fillet Weld[a] (in.)
To 1/4 inclusive	1/8
Over 1/4 to 1/2	3/16
Over 1/2 to 3/4	1/4
Over 3/4	5/16

[a]Leg dimension of fillet welds. Single-pass welds must be used.

(Courtesy of the American Institute of Steel Construction Inc.)

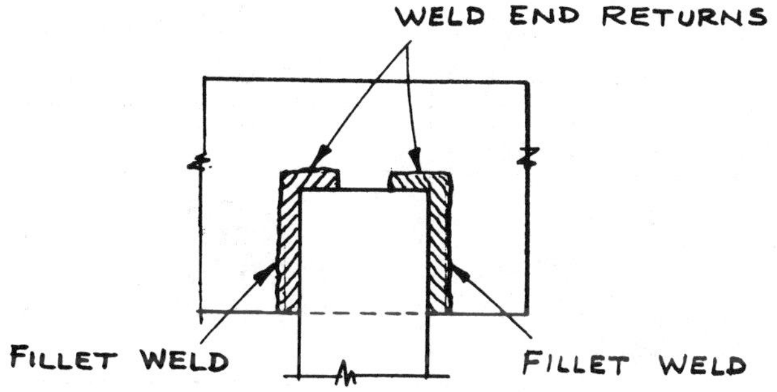

Figure 9–17 Weld end returns

be achieved by selecting fillet welds that require a minimum amount of metal and can thus be deposited in the least amount of time.

The strength of a fillet weld is directly proportional to its size. However, the *volume* of deposited metal, and thus the cost of the weld, increases as the square of the weld size. For example, a 5/8″ fillet weld contains four times the volume of weld metal required for a 5/16″ fillet weld, yet is only twice as strong. For this reason, many structural drafters tend to specify a smaller but longer weld rather than a larger, shorter, and more costly one whenever possible. On a pound basis, the cost of weld metal usually far exceeds the cost of any other part of the building structure.

Cost control for manual fillet welds is usually based on the 5/16″ fillet weld, which is generally considered the most economical size. Welds of this size and smaller can be deposited in a single pass of the electrode by the operator (welder). Thus, the majority of fillet welds in commercial and lighter industrial construction are either the 3/16″, 1/4″, or 5/16″ size. In fact, the welding tables used for designing shop-welded framing connections in the AISC manual list only the 3/16″, 1/4″, and 5/16″ sizes. The following example illustrates some of the basic principles discussed concerning the design of welded connections.

Example 9.1. A bar of A36 steel 4″ × 3/8″ in cross-section is to be welded to the back of a C12 × 25 channel with E70XX electrodes. The weld must be strong enough to develop the full tensile strength of the bar. Design the weld.

Solution. The area of the bar is:

$$4 \text{ in.} \times 0.375 \text{ in.} = 1.5 \text{ in.}^2$$

Because the allowable tensile stress of the bar is 22 ksi, the tensile strength of the A36 steel bar is:

$$F_t \times A = 22 \text{ ksi} \times 1.5 \text{ in.}^2 = 33 \text{ kips}$$

The weld must be large enough and long enough to resist a force of 33 kips.

Try a 5/16″ fillet weld because it is 1/16″ smaller than the 3/8″ thickness of the base metal and can be made in one pass of the weld electrode. Table 9–6 indicates that, using E70XX electrodes, a 5/16″ fillet weld has an allowable working strength of 4.6 kips per lineal inch. Thus, the required length of weld to develop the strength of the bar is:

$$33 \text{ kips}/4.6 \text{ kips/lin. inch} = 7.17 \text{ inches}$$

To compensate for irregularities in weld deposit and the tapered shape at the end caused by starting and stopping fillet welds, the length of weld called for on a structural steel drawing is usually increased by twice the weld size. Thus, the weld length actually called for in this example would be 7.17″ + 0.3125″ + 0.3125″ = 7.795″. Rounding to the nearest full inch, the weld shown on the drawing would be a 5/16″ × 8″-long fillet weld.

The actual position of the weld with respect to the bar might be shown as in either Figure 9–18*A* or *B*.

With 8″ of total weld length required, and using end returns in Figure 9–18*A* of not less than 2 × 5/16 = 10/16 = 5/8″ (assume 1″), the total length of vertical weld required would be:

8″ – 2″= 6″, or 3″ minimum on each side plus the 1″ end returns.

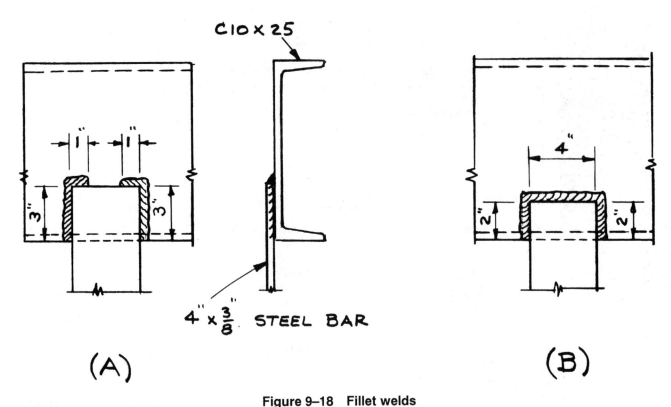

Figure 9–18   Fillet welds

Students desiring more information about welded connections than has been provided in this textbook may refer to the latest editions of the AISC *Manual of Steel Construction* and/or *Detailing for Steel Construction.*

9.6   SHOP-WELDED AND FIELD-BOLTED FRAMING CONNECTIONS

One of the most widely specified types of standard structural steel framing connections is the shop-welded/field-bolted connection. With this connection, the framing angles are welded to the webs of beams in the fabricator's shop and then field-bolted to connecting girders or columns by the ironworkers at the job site. Figure 9–19 illustrates a shop-welded/field-bolted connection.

Because shop-welded/field-bolted connections using A325-N high-strength bolts are so widely specified for commercial and industrial work, many structural design firms and structural steel fabricators have developed tables for their drafters and designers to use as guides in determining minimum shop-welded/field-bolted connection require-

ments. A typical example of one of these tables is illustrated in Table 9–8.

Notice first that Table 9–8 indicates shop-welded beam connection requirements in combination with A36 structural steel and ¼″ A325-N high-strength bolts, and fillet welds using E70XX electrodes. Notice also that the framing or connecting angles are 4 × 3½ unless the beam frames into a W8 column, at which time the outstanding leg of the angle may be reduced to 3″.

The procedure for designing a shop-welded/field-bolted framing connection is very similar to that for designing the shop-bolted/field-bolted connection previously discussed in section 9.3, except that it requires the use of Table III in the AISC manual for the welded part of the connection. This table, which is to be used with Table II connections, is reproduced in Table 9–9.

The procedure for selecting shop-welded/field-bolted standard flexible framed connections, using the tables in the AISC manual, are as follows:

1. Determine the minimum required length of the framing angles based upon the T-distance requirement of the connected beam.

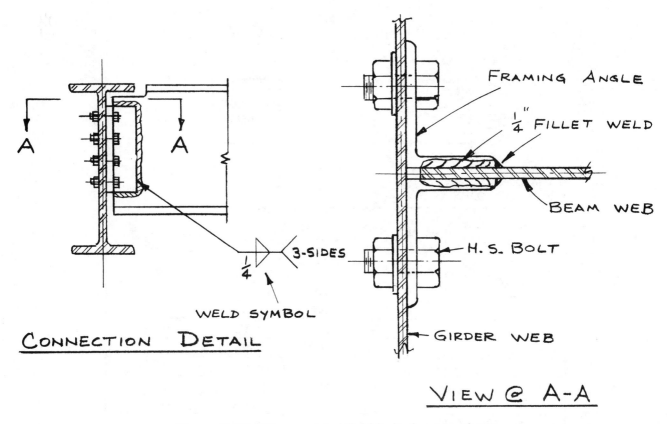

Figure 9–19 Shop-welded/field-bolted connections

2. From Table II-A in the AISC manual (Table 9–5 in this textbook) and keeping in mind whether the bolts are in single or double shear, determine the required quantity of bolts and the length and thickness of the outstanding legs of the framing angles according to the procedures previously discussed for bolted connections.

3. From Table III in the AISC manual (Table 9–9 in this textbook), and using the angle length already determined in steps 1 and 2, select the most economical weld size for the load.

4. Note the minimum web thickness required for weld A, the shop weld connecting the web leg of the framing angle to the beam web. If the beam web thickness is equal to or greater than the thickness required on the table, the connection is acceptable. If the beam web thickness is less than the minimum required, the tabulated capacity of the weld must be reduced proportionally.

The following example illustrates the procedure for selecting a shop-welded/field-bolted framing angle connection using ¾"-diameter A325-N bolts for the situation illustrated in Figure 9–20.

Example 9.2. Figure 9–20 shows a beam carrying a distributed load at a 16'-0" span between two W12 × 65 columns. The beam is to be connected to the webs of the columns with shop-welded/field-bolted connections. Because all the structural steel is A36, the shop welds will be made with E70XX electrodes. All high-strength bolts will be ¾"-diameter A325-N bolts. Design the connection.

Solution. Because the span is 16'-0", the allowable uniform load is listed as 107 kips (see Table 2–1). The beam itself weighs 60 lbs/lin. ft., so the beam weighs:

60 lbs/lin. ft. × 16' = 960 pounds (Round up to 1,000 lbs., or 1 kip)

TABLE 9–8
Minimum Welded Beam Connection Requirements

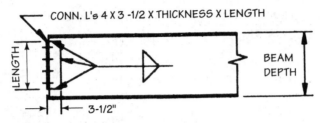

CONN. L's 4 X 3 -1/2 X THICKNESS X LENGTH

3-1/2"

SEE TABLE FOR NO. AND
SIZE OF H.S. A325N BOLTS

BEAM SIZE	NO. OF H.S. (A325N) BOLTS	CONN. ANGLES		3/4" DIA. BOLTS WELD SIZE
		THICKNESS	LENGTH	
W36	20	3/8"	2'-5-1/2"	1/4"
W33	18	3/8"	2'-2-1/2"	1/4"
W30	16	3/8"	1'-11-1/2"	1/4"
W27	14	3/8"	1'-8-1/2"	1/4"
W24	12	3/8"	1'-5-1/2"	1/4"
W21	10	3/8"	1'-2-1/2"	1/4"
W18	8	3/8"	0'-11-1/2"	1/4"
W16 X 45 & ABOVE	8	3/8"	0'-11-1/2"	1/4"
W16 X 36 & W16 X 40	8	5/16"	0'-11-1/2"	1/4"
W16 X 31 & BELOW	8	5/16"	0'-11-1/2"	3/16"
W14 & 34 & ABOVE	6	5/16"	0'-8-1/2"	1/4"
W14 X 30 & BELOW	6	5/16"	0'-8-1/2"	3/16"
W12 & 35 & ABOVE	6	5/16"	0'-8-1/2"	1/4"
W12 X 30 & BELOW	6	5/16"	0'-8-1/2"	3/16"
W10 X 30 & ABOVE	4	5/16"	0'-5-1/2"	1/4"
W10 X 26 & BELOW	4	5/16"	0'-5-1/2"	3/16"
W8 X 28 & ABOVE	4	5/16"	0'-5-1/2"	1/4"
W8 X 24 & BELOW	4	5/16"	0'-5-1/2"	3/16"
W6	2	5/16"	0'-4"	3/16"

NOTES: E70XX ELECTRODES MIN.
OUTSTANDING LEGS OF CONN. ANGLES MAY BE REDUCED TO 3" WHEN
BEAM FRAMES INTO W8 COLUMN WEB.

TABLE 9–9
Framed Beam Connections

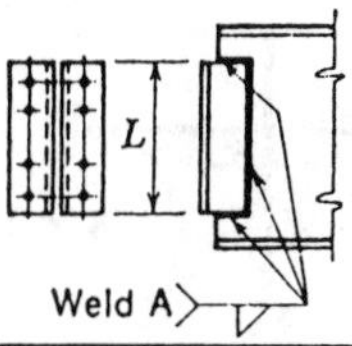

FRAMED BEAM CONNECTIONS
Welded—E70XX electrodes for combination with Table II connections
TABLE III Allowable loads in kips

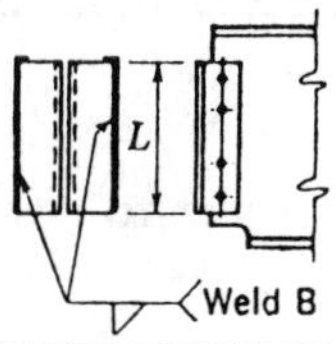

Weld A>

<L

Weld B

Weld A		Weld B		Angle Length L In.	[a]Minimum Web Thickness for Welds A		Maximum Number of Fasteners in One Vertical Row (Table II)
Capacity, Kips	[b]Size, In.	[c]Capacity, Kips	Size, In.		F_y = 36 ksi F_v = 14.5 ksi	F_y = 50 ksi F_v = 20 ksi	
266	5/16	296	3/8	29½	.64	.46	
213	1/4	247	5/16	29½	.51	.37	10
160	3/16	197	1/4	29½	.38	.28	
245	5/16	261	3/8	26½	.64	.46	
196	1/4	217	5/16	26½	.51	.37	9
147	3/16	173	1/4	26½	.38	.28	
222	5/16	223	3/8	23½	.64	.46	
178	1/4	186	5/16	23½	.51	.37	8
133	3/16	149	1/4	23½	.38	.28	
198	5/16	187	3/8	20½	.64	.46	
158	1/4	156	5/16	20½	.51	.37	7
119	3/16	125	1/4	20½	.38	.28	
174	5/16	152	3/8	17½	.64	.46	
139	1/4	126	5/16	17½	.51	.37	6
104	3/16	101	1/4	17½	.38	.28	
148	5/16	115	3/8	14½	.64	.46	
118	1/4	95.7	5/16	14½	.51	.37	5
88.7	3/16	76.6	1/4	14½	.38	.28	
121	5/16	80.1	3/8	11½	.64	.46	
97.0	1/4	66.9	5/16	11½	.51	.37	4
72.7	3/16	53.4	1/4	11½	.38	.28	
92.1	5/16	48.2	3/8	8½	.64	.46	
73.7	1/4	40.3	5/16	8½	.51	.37	3
55.3	3/16	32.2	1/4	8½	.38	.28	
61.8	5/16	21.9	3/8	5½	.64	.46	
49.5	1/4	18.3	5/16	5½	.51	.37	2
37.1	3/16	14.6	1/4	5½	.38	.28	

[a]When the beam web thickness is less than the minimum, multiply the connection capacity furnished by Weld A by the ratio of the actual web thickness to the tabulated minimum thickness. Thus, if 5/16 in. Weld A, with a connection capacity of 148 kips and a 14½-in. long angle, is considered for a beam of web thickness of 0.375 in. with F_y = 36 ksi, the connection capacity must be multiplied by 0.375/0.64, giving 86.7 kips.

[b]Should the thickness of material to which connection angles are welded exceed the limits set by AISC ASD Specification Sects. J2.1b or J2.2b for weld sizes specified, increase the weld size as required, but not to exceed the angle thickness.

[c]When welds are used on outstanding legs, connection capacity may be limited by the shear capacity of the supporting member as stipulated by AISC ASD Specification Sect. F4. See Ex. 13 and 14 for Table IV.

Note 1: Connection angles: Two L 4 × 3½ × thickness × L; F_y = 36 ksi. See discussion preceding examples for Table III for limiting values of thickness and optional width of legs.

Note 2: Capacities shown in this table apply only when the material welded is F_y = 36 ksi or F_y = 50 ksi steel.

(Courtesy of the American Institute of Steel Construction Inc.)

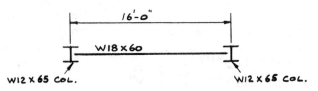

Figure 9–20 Beam-to-column connection

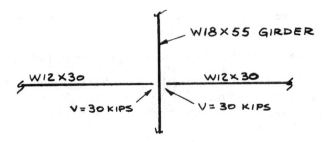

Figure 9–21 Beam-to-girder connection

This gives a total load of 108 kips, thus each reaction would be:

$$108 \text{ kips}/2 = 54 \text{ kips}$$

The T-distance of a W18 × 60 beam is 15½", so the minimum AISC recommended angle length is:

$$15.5''/2 = 7.75''$$

This means the minimum connection must be at least a 3-row connection with a 8½"-long angle.

Table 9–5 shows that a 3-row connection using ¼"-diameter A325-N high-strength bolts is rated at 55.7 kips, and the framing angles are recommended to be 8½" long and ⁵⁄₁₆" thick:

$$55.7 \text{ kips} > 54 \text{ kips} \text{ (OK)}$$

Since the web thickness of the W12 × 65 column (0.390") is greater than the ⁵⁄₁₆" (0.3125") angle thickness, a bearing check is not required.

With the framing angle length and thickness already established, the drafter or designer then turns to AISC Table III (Table 9–9) and looks in the column for angle lengths of 8½". There, for the weld connecting the framing angle to the web of the beam (weld A), a ³⁄₁₆ fillet weld is rated at a capacity of 55.3 kips.

$$55.3 \text{ kips} > 54 \text{ kips} \text{ (OK)}$$

The minimum web thickness recommended for weld A is .38". The web thickness of the W18 × 60 is 0.415" (see Table 2–1). Table 9–7 shows that a ³⁄₁₆ fillet weld can be used with material up to ½" (0.500") thick. Because the web thickness of the W18 × 60 is greater than the minimum recommended web thickness of .38" but less than the maximum recommended web thickness of .500", ³⁄₁₆ fillet weld is acceptable.

The last check required for this connection would be to see if the framing angles will fit into the web of the W12 × 65 column. The T-distance of a W12 × 65 is 9½". If the outstanding legs of the framing angles were 4" legs, the required total length for the connection angles would be:

$$4'' + 4'' + 0.415'' = 8.415''$$
$$9.500'' > 8.415'' \text{ (OK)}$$

The shop-welded/field-bolted connection at each end of the W18 × 60 beam would require (2) 4" 3½" × ⁵⁄₁₆" × 0'–8½"-long framing angles. The 3½" leg is the web leg that would be fastened to the web of the W18 × 60 with ³⁄₁₆ fillet welds (shop welds). This connection would also require (6) ¼"-diameter A325-N field bolts.

It is interesting to note that the minimum shop-welded/field-bolted requirements shown in Table 9–8 for a W18 beam would require (8) ¼"-diameter A325-N field bolts (which indicates a 4-row connection and a 0'–11½" angle length), ⅜"-thick connecting angles, and ¼" fillet welds. In other words, if a structural drafter or designer were to use this table directly, the connection would be quite over-designed for the situation. Thus, these types of tables tend to be conservative on the side of safety.

Although it is not possible in this chapter to cover a large variety of situations encountered in the design of shop-welded/field-bolted connections, the following example will illustrate another fundamental concept.

Example 9.3. Select shop-welded/field-bolted connections for the W12 × 30 beams framing into either side of the W18 × 55 girder as shown in Figure 9–21. The reaction for each beam is 30 kips.

Solution. The T-distance of a W12 × 30 beam is 10½", so the minimum AISC recommended angle length is:

$$10.5''/2 = 5.25'',$$

which means that the minimum connection must be at least a 2-row connection with a 5½" angle length.

Table 9–5 shows that a 2-row connection using ¼"-diameter A325-N high-strength bolts is rated at 37.1 kips, and the framing angles called for are 5½" long and ⁵⁄₁₆" thick:

$$37.1 \text{ kips} > 30 \text{ kips} \text{ (OK)}$$

The W12 × 30 beams framing into the web of the W18 × 55 girder would create a total load of 30 kips × 2 = 60 kips in double shear. Table 9–1 shows that a 3/4″ A325-N bolt in double shear has a rated capacity of 18.6 kips/bolt.

$$18.6 \text{ kips/bolts} \times 4 \text{ bolts} = 74.4 \text{ kips} > 60 \text{ kips} \quad \text{OK}$$

Checking for bearing stress on the girder web, we see that the web thickness of the W18 × 55 is 0.390″ (Table 2–1). Table 9–2 shows that the allowable bearing capacity of the W18 × 55 is:

$$4 \text{ bolts} \times 52.2 \text{ kips} \times 0.390'' = 81.4 \text{ kips}$$
$$81.4 \text{ kips} > 60 \text{ kips} \quad \text{OK}$$

Checking the required weld on the web legs of the framing angles, Table 9–9 shows that, for 5½″-long angles, a ³⁄₁₆″ fillet weld is rated for an allowable load of 37.1 kips:

$$37.1 \text{ kips} > 30 \text{ kips} \quad \text{OK}$$

However, a check of the AISC tables reveals that the web thickness of a W12 × 30 is 0.260″, and Table 9–9 indicates that the minimum web thickness for a load of 37.1 kips and a ³⁄₁₆″ fillet weld is .38″. This means, if the web thickness of the beam is not .38″, the capacity of the weld must be proportionally reduced. This calculation shows:

$$.260''/.38'' \times 37.1 = 25.4 \text{ kips}$$
$$25.4 \text{ kips} < 30 \text{ kips} \quad \text{N.G.}$$

The welded part of the connection in this situation is inadequate, or N.G. (No Good).

At this point, the structural drafter or designer has two options. First, he or she could increase the angle length to 8½″ by going to a 3-row connection. An 8½″-long angle, using a ³⁄₁₆″ fillet weld is rated for a capacity of 55.3 kips. Again, reducing the weld proportionally:

$$.260''/.38'' \times 55.3 = 37.8 \text{ kips}$$
$$37.8 \text{ kips} > 30 \text{ kips} \quad \text{OK}$$

The other possibility would be to stay with the 2-row connection and the 5½″ angle length and increase the size of the fillet weld from ³⁄₁₆″ to ¼″. A ¼″ fillet weld is rated at 49.5 kips for a 5½″-long angle in Table 9–9:

$$.260''/.38'' \times 49.5 = 33.8 \text{ kips}$$
$$33.8 \text{ kips} > 30 \text{ kips} \quad \text{OK}$$

Thus, in this case, increasing either the angle length or the weld size would be acceptable. It is interesting to note that Table 9–8 shows a 3-row connection with a 3-row connection and a ³⁄₁₆″ weld size recommended for a W12 × 30 beam size. Again, this is obviously the most conservative (safest) of the two possibilities checked for this connection.

The previous example has illustrated that the welded part of the connection sometimes governs. It has also shown how to proportionally reduce the capacity of the weld to fit

a specific situation. These are the types of concepts most structural engineering firms and structural fabricators want their structural drafters to be familiar with so that they can perform the investigations and calculations required in their day-to-day work.

9.7 SEATED CONNECTIONS

Chapter 7 pointed out that seated beam connections are very economical and widely specified. Figure 7–24 illustrated a welded seated beam connection in which a W18 × 46 beam was connected to the face of a W8 × 31 column. This type of unstiffened seated connection is one of the most frequently used connections in structural steel-frame commercial and industrial buildings. Using the proper tables in the AISC manual, this type of connection is remarkably easy to select for most standard situations. Table 9–10, a reproduction of Table VI from the AISC manual, is used to select unstiffened welded seated beam connections.

Notice first that Table 9–10 is comprised of three separate tables. Table VI-A and VI-B show the thickness of seat angle required to support various loads for various beam web thicknesses and seat angle lengths using either grade 36 or grade 50 steel and E70XX electrodes. After the proper seat angle length and thickness have been selected for a given load, the structural drafter or designer looks down to Table VI-C to select the proper weld size and length of vertical leg. The tables assume the OSL (outstanding leg or horizontal leg of the angle) will always be either a 3½″ or 4″ leg. Thus, the beam web thickness is critical. With the beam bearing only 3″ to 3½″ on the horizontal leg, the beam web could buckle if the recommended loads were exceeded. The usual practice is to run the fillet welds along both ends of the entire length of the vertical leg of the angle and then continue the welds to make end returns a distance equal to at least twice the weld length along the horizontal leg at the heel of the angle. At this point, we will illustrate the procedure for using the AISC tables to determine the requirements for a welded unstiffened seat angle connection.

Example 9.4. Select an unstiffened welded seated beam connection for a W21 × 57 beam (web thickness ⅜″ and flange width 6½″) to support a beam reaction of 32 kips. The beam is to connect to the web of a W14 × 74 column (web thickness = 0.450″) with the seated connection. All structural steel is A36, and E70XX electrodes are to be used.

Solution. First check the T-distance of the W14 × 74 column and see that it is 11″. This means the column web is long enough that an 8″-long seat angle can be used.

TABLE 9–10
Seated Beam Connection Tables

SEATED BEAM CONNECTIONS
Welded—E70XX electrodes
TABLE VI Allowable loads in kips

TABLE VI-A. Outstanding Leg Capacity, kips (based on OSL = 3½ or 4 in.)

		Angle Length	6 In.					8 In.				
		Angle Thickness, In.	⅜	½	⅝	¾	1	⅜	½	⅝	¾	1
F_y, ksi	36	**Beam Web Thickness, In.** 3/16	8.23	11.5	14.1	16.7	19.7	9.50	12.7	15.5	18.4	19.7
		¼	9.50	14.2	18.9	21.9	28.1	11.0	16.2	20.4	23.8	28.6
		5/16	10.6	16.1	21.8	27.5	36.3	12.3	18.3	24.5	30.7	39.0
		⅜	11.6	17.9	24.3	30.9	41.1	13.4	20.3	27.3	34.4	47.6
		7/16	12.6	19.5	26.8	34.1	48.9	14.5	22.2	29.9	37.8	53.8
		½	13.4	21.1	29.1	37.2	53.6	15.5	23.9	32.5	41.2	58.8
		9/16	14.2	22.6	31.4	40.2	58.2	16.5	25.5	34.9	44.4	63.6

Note: Values above heavy lines apply only for 4-in. outstanding legs.

TABLE VI-B. Outstanding Leg Capacity, kips (based on OSL = 3½ or 4 in.)

		Angle Length	6 In.					8 In.				
		Angle Thickness, In.	⅜	½	⅝	¾	1	⅜	½	⅝	¾	1
F_y, ksi	50	**Beam Web Thickness, In.** 3/16	9.69	14.5	17.7	20.9	27.2	11.2	15.8	19.3	22.9	27.4
		¼	11.2	17.1	23.2	28.0	35.7	12.9	19.4	26.0	30.2	38.6
		5/16	12.5	19.5	26.6	33.9	46.9	14.5	22.0	29.8	37.6	49.9
		⅜	13.7	21.6	29.9	38.2	55.2	15.8	24.4	33.3	42.3	60.4
		7/16	14.8	23.7	32.9	42.4	61.5	17.1	26.7	36.6	46.7	67.1
		½	15.8	25.6	35.9	46.4	67.6	18.3	28.8	39.8	51.0	73.6
		9/16	16.8	27.5	38.8	50.3	73.6	19.4	30.9	42.9	55.1	80.5

Note: Values above heavy lines apply only for 4-in. outstanding legs.

TABLE VI-C. Weld Capacity, kips

Weld Size In.	E70XX Electrodes				
	Seat Angle Size (long-leg vertical)				
	4×3½	5×3½	6×4	7×4	8×4
¼	11.5	17.2	21.8	28.5	35.6
5/16	14.3	21.5	27.3	35.6	44.5
⅜	17.2	25.8	32.7	42.7	53.4
7/16	20.1	30.1	38.2	49.8	62.3
½	—	34.4	43.6	56.9	71.2
⅝	—	43.0	54.5	71.2	89.0
11/16	—	47.3	60.0	78.3	—
¾	—	—	—	—	—
Range of Available Seat Angle Thicknesses					
Minimum	⅜	⅜	⅜	⅜	½
Maximum	½	¾	¾	¾	1

(Courtesy of the American Institute of Steel Construction Inc.)

Enter Table 9–10 (AISC Table VI) and see in Table VI-A that, for a beam web thickness of ⅜″ and an angle length of 8″, a ¾″-thick seat angle length is rated at a capacity of 34.4 kips.

$$34.4 \text{ kips} > 32 \text{ kips} \quad \text{OK}$$

Going down to Table VI-C, it can be seen that, using E70XX electrodes and a ⁵⁄₁₆″ weld size, a 7″ × 4″ angle is good for an allowable load of 35.6 kips.

$$35.6 \text{ kips} > 32 \text{ kips} \quad \text{OK}$$

Thus, the seat angle should be 7 × 4 × ¾ × 0′–8″ long and welded to the web of the column with the long leg in the vertical position. The beam would probably be mounted in place first with (2) ¾″-diameter bolts through the bottom flange of the beam and the outstanding leg of the seat angle as shown in Figure 7–24, then welded to the seat angle with ¼″ or ⁵⁄₁₆ fillet welds, which are compatible with both the ¹¹⁄₁₆″ beam flange and ¾″ seat angle thicknesses.

9.8   SUMMARY

This chapter has discussed some of the more commonly used structural steel connections in commercial and industrial steel-frame buildings. Structural steel design and structural steel fabrication offices employing entry-level structural drafters assume that the drafter has a fundamental knowledge of these connections as well as commonly used connection methods such as bolting with high-strength bolts and welding with fillet welds.

The structural drafter must know how to use weld symbols and how to show them on structural design or fabrication drawings. As we shall see in Chapters 10 and 11, the structural drafter working in a structural fabrication office must know how to select standard connections because that is often one of the first tasks required in the preparation of shop fabrication detail drawings of structural steel beams and columns. With a background of the material presented in this chapter, the student is now prepared to move on to learn how to make the type of shop or detail drawings prepared by the structural steel detailer.

STUDY QUESTIONS

1. AISC specifications define three types of steel-frame construction based upon the type of framing connection used. Unless the structural design plans and specifications clearly specify otherwise, which type of connection will the structural steel fabricator provide?

2. Currently, the two primary methods of fastening together the major components of a steel-frame structure are _____ and _____.

3. Because of their low carbon content, bolts referred to as _____ are *not* recommended for major connections.

4. High-strength bolts used for the vast majority of field connections are furnished in what two grades?

5. In recent years, a new and efficient type of high-strength bolt has been introduced. What is the bolt called?

6. What is the difference between bearing-type and slip-critical bolted connections?

7. What is the difference between single shear and double shear when referring to bolted connections? Use a sketch if necessary to help answer this question.

8. What is the difference between shop bolts and field bolts?

9. What is the difference between bolts designated as A325-N and A325-X?

10. Which of the two bolts referred to in question 9 is the most commonly used in commercial and industrial construction?

11. What is the allowable load in kips for a ¾″ A325-N bolt in single shear? What is the allowable load for the same bolt in double shear?

12. Why should the mating surfaces of slip-critical connections not be painted?

13. What is meant by the term *block shear*?

14. The spacing between bolts, or *pitch*, in standard bolted connections is usually _____.

15. What angle length is required for a 5-row bolted connection?

16. Sketch the commonly used symbol to designate shop bolts on shop drawings.

17. What does the symbol ● indicate on a shop drawing?

18. The tabulated values shown in Table 9–5 (AISC Table IIA) are based upon the _____ condition of the bolts.

19. If beam reactions are not shown on the structural design drawings, what assumption does the structural detailer make concerning the value of the reactions?

20. Generally, three basic types or classifications of welded joints are used for the majority of steel-frame construction. What are they?

21. Sketch the symbol for a ¼″ fillet weld, 2″ in length, to be made by the ironworkers at the job site.

22. When drawing a fillet weld symbol, the size of the fillet weld is always called out on which side of the symbol?

23. When showing the standard triangular fillet weld symbol, the vertical leg of the triangle is always shown on the _____ side of the symbol.

24. Sketch the symbol for a square butt weld.

25. What is the purpose of a circle located at the bend in the reference line on a weld symbol?

26. The two most commonly used electrodes for structural steel welded connections are the _____ and _____.

27. The allowable working strength of fillet welds is usually given in _____.

28. In welded construction, what is meant by the term *end return*? Use a sketch if necessary.

29. What is the maximum size of fillet weld that can be deposited in a single pass of the electrode?

30. What is meant by a shop-welded/field-bolted framing connection?

31. Select a shop-bolted/field-bolted standard framing connection in which a W18 × 46 beam frames into the web of a W21 × 68 girder with a reaction of 38 kips. Assume A36 steel and ¾″-diameter A325-N bolts.

32. Select a shop-welded/field-bolted connection for a W16 × 40 beam framing into a W21 × 50 girder with a reaction of 26 kips. Assume A36 steel, E70XX electrodes, and ¾″-diameter A325-N field bolts.

Chapter 10

Structural Steel Column Detailing

OBJECTIVES

Upon completion of this chapter, you should be able to:

- Prepare shop detail drawings for a variety of structural steel columns in accordance with generally accepted practices of structural steel fabrication.

10.1 INTRODUCTION

Structural steel drafters employed in the offices of structural steel fabricators are known throughout the construction industry as structural detailers. A structural detailer begins his or her work after the steel fabricator has been awarded a contract to fabricate and deliver the required structural steel for a building project to the job site. The building might be anything from a gigantic shopping mall or tall office building to a small, one-story fast-food restaurant or branch bank. A large project requires the efforts of a team of structural detailers, while a small project can often be adequately detailed by one person.

10.2 INITIAL STEPS IN SHOP DETAILING

In most fabricators' offices, work assignments come from the chief structural drafter. After assigning a project to a structural detailer, the chief drafter usually reviews it with him or her by carefully going through the set of structural design drawings that have been prepared by the structural engineering design office. Many times sketches prepared by the estimator in the structural fabricator's office who put together the successful estimate are also used to acquaint the structural detailer with the project.

Anchor Bolt Plans and Details

Usually the first ask of the structural detailer is to prepare an anchor bolt plan and the required anchor bolt details. An *anchor bolt plan* is a plan view showing the location of all of the anchor bolts required to fasten the structural steel columns to their footings, piers, tops of foundation walls, and other connection points. *Anchor bolt details* are detail drawings of the various sizes and lengths of anchor bolts required for the fabricator's shop to produce them. Anchor bolt plans and details are prepared first because the anchor bolts must be set in place at the job site before the concrete footings and walls are poured. Anchor bolt plans and details will be covered in Chapter 12 after the fundamentals of column detailing have been introduced.

Column Details

Typically, after the anchor bolts, the next components of the structural support frame to be detailed are the structural steel columns. Columns are drawn next because they must show, or at least help to show, the top-of-steel elevations for the floor and roof beams and girders connecting to them as well as information about the sizes and types of connections required. For example, if a beam or girder were to connect to a structural steel column with a seated beam connection (as illustrated in Figures 7–25 and 7–26), the beam seat angle or plates would have to be properly sized and then located and welded into place on the flange face or web of the column before the column was shipped to the job site from the fabricator's shop. Likewise, mounting holes for standard framing angle connections (as illustrated in Figures 7–20, 7–21, and 7–22) must be located and then punched or drilled through column webs or flanges in the fabricating shop. Single- or thru-plate shear connection plates (such as those illustrated in Figures 7–28, 7–29, and 7–30) must be drilled or punched as required and then located and welded in place on the column before it leaves the shop. The same is true for steel joist seat angle connections, as illustrated in Figure 7–41.

Thus, before the steel detailer can begin to detail the columns, he or she must know the top-of-steel beam and girder elevations and the bottom-of-column baseplate elevations and baseplate sizes. The detailer also needs to have located, sized, and selected all connections pertaining to that particular column. It is not surprising, therefore, that the shop detail drawings of the structural steel columns for a commercial or industrial building can often be time-consuming to prepare. The structural detailer must refer many times to the structural design drawings to find such information as sizes of connecting beams and girders, top-of-steel elevations, and sizes of baseplates. And, as discussed in Chapter 9, he or she must design the connections if that information is not included in the structural design drawings. However, once the time-consuming column shop drawings are completed, the rest of the project (i.e., detailing of beams, girders, etc.) should go fairly quickly because the connections and top-of-steel and bottom-of-baseplate elevations have been established. The time spent to do an accurate and complete job on the columns will more than pay for itself down the line.

10.3   STANDARD PROCEDURES IN STRUCTURAL STEEL COLUMN DETAILING

The structural detailer will have at his or her workstation the set of structural design drawings previously prepared by the structural engineering design firm. These drawings, along with the structural steel specifications, will be the detailer's primary reference while preparing the detail or shop fabrication drawings. The initial phase of the steel detailer's work, after he or she has become familiar with the structural requirements, is usually to make sketches of the structural support columns. As previously mentioned, by referring to the design framing plans, sections, details, and column schedule, the detailer can determine such requirements as column, beam, girder, and baseplate sizes, top-of-steel elevations, and connection requirements. From that information, he or she then develops freehand sketches showing what each individual column will look like. This includes the locations of connection holes through beam webs, flanges, cap plates and baseplates, top of cap plate, bottom of baseplate, and top of steel, as well as the location of the column itself, as shown in Figure 10–1.

Notice first in Figure 10–1 that the column on the left is a 3″ standard steel pipe column and the column on the right is a W-shape, or wide-flange, column. The columns are identified as columns A1 and B4. On preliminary sketches, steel detailers often identify columns in reference to their locations on the building's grid system. This is not how they will be primarily identified on the shop drawings, although sometimes column grid locations are shown on shop drawings as a secondary reference. Notice also that the columns are sketched in only the principal view. The top-of-cap-plate and bottom-of-baseplate elevations are shown, as are the cap plate size, baseplate size, and length of steel pipe required to fabricate the column. There is no need to show such features as hole sizes and weld symbols because that will be shown on the column detail drawing.

Column B4 shows routine information such as the length of W8 × 31 W-shape required, the baseplate size, and the locations of connection holes in both the column web and flanges. However, column B4 also shows dimensions and elevations that may, at first glance, appear confusing—the +8′–0 elevations at about mid-height on the sketch and the +15′–0 elevation above the top of the column. These dimensions are the top-of-steel elevations for the beams that will be framing into the columns or, in other words, elevations that must be maintained when the columns are installed by the ironworkers at the job site.

Notice also that "mark 'south'" is written on column A1 and "mark 'west'" is written on column B4. These noes are actually painted on the columns in the places indicated while the columns are still in the fabricator's shop. Their purpose is to show the ironworkers the proper column orientation when the column is set in place at the job site.

It is important to remember that the structural detailer must be able to determine all the information shown in Figure 10–1 either from reviewing the design drawings or by selecting the proper connection, as discussed in Chapter 9. It is also important to realize that the structural detailer will make as many sketches as required, like those in Figure 10–1, to ensure that he or she has a good grasp of how the structural steel components will fit together before making the first shop fabrication drawing.

The preparation of shop detail drawings for columns will be discussed in greater detail in this chapter, but before going into that, we will first examine how the structural detailer may have found the information required to make the sketch for the pipe column in Figure 10–1.

Where to Find Information

Suppose the structural design drawings show that, at grid line A1, a 3″ standard steel pipe column sits on top of a foundation wall as shown in Figure 10–2. The structural

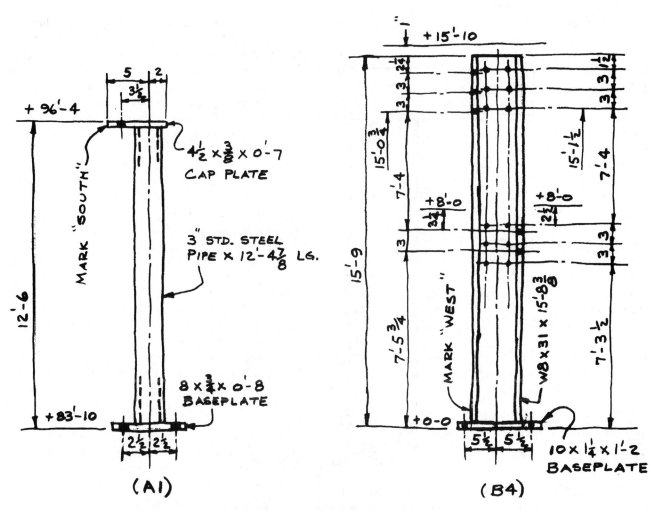

Figure 10–1   Column detail sketches

design drawings, on a section or detail, should also have shown the top-of-steel elevation for the M14 × 18 beam and the top-of-foundation-wall elevation, grout thickness, bottom-of-baseplate elevation and baseplate size—in other words, the information indicated with an asterisk in Figure 10–2. By referring to Part 1 of the AISC manual, the structural detailer would find that the actual depth of a M14 × 18 (a very light, wide-flange beam) is exactly 14″. With the top-of-beam elevation given as 97′–6″, the top-of-column cap plate elevation must be:

$$(97'–6'') – (1'–2'') = 96'–4''$$

The top-of-foundation wall elevation is listed as +83′–9″, and the designer has called for a 1″ thickness of grout, which makes the bottom-of-baseplate elevation (83′–9″) + (1″) = 83′–10″. The overall length of the column from top of cap plate to bottom of baseplate must then be:

$$(96'–4'') – (83'–10'') = 12'–6''$$

The size of the baseplate, 8″ × ¼″ × 0′–8″, should have been shown on the design drawings on a detail, the column schedule, or both. The actual dimensions of the cap plate that supports the beam might be shown on a design drawing detail, but this is often left to the structural detailer. Assuming in this case that the structural detailer is to select the cap plate, he or she would use the following process.

Looking in Part 1 of the AISC manual, the structural detailer would see that the bottom flange of the M14 × 18 beam is 4″ wide and ¼″ thick. The column cap plate should then be at least 4″ wide and at least as thick, or slightly thicker, than the bottom flange of the beam. The standard procedure with this type of column is to connect both the cap plate and the baseplate to the pipe with a fillet weld all around. A cap plate 7″ in length, set on top of the steel pipe so it extends 2″ to the right and 5″ to the left of the center line of the pipe, would provide enough room for a fillet

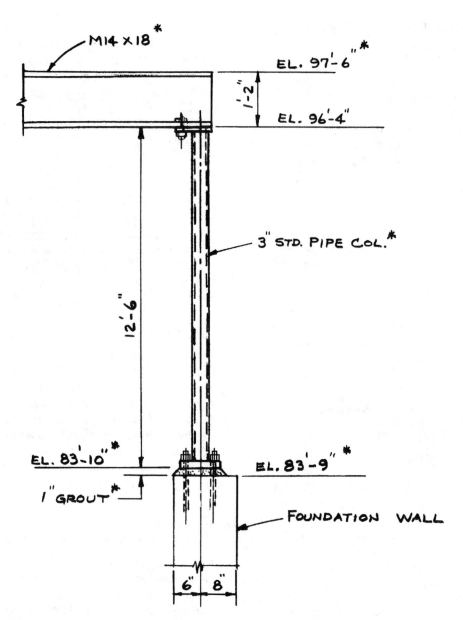

Figure 10–2   A structural design column detail

weld all around and make it possible to locate the connecting holes 3½″ away from the center line of the pipe. This is far enough to ensure that the hex. nuts of the bolted connection will not interfere with the fillet weld and still allow the AISC recommended edge distance of 1½″. Thus, a 4½″ × ⅜″ × 0′–7″-long cap plate is a logical choice.

The length of 3″ standard steel pipe required to fabricate the column is found by adding the combined thicknesses of the cap plate and baseplate and subtracting the sum from the 12–6″ overall length of the column:

$$(12'-6'') - (\tfrac{3}{8}'' + \tfrac{1}{4}'') = 12'-4\tfrac{7}{8}''$$

As emphasized previously, it is the responsibility of the structural detailer to use the information given in the design drawings and the standards provided by the AISC in the *Manual of Steel Construction* to prepare shop drawings showing exactly how the components of the structural framework must be fabricated in order to fit together at the job site the way the engineer envisioned they should. Also, it is the responsibility of the structural drafter in the design office to provide the necessary information on the design drawings so that the structural detailer can fulfill his or her responsibilities. With this in mind, we will examine some shop fabrication drawings of structural steel columns.

10.4 FEATURES OF A TYPICAL COLUMN DETAIL

You may recall from Chapter 8 that shop detail drawings have certain unique characteristics. For instance, we discussed the fact that the lines should be dark and the lettering clear so that fabricators or ironworkers can read the plans under less than ideal circumstances. We also mentioned that inch marks (″) are never shown on shop drawings and that column details might be drawn to a scale in depth, but are rarely drawn to scale in the longitudinal or length dimension. We will now discuss some other unique features of shop details, including the drawing number, shipping marks, material list, assembling marks, bolt symbols, and gage dimension.

Drawing Number

Figure 10–3 is a detail drawing of two different types of structural steel columns. One column, of which four are required, is a 3″ standard pipe column often used in one-story structures such as small office buildings, branch banks, or restaurants. The second column might be for a similar application, but it is a 4″ × 4″ structural tube column, and the drawing calls for twelve of them to be fabricated.

Notice first that the drawing number at the bottom of the page is simply a number 1. It is not A-1, S-1, or P-1 such as would be found on drawings in a design office where all architectural drawings are "A" numbers, structural drawings are "S" numbers, and plumbing drawings are "P" numbers. In fabricating offices, drawings are usually either numbered with "E" numbers (i.e., E-1, E-2, E-3 . . .) to designate erection drawings (see Chapter 12) or simply with 1, 2, 3, 4 . . . , which are the numbers assigned to detail sheets. These detail sheets are usually 24″ × 36″ sheets upon which many beams, columns, or other structural components are detailed for shop fabrication.

Shipping Marks

Notice also that the columns in Figure 10–3 are identified as columns 1C1, 1C2, and 1C3. Every structural component is given an identification mark, or *shipping mark*, as it is drawn. That mark is then shown on the erection plan in the proper location and painted in bright yellow on the beam or column in a certain place so the ironworkers out on the job can find it easily when they erect the structural steel.

A shipping mark number has three parts. First is the detail sheet number. Thus, as in Figure 10–3, all of the details on the first detail sheet (beams, columns, leveling plates, bearing plates, etc.) would have the number "1" as the first number. The details on the second detail sheet would have the sheet number "2" as the first number, and so on.

The second part of the number identifies the type of component shown by the drawing. Fabricators have their own identification systems, but most of them follow the convention of identifying all columns with a "C", all beams with a "B", and all anchor bolts with "AB."

The third part of the shipping mark number indicates whether the drawing is the first column on the sheet, the second column, the third beam, the fifth support angle, or however many drawings are on the same sheet. For example, notice on Figure 10–3 that the detail sheet is identified as sheet 1. This means every shipping mark on this sheet will begin with number 1. Since the first detail drawn on this sheet is the pipe column, the first part of the shipping mark is 1C. If it were a beam, it would have been 1B. Then, of course, the first column to be detailed on sheet 1 would be 1C1, the second column 1C2, and the third column 1C3. If the next item detailed on the sheet were to be a beam, it would be identified as 1B1.

Material List

Another important part of a shop detail drawing is the material list. Material lists are usually located in the upper right-hand corner of the sheet as shown in Figure 10–3. From this list, the fabricating shop determines the sizes and quantities of the steel stock required to fabricate all of the beams, columns, support angles, leveling plates, or whatever components are shown on that sheet. Thus, to fabricate the four pipe columns identified as 1C1 on the detail sheet, the ironworkers in the fabrication shop will need:

- (4) 3″ standard steel pipes × 14′–4⅞″ long
- (4) steel plates 4″ × ½″ × 0′–10″ long
- (4) steel plates 8″ × ¼″ × 0′–8″ long

Although the columns in Figure 10–3 are drawn with only the principal view, more than one view might be required for more complex columns. For example, the detailer would make additional drawings if items were to be welded in several different places on the flange faces of W-shaped columns or if many different nonstandard holes were to be drilled or punched through column flanges. However, the majority of simple columns, manually drawn, can be adequately detailed with only the principal view. On the other hand, CAD-generated column details often show two or even three views of a column and a top view automatically because of the way the software is set up rather than because all the views are required.

Assembling Marks

Notice the cap plate and baseplate on column 1C1. Each of these plates is identified on the drawing by a lower-case letter. The cap plate is identified with an *a* and the baseplate with a *b*. These identification marks are called

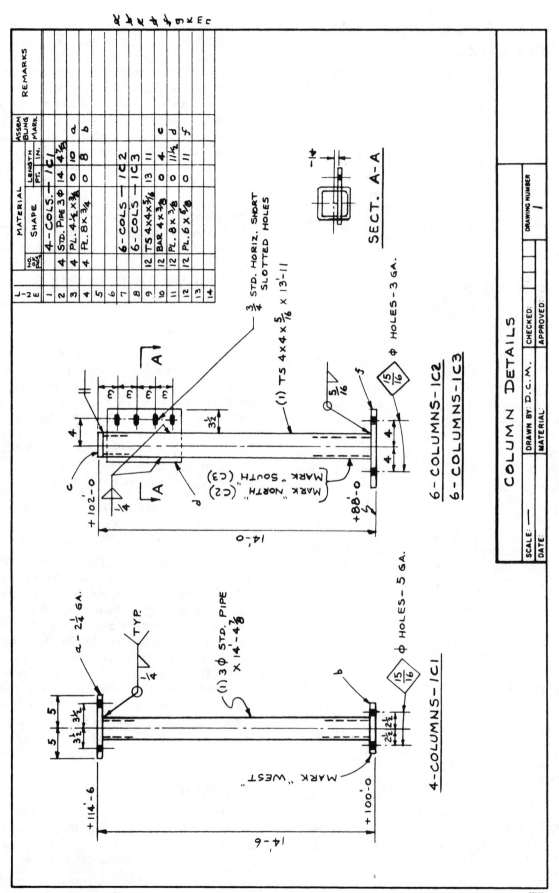

Figure 10-3 Column details

Table 10–1
Nominal Hole Dimensions

Nominal Hole Dimensions

Bolt Dia.	Hole Dimensions			
	Standard (Dia.)	Oversize (Dia.)	Short-slot (Width × length)	Long-slot (Width × length)
½	$^9/_{16}$	$^5/_8$	$^9/_{16}$ × $^{11}/_{16}$	$^9/_{16}$ × 1¼
$^5/_8$	$^{11}/_{16}$	$^{13}/_{16}$	$^{11}/_{16}$ × $^7/_8$	$^{11}/_{16}$ × 1$^9/_{16}$
¾	$^{13}/_{16}$	$^{15}/_{16}$	$^{13}/_{16}$ × 1	$^{13}/_{16}$ × 1$^7/_8$
$^7/_8$	$^{15}/_{16}$	1$^1/_{16}$	$^{15}/_{16}$ × 1$^1/_8$	$^{15}/_{16}$ × 2$^3/_{16}$
1	1$^1/_{16}$	1¼	1$^1/_{16}$ × 1$^5/_{16}$	1$^1/_{16}$ × 2½
≥1$^1/_8$	$d + ^1/_{16}$	$d + ^5/_{16}$	$(d + ^1/_{16}) × (d + ^3/_8)$	$(d + ^1/_{16}) × (2.5 × d)$

(Courtesy of the American Institute of Steel Construction Inc.)

assembling marks, or *piecemarks* and each plate is identified by that mark on the material list. Also notice to the right of the material list a row of lowercase letters (a, b, c . . .) running vertically. Whenever the drafter uses one of those letters to identify a piecemark in the material list, he or she immediately crosses it off the column of letters on the right-hand side of the sheet so that the letter will not be used again. At the same time the piecemark is used on the material list, it must also be shown in the proper location on the detail drawing.

Because the cap plate and baseplate are drawn with only one view, enough dimensions must be given that there will be no mistake in the shop as to how the plates are to be oriented on the column. For example, the cap plate on column 1C1 is listed as plate *a*. On the material list, plate *a* is shown to be 4 ½″ wide, ⅜″ thick, and 10″ long. The way the cap plate is drawn on the detail indicates that the plate is to be set on top of the column so that it extends 5″ to the left and right of the center line of the pipe. Thus, we must be looking at the 10″ length in the view shown, which means the 4 ½″ width dimension is the dimension that cannot be seen. However, even without a plan view to actually show the width dimension, the fabricating shop will know how to set the cap plate on top of the column.

Bolt Hole Symbols

The solid black areas shown on the cap plate of column 1C1, on the thru-plate at the top of columns 1C2 and 1C3, and on the baseplates of columns 1C1, 1C2, and 1C3 indicate open holes through the plates. The holes through the cap plate of column 1C1 would be used for bolted connections that run through the cap plate and the bottom flange of a W-shape beam. (Additional examples can be viewed in Figures 7–15 through 7–19.) The horizontal slots through the thru—plate at the top of columns 1C2, and 1C3 are for bolts that will connect the thru-plate to the web of a beam (see Figures 7–29 and 7–31). The holes through the

column baseplates are for anchor bolts coming up out of a footing, pedestal, pier, or foundation wall similar to those shown in Figures 7–1 through 7–7. The baseplate for columns 1C2 and 1C3 would be very similar to the detail shown in Figure 7–3.

Notice that the sizes of the holes through the cap plate of column 1C1 are not given. As a standard practice on shop drawings, $^{13}/_{16}$″-diameter holes for ¾″-diameter bolts are assumed unless specified otherwise. This is because the use of ¾″-diameter high-strength bolts for commercial and industrial buildings is standard in the construction industry, and the standard hole diameter is always $^1/_{16}$″ greater than the bolt diameter. For example, if a project calls for ⅞″-diameter bolts, a general note saying "All bolt holes to be $^{15}/_{16}$″ diameter U.N. (Unless Noted)" will be placed conspicuously on each detail sheet by the structural detailer.

In Figure 10–3, the holes through the baseplates are larger than $^{13}/_{16}$″ in diameter so their sizes are given. Any holes larger or smaller than $^{13}/_{16}$″ in diameter must be shown if the majority of holes are $^{13}/_{16}$″. Actually, the $^{15}/_{16}$″-diameter holes called for through the column baseplates are for ¾″-diameter anchor bolts. Anchor bolt holes are usually oversized to allow for slight errors in placing the bolts in the field and to provide the ironworkers room for adjustments as required during field erection. Thus, a ¾″-diameter anchor bolt would require a $^{15}/_{16}$″-diameter hole through the baseplate, while a 1″-diameter anchor bolt would require a 1¼″ oversized hole. Table 10–1, reproduced here from the AISC manual, shows nominal standard and oversized hole and short and long-slot dimensions for various bolt diameters.

Gage Dimension

Figure 10–3 also shows holes through the cap plate at the top of column 1C1 located 3½″ on either side of the center line of the column. However, there is no plan view—that is, no view looking down at the cap plate. How many holes are required? If more than two, where are they located? Here

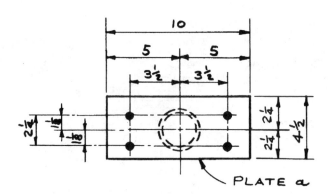

Figure 10–4   A column cap plate detail.

the structural detailer applies a common practice that is, again, quite unique to structural shop drafting. Notice that the cap plate for column 1C1 is identified by an arrow leading from piecemark *a* to the edge of the cap plate. The letter *a* is then followed by "2¼ GA." GA. stands for *gage,* which is the total distance between the holes *and* assumes the holes are of equal distance on either side of a center line in the view that is *not shown.* Thus, the information given on the drawing for the cap plate on top of column 1C1 in Figure 10–3 would indicate to the structural fabrication shop that, in plan view, the top of the column would look as shown in Figure 10–4.

The gage dimension of 2¼″ shown in Figure 10–3 and in plan view in Figure 10–4 is determined by the flange gage of the W-shape that will be supported by the column cap plate. By definition, the usual *flange gage* is the AISC recommended distance between rows of connecting holes through the flanges of W-shapes, assuming that the rows of holes are equally spaced on either side of the center line of the web as shown in Figure 10–5.

Standard flange gages depend upon and vary directly with the width of the flange, according to the following guidelines:

* For flange widths 8 inches and over, the usual flange gage is 5½ inches.
* For flange widths from 6 inches up to (but not including) 8 inches, the usual flange gage is 3½ inches.
* For flange widths from 5 inches to 5¾ inches, the usual flange gage is 2¾ inches.
* For flange widths below 5 inches, the flange gage is 2¼ inches.

Thus, the flange gage for the M14 × 18 shown in Figure 10–2 would be 2¼″ because the flange width of a M14 × 18 is 4″. The flange gage of the W16 × 40 beams in Figures 7–15 and 7–18 are 3½″ because the flange width of a W16 × 40 is 7″ (see Table 2–1). What is the recommended flange gage dimension, and thus the gage dimen-

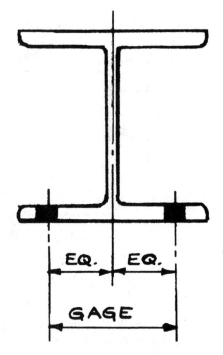

Figure 10–5   Flange gage dimensions.

sion through the cap plate, for the W18 × 65 beams shown in Figure 7–16?

The baseplate for column 1C1 in Figure 10–3 is shown as 8″ × ¾″ × 0′–8″ long. The holes through the baseplate are shown to be ⟨15/16⟩ diameter, which is standard for ¾″-diameter anchor bolts, and the holes are shown to be 5″ apart in the principal view of the column. The 5-GA. noted on the column detail would tell the fabricating shop to locate the holes through the baseplate as shown in Figure 10–6A.

Baseplate dimensions are selected by the structural designer to distribute the column load over a sufficient area of a concrete foundation wall or independent column footing so that the allowable compressive strength of the concrete will not be exceeded (see Chapter 5). Beyond that, there are two main considerations: (1) The anchor bolt holes must be far enough away from the column itself so that the hold-down nuts of the anchor bolts do not interfere with the column, and (2) the baseplate must be large enough that the minimum recommended edge distance from the center of the holes to the edge of the baseplate is maintained. Table 10–2, taken from the AISC manual, lists the minimum recommended edge distance for a ¾″-diameter bolt as 1¼″, so the 1½″ edge distance for the anchor bolts on column 1C1 is sufficient.

Notice that the baseplate for columns 1C2 and 1C3 in Figure 10–3 is 11″ long, with ⟨15/16⟩ diameter anchor bolt holes shown to be 8″ center-to-center in the principal view and calling for the holes to be 3 GA. This tells the fabricating shop to locate the holes through the baseplate as shown

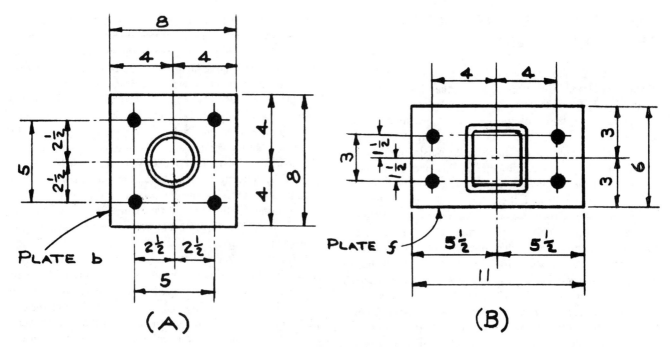

Figure 10–6   Column baseplate details.

Table 10–2
Minimum Edge Distances for Steel Plates

Minimum Edge Distance, in.
(Center of Standard Hole[a] to Edge of Connected Part)

Nominal Bolt or Rivet Dia. (in.)	At Sheared Edges	At Rolled Edges of Plates, Shapes or Bars, Gas Cut or Saw-cut Edges[b]
½	⅞	¾
⅝	1⅛	⅞
¾	1¼	1
⅞	1½[c]	1⅛
1	1¾[c]	1¼
1⅛	2	1½
1¼	2¼	1⅝
Over 1¼	1¾ × Dia.	1¼ × Dia.

[a]For oversized or slotted holes, see Table J3.6.
[b]All edge distances in this column may be reduced ⅛-in. when the hole is at a point where stress does not exceed 25% of the maximum design strength in the element.
[c]These may be 1¼ in. at the ends of beam connection angles.

(Courtesy of the American Institute of Steel Construction Inc.)

in Figure 10–6B. The 6″-wide baseplate provides more than enough width for the 5⁄16 fillet weld all around the ST column, and again, the minimum 1½″ edge distance for the anchor bolts has been maintained.

Two final points can be made about Figure 10–3. First, between columns 1C2 and 1C3 is the orientation of "north" and "south" when the columns are set in place at the job

site. Second, the thru-plate (d) is set ¼″ off-center of the column. Thus, assuming the web of the column connecting to the plate is ½″ thick, when the beam is connected to the thru-plate, the center line of the 4 × 4 structural tube column and the center of the web of the beam connecting to it will be in line.

10.5 EXAMPLES OF COLUMN DETAIL DRAWINGS

Figure 10–7 shows a detail sheet for two structural steel W-shape columns. Since this is detail sheet 2, the columns are labeled 2C1 and 2C2.

Notice first how column 2C1 is dimensioned. The dimensions are not crowded, and the lettering is dark and easy to read. The overall dimension is the outside dimension on the far left. Closer to the column, the inner dimension string locates the bottom bolt holes for the 5-row connection that will be made through the column web. Obviously, the 1′–2½ dimension shown from the bottom of the baseplate to the first connection hole is drawn completely out of scale, but this is permissible in the structural steel shop detail drawings. Because a standard 5-row connection will have four spaces, the connection holes in the vertical direction are simply called out as 4 × 3 = 1′–0, which means four spaces at 3″ pitch between rows of holes = 1′–0. The 10′–6¾ dimension from the top row of bolt holes to the top of the column cap plate completes the dimension string and verifies the location of the top of the cap plate from the top row of bolt holes. It also equates the dimensions in the dimension string to the overall dimension of 12′–8¾:

$$1′–2½ + 1′–0 + 10′–6¾ = 12′–8¾$$

Looking just to the right of the 10′–6¾ dimension, we see a 3″ dimension from the top of the web connection holes to a short horizontal line. This signifies that the top-of-steel elevation for the beam connecting to the column web at that point is +13′–0½, and that the top of the beam is 3″ above the top of the connection holes through the web of the column.

The principal view of the column also shows that the connection holes through the column web are 3½″ center-to-center, and since not stated otherwise, the holes are assumed to be spaced at equal distances on either side of the center line of the column web as shown. The standard gage hole spacing dimension through column webs is 3½″ for W8 and W10 columns and 5½″ for W12s and larger, as shown in Figure 10–8.

For W8 and W10 columns, when a beam is fastened to the column web with framing angles, the angles are usually 3 × 3 angles, with the gage distance of the outstanding legs of the angles 3½″. For W12 and larger columns using standard framing connections, the angles are usually 4 × 3 or 4 × 3½, with the 4-inch legs outstanding. Of course, other angle sizes and gages are also used, depending on the situation and the standards of various structural steel fabricators.

Holes for a 3-row connection are shown through the right flange of column 2C1 and the left flange of column 2C2. Notice in both cases that the flange gage of the holes is 3½″, which is in keeping with our previous discussion of flange widths and gages. However, on both columns, notice that the 3½ GA. is pointing not to the holes or their center lines, but to the flange face. This indicates that any holes on the flange face indicated would be 3½ gage.

On column 2C1, the holes on the flange face are located so that they are staggered (not in line with) the holes through the web of the column. This is often done with W8 columns to ensure that the connecting bolts through the web and the flange will not interfere with each other. It is usually accomplished by locating the first hole at some dimension other than the standard 3″ below the top of the beam. In this example, locating the first hole in the flange face connection 4″ below the +13′–3 top-of-beam dimension will stagger the holes as required.

Column 2C2 is a detail for a W8 × 24 column. This detail illustrates a situation where the holes through the column cap plate are not centered on a gage, requiring a top view to be drawn showing that the two holes through the cap plate are located 5½″ off the center line of the column web.

A final point about this column detail sheet is that the structural drafter has specified that column 2C1 is to be located at grid line intersection B2 and column 2C2 is to be located and set in place at grid D4.

Figure 10–9 is a structural steel shop fabrication drawing of two steel columns and their leveling plates. You may recall from Chapter 7 that leveling plates, also called setting plates, are ¼″-thick plates exactly the same dimension as the column baseplate they are designed for.

We will first examine column 3C1 in Figure 10–9. Note that it does not have a cap plate and that the overall dimension of 28′–0 is from the bottom of the baseplate to 1″ above the top of the column. This indicates that the top-of-steel elevation for the beams connecting to the upper end of the column web will be 1″ above the top of the column itself. This is a fairly common occurrence, and when the dimension of 1 inch is used, most structural fabricators will want the inch mark (″) included for clarity.

The dimensioning of the holes through the column web is done a little differently on column 3C1 than on previous examples because the bottom row of connection holes must always be located from the bottom of the baseplate. The reason is that, as a W-shape rolls down the beam line in the shop to have standard holes or slots punched or drilled in it, the location of the first holes for a connection must be referenced from one end of the beam or column. On beam 3C1, the lower connection holes start 13′–0 from the bottom of the baseplate and are dimensioned like the web holes through column 2C1 in Figure 10–7. However, the upper row of connection holes must also be referenced from the bottom of the baseplate, and this is accomplished by a

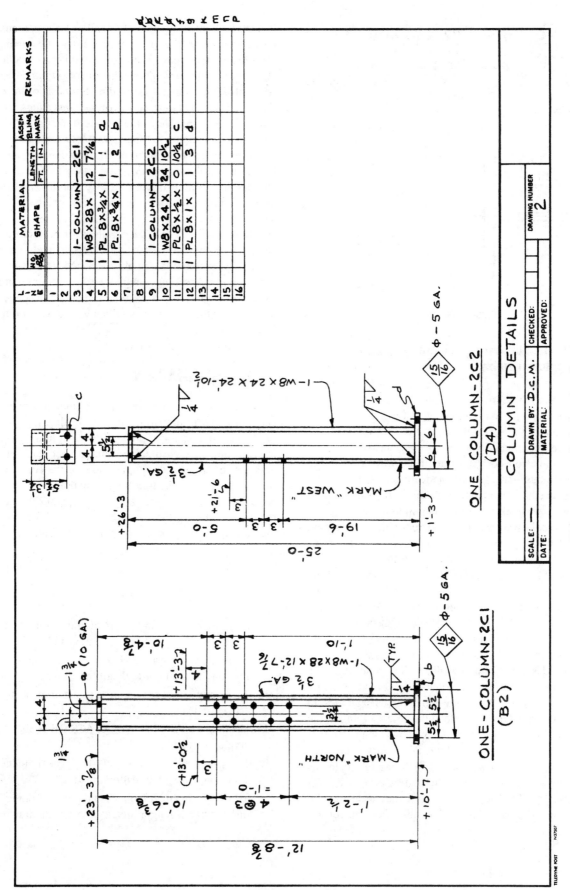

Figure 10-7 Column details

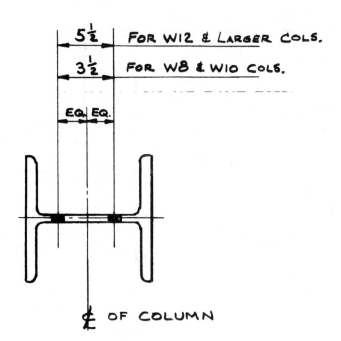

Figure 10–8   Standard gage hole spacing through column webs.

running dimension. The running dimension of 27′–0 on column 3C1 locates the first holes of that connection from the bottom of the baseplate. Notice that the dimension line for the 27′–0 dimension does not run all the way down to the bottom of the baseplate, but the fabricating shop knows that is where it is taken from. The discussion of beam detailing in Chapter 11 will call for many running dimensions.

Column 3C1 also has a seat angle welded in place on the face of the left flange. The structural drafter has indicated that the angle is to be centered on the flange by the note "*b* (on ℄)." The note "G.O.L. = 3½" indicates that the holes are to be spaced 3½″ apart on the outstanding leg (the 4″ leg) of the seat angle, similar to the situation shown in Figure 7–25. G.O.L. stands for Gage Outstanding Leg. Notice also that the structural drafter is calling for ⁵⁄₁₆ fillet welds on the vertical legs at both ends of the seat angle.

The weld symbols for the fillet welds connecting the baseplates to the bottom of columns 3C1 and 3C2 are slightly different from those shown in previous examples. In addition to the standard weld along the column flanges, the structural drafter has indicated a fillet weld 3″ long on either side of the column web. Structural fabricators commonly specify this type of web weld on all W10 or larger columns as well as on smaller columns if dictated by design requirements. Also see that the holes through the baseplates are specified as ⟨1¼⟩ ϕ, which means that they are for 1″-

diameter anchor bolts (see Table 10–1) and the edge distance from the center of the holes to the ends of the plates exceeds the minimum 1¼″ distance specified in Table 10–2.

A note on the principal view of column 3C2 shows two plates (*f* and *g*) located on the center line of the column flange. These plates are called *connection plates,* or *gusset plates.* In this case, they could be used to connect diagonal cross-bracing angles between two columns. The locating elevation dimensions on the principal view are taken between established work points (w.p.). A ₆L̲₁₂ slope or pitch line is shown from the work points, and the connection hole locations—from the work points—are shown along this line.

Details 3LP1 and 3LP2 show the leveling plates required for columns 3C1 and 3C2, respectively. These leveling plates would be set in place, leveled, and secured at the job site by the erecting crew before the columns themselves were installed. This is common practice when erecting heavy columns such as 3C1, which would weigh more than 1,700 pounds.

10.6   SUMMARY

This chapter has covered many of the fundamental concepts involved in preparing shop detail drawings for structural steel columns. The preparation of column shop drawings is undoubtedly time-consuming, but the time spent doing an accurate and complete job on the column details will greatly reduce the time spent detailing beams and girders later. Thus, the structural drafter in either the engineering design office or the structural fabricator's drafting room must know how to properly prepare shop detail drawings for structural steel columns.

STUDY QUESTIONS

1. Why are the structural steel columns the first part of the structural support frame to be detailed?

2. Where does the structural detailer obtain the information required to detail the columns?

3. How many views are required to detail most structural steel columns?

4. What does a shipping mark of 6C3 tell about a structural steel column?

5. How is a dimension of twenty-two feet six inches shown on a shop fabrication drawing?

6. To what scale are columns usually drawn in the longitudinal or length dimension on shop drawings?

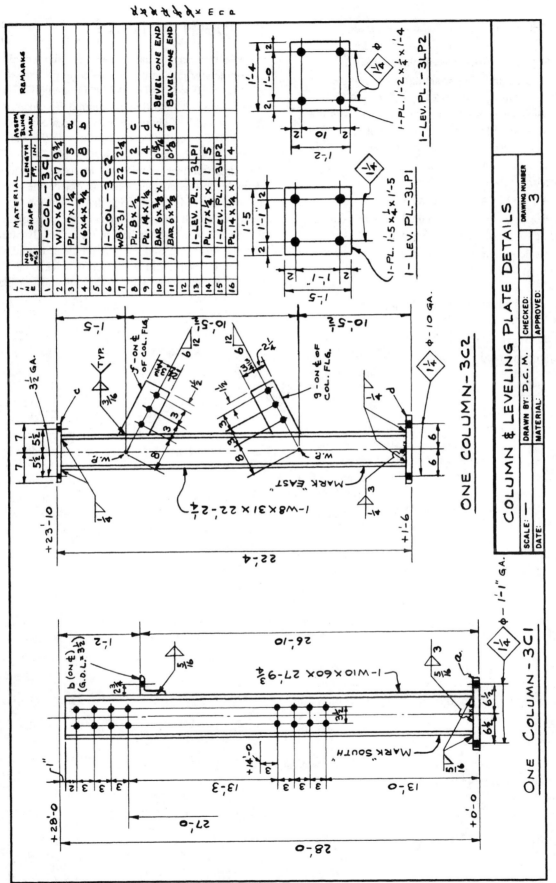

Figure 10–9   Column and leveling plate details

7. What depth scale is usually used for most steel columns?

8. What is a piece mark, or assembling mark?

9. Why is it especially important that structural steel, whether prepared manually or by CAD methods, always have clear, sharp, dark linework and lettering?

10. Why is it standard practice on shop detail drawings not to show sizes for $^{13}\!/_{16}''$-diameter holes?

11. Why are anchor bolt holes through baseplates oversized?

12. What is meant by a GAGE dimension?

13. What size holes would be required through a column baseplate for $^{7}\!/_{8}''$-diameter anchor bolts?

14. What would be the standard flange gage for a W8 × 31 column? (Note: A W8 × 31 has a flange width of 8 inches.)

15. What is the standard web gage through the web of a W8 × 31 column?

16. What do the letters G.O.L. indicate?

17. It is common practice among structural fabricators to specify a web weld on the baseplate connection for _____ or larger columns.

STUDENT ACTIVITY

Using either manual or CAD drafting technique, make a shop detail drawing of the two columns shown in Figure 10–1 on a 12″ × 18″ sheet. Make a material list on the right side of the sheet and fill in as required. The material list will require a width of about 5½ inches and a depth of about 3½ inches. The space between most lines on the material list is ¼ inch. Draw the columns to a scale of 1″ = 1′–0 in depth. The length dimension need not be to scale. Following the examples in this chapter, make all object lines heavy, dark, and easy to read, all extension and dimension lines thin but clear, and all lettering dark and about ⅛″ in height, except for titles, which can be ¼″ in height. Be especially careful not to crowd dimensions. Show required hole sizes and weld symbols. Assume all framing connection holes to be $^{13}\!/_{16}''$ diameter and all anchor bolt holes through the baseplate to be $^{15}\!/_{16}''$ diameter.

Chapter 11

Structural Steel Beam and Miscellaneous Steel Detailing

OBJECTIVES

Upon completion of this chapter, you should be able to:

- prepare shop detail drawings for a variety of structural steel beams in accordance with generally accepted practices of structural steel fabrication.
- prepare shop drawings for a variety of miscellaneous structural steel components common to commercial and industrial building construction.

11.1 INTRODUCTION

Structural steel beams are the most common components of structural steel framing systems. They can be broadly defined as horizontal members subjected to gravity or vertical loads, meaning loads perpendicular to their length. The roofs and floors of commercial and industrial buildings of most any size are usually supported by beams, and the beams are supported by girders, which are simply large beams into which the smaller beams are connected.

Beams are not always perfectly horizontal. For example, roof beams are generally sloped or pitched downward to some degree, which allows rainwater to run toward a low point in the roof for removal by the roof drainage system. A structural steel framing plan identifying beams and girders is shown in Figure 5–3, (p. 53) and a sloped beam connection detail is shown in Figure 7–33 (p. 137).

Because beams are so widely used in almost any type of commercial or industrial building construction, the proper method of preparing shop drawings for structural steel beams is one of the skills every structural drafting technician must know. This is true not only for the structural drafter working in the structural fabricator's office, but also for the structural drafter working in the design office, who will check and approve shop detail drawings made from the design drawings he or she previously prepared.

In addition to knowing how to prepare shop drawings of columns, beams, and girders, the structural steel detailer must also prepare shop drawings for a variety of miscellaneous structural steel items needed for commercial or industrial buildings. Examples include roof frames, steel joist bearing angles or plates, steel handrails, steel stair stringers, steel ladders, deck support angles, and beam bearing plates. Although the limitations of a beginning textbook make it impossible to do more than scratch the surface of the subject of miscellaneous steel, this chapter will introduce some of the simple and more commonly used components classified as miscellaneous steel.

But first, we will discuss the generally accepted rules for preparing structural steel beam shop fabrication drawings. While these rules may vary slightly from office to office, they will generally be very close to the same nationwide and will be based on information in *Detailing for Steel Construction,* a publication of the AISC.

11.2 BEAM DETAILING

Shop fabrication drawings of structural steel beams are usually not drawn until after the column details have been prepared. With the column details complete, the steel detailer can use both the structural design drawings and the column shop drawings for reference when preparing the beam details. From the design drawings, he or she will find the grid system dimensions required to determine approximate lengths of beams and connection points between beams and supporting girders. Most top-of-steel elevations will be found on either the design drawings and/or the column details. The connection requirements (i.e., seated beam connection or 2-row, 3-row, 4-row framed connection) will have been determined as the columns were being detailed.

Most beam details, like column details, are drawn in only one view, the principal view. The same rules apply for beam details as for column details: All object lines must be heavy and dark; arrowheads must be dark and well proportioned; lettering must be dark, large enough to be easily read, and not crowded; and all extension lines must be thin enough to be easily distinguished from object lines, but dark enough to be easily seen. Also, as with column details, no attempt should be made to draw beams to scale in the longitudinal (length) dimension. Overall lengths are usually foreshortened, although it is common to exaggerate very small distances to a length where the view and dimensions are clear. Connection locations and the spaces between open holes are drawn approximately to scale, but again, large enough so that dimensions are not crowded.

Connection angle and beam flange thicknesses are not drawn to scale. For example, if a structural steel detailer were drawing a detail of a W12 × 22 beam, he or she would most likely draw the beam depth to a scale of $1'' = 1'-0$ in the principal view. However, it would be very impractical, and certainly unnecessary, to check in the AISC manual and find that the thickness of the flanges for a W12 × 22 is .335″ (less than ⅜″) thick and then attempt to draw the flanges to that thickness at a scale of $1'' = 1'-0$. The usual practice is to draw the lines indicating the thickness of the flanges about ¹⁄₁₆″ apart. A drafter soon learns to judge the distance closely enough without measuring. Of course, the drafter should not show the flanges as if they were 3 or 4 inches thick either. The same holds true for the drawn thickness of connection angles, which are usually shown on the shop detail drawing to be as thick or very slightly thicker than the beam flanges.

Dimensioning

One of the first and most important considerations when preparing shop detail drawings for beams is that there is a definite way beams must be dimensioned. This dimensioning concept takes time to learn, but once learned, it becomes quite easy and automatic. The steel detailer knows that every beam detail must have certain dimensions located in certain places and showing certain things. By far the most important consideration in dimensioning a beam is that the dimensions must be *correct*. That is, the number must add up, because when the beam leaves the fabrication shop and

goes out on the job, it must fit where it was intended to go. To that end, the dimensions must be right, their intent unmistakably clear, and all numbers and letters legible.

A very basic rule of dimensioning is that the dimensions must be arranged in a manner most convenient to the workers who will use the drawing. They should not be crowded. Extension lines should cross the fewest possible number of other lines. Overall dimensions and long dimensions should be located farthest away from the views to which they apply. Dimensions are usually referenced to the center lines of beams but to the backs of channels and angles. Vertical dimensions are given to either the top or bottom of beams and channels (whichever elevation is to be held), but never to both. Usually this means vertical dimensions will be located from the top of the beam because the top-of-steel elevation is the important reference for floor and roof systems and for most structural support frames used to support mechanical equipment (see Chapters 6 and 7). However, this is not always true. For example, bottom-of-steel elevation is the most important consideration for beams used as lintels over door or window openings. Thus, on lintels, it is usually the bottom-of-steel elevation that must be held. Perhaps the best way to introduce how shop drawings are drawn and dimensioned would be to trace the process of preparing such a drawing for a simple beam.

Procedure for Detailing a Simple Beam

Figure 11–1A is a small part of a framing plan as it might look on a structural design drawing. It shows that a W16 × 57 beam connects to the web of a W24 × 76 girder on grid line ① and to the flange face of a W8 × 31 column on grid line ②. The distance from grid line ① to grid line ② is $18'-0$. Notice that the top-of-steel elevation for both the W16 × 57 beam and the W24 × 76 girder is $+99'-8$.

From the information given in Figure 11–1A, the steel detailer would visualize connections as shown in Figure 11–1B, which may or may not be shown on a design drawing connection detail. Because the top-of-steel elevation for both the W24 × 76 and W16 × 57 is shown as $+99'-8$ on the framing plan, the steel detailer knows that the top flange of the W16 × 57 must be cut out, or coped, as shown in Figure 11–1B to avoid interference with the top flange of the W24 × 76 girder. He or she also knows that, at the connection between the end of the beam and the flange face of the column, the W16 × 57 can be cut off square. First the steel detailer sizes the framing angle connections, using methods described in Chapter 9, and details the W8 × 31 column as shown in Chapter 10. Then he or she is ready to detail the shop fabrication drawing for the beam as shown in Figure 11–1C.

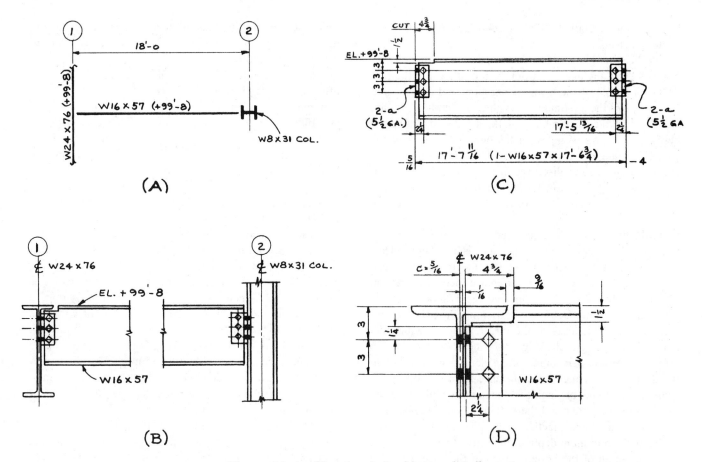

Figure 11–1   Structural steel beam detail

Figure Setback Dimension. Notice first in Figure 11–1C that the overall dimension, which is the end-to-end of connection angles dimension, is shown to be 17′–7¹¹⁄₁₆. Notice also that, outside and directly opposite the dimension line is a –⁵⁄₁₆ on the left end and a –4 on the right end. These dimensions, called setbacks, are always shown on beam details with simple (end-supported) framing angle connections. *Setbacks* are distances from the center lines of the supporting structural steel girders or columns to the backs of the connection angles. Because the beam, when installed, must fit into place on the grid system as shown in Figure 11–1A, the sum of the overall back-to-back of framing angle dimension plus the setbacks on either end must equal the 18′–0 dimension between grid lines ① and ②.

$$17′-7\tfrac{7}{16} + \tfrac{5}{16} + 4 = 18′-0$$

For beams framing into girders, the setback dimension is found by taking one-half of the web thickness of the girder plus ¹⁄₁₆ inch. This is commonly referred to as the C-dimension. Part 1 of the AISC manual lists one-half of the web thickness (t_w) of a W24 × 76 (t_w) as ¼ inch. Thus, ¼ + ¹⁄₁₆ = –⁵⁄₁₆ setback. Ironworkers have standard shims at the job site to fill in spaces as required during the erection of the steel frame.

The setback for the right side of the W16 × 57 that frames into the flange face of the column is simply one-half of the column depth. Since a W8 × 31 is 8 inches deep, the setback is –4. Notice that the setbacks are given as minus (–) dimensions. That is because, in this case, the overall end-to-end of beam dimension is short of the grid line distance between supporting members. It is important to be aware of this factor because it will relate to our discussion of beams set on foundation walls and overhanging or cantilever beams later in this chapter.

Calculating Beam Length. The length of the beam itself (in this case, the W16 × 57) is usually calculated to be approximately 1 inch shorter than the end-to-end of framing angles dimension for a simply supported beam as shown in Figure 11–1. In other words, if the overall dimension were 15′–0, the W16 × 57 would be cut to a length of 14′–11. Many fabricators like to keep beam lengths in ¼-inch increments, which is why the beam length in Figure 11–1C is shown as 17′–6¾. The actual dimension from the face of the outstanding leg of the connection angles to the end of the beam is usually not shown. Thus, in Figure 11–1C, it is assumed that the fabricating shop would center the beam between the overall 17′–7¹¹⁄₁₆ face-to-face of connection angle dimension.

Notice that Figure 11–1C shows the top-of-steel elevation (+99′–8), the distance from the top of steel to the first connection hole of the 3-row connection, and the required cut or cope of the top flange. Also, the connection angles are identified by their assembling mark, and the gage distance between the rows of holes on the outstanding legs of the connection angles is shown as 5½ GA. Connection angle considerations will be discussed in greater depth shortly, but first we will focus on how the structural steel detailer calculates for the required cut or cope.

Calculating Cope. The required cope, which is cut into the top of the W16 × 57 in this example, is based upon clearance requirements necessary so that the ironworkers can install the beam without interference between the top flanges of the W16 × 57 beam and the W24 × 76 girder. Thus, the governing factors in determining the cope requirements are the thickness of the girder web, the width and thickness of the girder flange, and the k-distance of the girder. The *k-distance,* an important element in determining the depth of the cope, is the distance from the top of the girder flange to the beginning of the radius or fillet between the web of the girder and the bottom of the girder flange. Thus, it is the usual practice to use the k-distance of the girder as the minimum depth of the vertical cut of the cope.

The length of the horizontal cut of the cope, which is measured from the back of the outstanding legs of the connection angles, should be long enough to provide ½ to ¾ inch of clearance between the toe of the girder flange and the end of the beam flange. The AISC recommended formula for determining the required length of the cope cut is:

$$\text{Length of cope cut} = \frac{b_f - t_w}{2} + \frac{1}{2} \text{ in.}$$

For the cope shown in Figure 11–1C and D, the steel detailer would look in Part 1 of the AISC manual for the flange width and thickness and the web thickness of the W24 × 76 before determining the required length of horizontal cone cut to be:

$$\text{Length of cut} = \frac{8.990 - 0.440}{2} + 0.500 = 4.775 \text{ in.}$$

From this formula and the detail in Figure 11–1D, it can be seen that a horizontal cut length of 4¾ would result in a clearance of approximately ⁹⁄₁₆ between the toe of the girder flange and the end of the beam flange.

The depth of beam cope should be equal to or greater than the k-dimension of the girder. The k-dimension for a W24 × 76 is listed as 1⁷⁄₁₆. The depth and length of coping cuts are usually specified by rounding up to the next ¼ inch.

Thus, the cope specified on the steel detailer's shop drawing for the W16 × 57 would be 1½ deep by 4¾ long as shown in details 11–1C and D.

Many structural fabricators have developed their own office standards by which their steel detailers can quickly determine cope requirements. Figure 11–2 is an example of an office *standard,* a reference that might be given to an entry-level structural steel detailer to help him or her determine required copes for steel beams. This example will give a slightly larger horizontal cut than the method previously described because it does not subtract anything for the thickness of the girder web, but it is accurate enough to meet modern structural steel fabrication requirements.

The intersection of the horizontal and vertical cope cuts is called a *re-entrant corner.* The AISC manual specifies that all re-entrant corners should be shaped, notch-free, to a radius. For most commercial building work, this radius is usually about ½ inch. Some structural fabricators require their steel detailers to show this radius on their drawings, as in Figure 11–1D. Others prefer re-entrant cuts on drawings to be shown as sharp corners, as in Figure 11–1C, because fabricating shop workers are trained to provide the corner fillet as standard procedure whether it is shown on the drawing or not. In either case, the re-entrant cut fillet size is *not* shown on the steel detailer's drawing.

The last point to consider about coping cuts is how they are dimensioned. Many fabricators prefer the vertical and horizontal cuts dimensioned as shown in Figure 11–1C. Some prefer to show a standard cope mark or symbol such as 1½ ⌐4¾ or 1½ × 4¾. Still others simply note on the detail drawing "Cope to W24 × 76″, and the fabrication shop provides the proper cope according to shop standards.

Detailing Framing Angle Connections. The connection angles shown in Figure 11–1C for the W16 × 57 are shop- or web-bolted and field-bolted connections at either end of the beam. The angles, which would be called for on the material list, are 4 × 3½ × ⁵⁄₁₆ × 0′–8½. The shop or web bolts, shown on the beam detail by the standard diamond ◇ symbol, are on the 3½ web leg of the framing angles. The 2¼-inch gage distance from the back of the connection angles to the center line of the web bolts on the 3½-inch leg meets the AISC specified minimum edge distance of 1¼ inch from the center line of the bolt holes to the rolled edge of the angle.

Notice in Figure 11–1C the dimension of 17′–5¹³⁄₁₆ locating the web bolts for the framing angles at the right end of the beam. This is a running dimension similar to the running dimensions in Chapter 10 that located column bolt holes from the bottom of the baseplate. When detailing structural steel beams, these running dimensions always

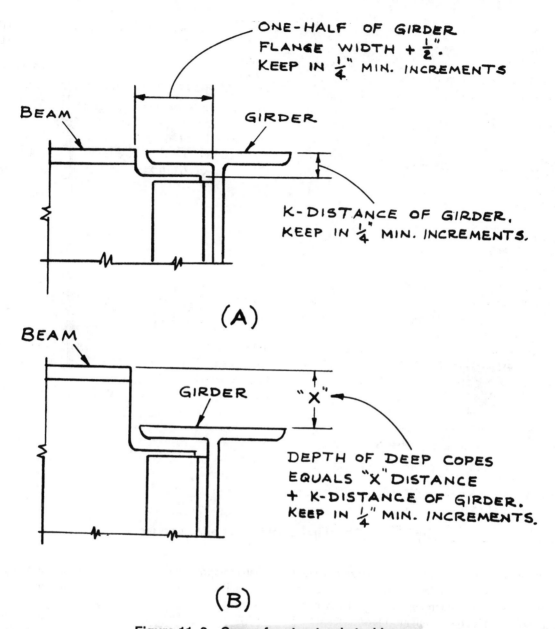

ONE-HALF OF GIRDER
FLANGE WIDTH + $\frac{1}{2}$".
KEEP IN $\frac{1}{4}$" MIN. INCREMENTS

BEAM

GIRDER

K-DISTANCE OF GIRDER.
KEEP IN $\frac{1}{4}$" MIN. INCREMENTS.

(A)

BEAM

GIRDER

"X"

DEPTH OF DEEP COPES
EQUALS "X" DISTANCE
+ K-DISTANCE OF GIRDER.
KEEP IN $\frac{1}{4}$" MIN. INCREMENTS.

(B)

Figure 11–2 Copes for structural steel beams

locate holes or sets of holes from the *left* end of the beam as it appears on the detail sheet. This practice helps the fabricating shop set up automated hole-punching and drilling equipment on beam lines, which significantly reduces the cost of fabricating structural steel beams and columns.

Before discussing examples of shop detail drawings for structural steel beams, one last point should be made about framing angle connection details. The 5½-inch gage dimension shown in Figure 11–1C for the bolt holes through the outstanding legs of the beam connection angles, and the web of the W24 × 76 girder, were possible because no beam was framing into the girder at that point from the opposite side

as illustrated in Figure 11–3A. If beams had been directly opposite each other, the beam-to-girder connection could not have been made as shown in Figure 11–1B. That is, interference between the shop bolts and the field bolts would have made it impossible for the ironworkers to insert and tighten the field bolts during the erection of the structural steel.

The competent structural steel detailer will recognize the bolt interference problem and solve it in one of two ways. One choice is to stagger the shop bolts and field bolts as shown in Figure 11–3B, which would leave ample clearance between the shop and field bolts (see also

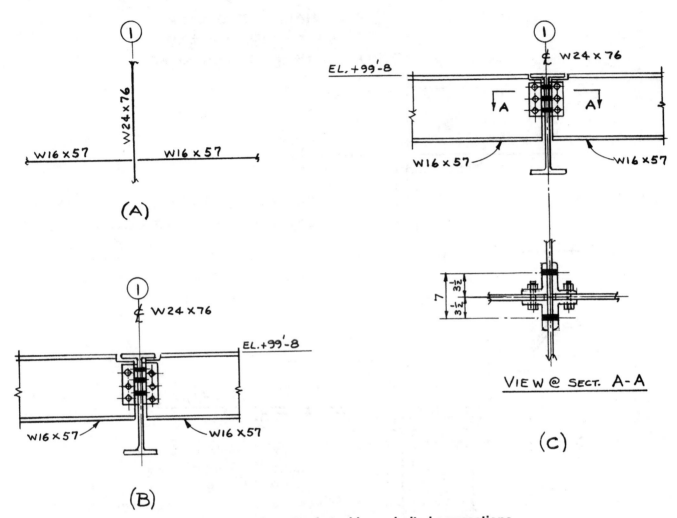

Figure 11–3   Structural steel beam bolted connections

Figure 9–9B). A second possibility is to increase the angle size to 5 × 3½, making it possible to increase the gage distance for the holes in the 5-inch outstanding legs from 5½ inches up to 7 inches as shown in Figure 11–3C. This would solve the interference problem and eliminate the need to stagger the holes in the connection angles.

A similar situation occurs when connection holes are located in both the webs and flanges of W8 columns, as was described in an example in Chapter 10. This is why the open holes through the webs and flanges of the W8 columns in Figure 10–1 and Column 2C1 in Figure 10–7 were shown staggered.

11.3   DETAILING PRACTICES FOR VARIOUS BEAM TYPES

Having discussed some of the fundamental concepts involved in preparing shop detail drawings for structural steel beams, we now have the background to concentrate on

detailing practices for a variety of beam types found in commercial and industrial building construction.

Simple Beams

The discussion from this point will no longer differentiate between beams and girders, but will refer to all the examples simply as beam details. The first type examined will be simple beams, or beams supported at their ends.

Figure 11–4 illustrates a simple structural steel beam. Figure 11–4A shows that the beam, a W18 × 46, is fastened to the flange face of a W8 × 31 column on one end and set in a beam pocket near the top of a foundation wall at the other end.

A similar example of a simple beam was shown in Chapter 7. Notice in Figure 7–52 that the 6 × ¾ × 1′–2 beam bearing plate is welded to the bottom of the W18 × 46 with a ³⁄₁₆ field fillet weld 4 inches long at the near and far side of the toe of the bottom flange. Thus, the bearing plate would be detailed and identified separately and not shown on the beam detail. With a beam bearing plate 1′–2 in length, the

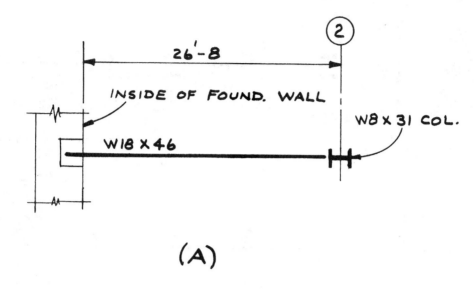

(A)

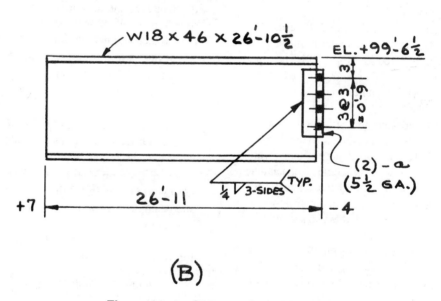

(B)

Figure 11–4 Structural steel beam detail

¾-inch-diameter anchor bolts would be spaced 11 inches apart. Thus, since the flange width of a W18 × 46 is only 6 inches, there would be no holes through the bottom flange of the beam for the anchor bolts. Figure 7–52 also indicates that the W18 × 46 beam sets 7 inches back onto the wall. Because the beam runs 7 inches beyond the face of the support, the beam detail drawing (Figure 11–4*B,* shows a +7 at the left end of the beam just beyond the 26′–11 overall dimension.

Figure 11–4*A* shows that grid line ② at the center line of the W8 × 31 column is 26′–8 from the inside of the foundation wall. It can also be seen that the connection is made

at the column flange, and because the W8 × 31 column is 8 inches deep, the face of the outstanding legs of the connection angles would be 4 inches short of the grid line. Thus, – 4 is shown to the right of the overall dimension of 26′–11. The dimension from the inside of the foundation wall to the end of the connection angles is (26′–8) – (4) = 26′–4, and 26′– 4 + 7 = 26′–11, the overall dimension shown in Figure 11– 4*B*.

Because the flange width of a W8 × 31 column is 8 inches, the flange gage used by most structural steel fabricators would be 5½ inches. Thus, the framing angles shown on the right end of the beam would be (2) – 4 ×

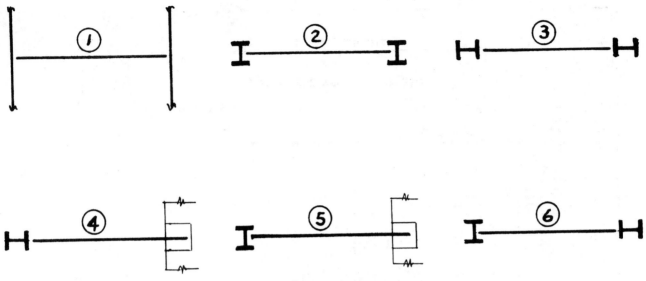

Figure 11–5   Typical structural connections

$3 \times \frac{5}{16} \times 0'–11\frac{1}{2}$, and the holes through the outstanding legs would be designated 5½ GA. as shown.

With an overall dimension of 26'–11 and the usual ½" setback from the face of the connection angles to the end of the beam, ½ the beam length called for is 26'–10½. The top-of-steel elevation (+99'–6½) and proper weld symbol to connect the web legs of the framing angles to the web of the W18 × 46 complete the detail.

Beams may be fastened into the structural support frame in many ways—to the webs of columns, to the flanges of columns, or to the web of a column on one end and a wall on the other end, and so on. Thus, the structural steel detailer must be able to dimension beams for all those situations. Figure 11–5 shows how to determine overall dimensions for shop beam details by explaining what the setback dimensions should be for various situations.

In the example of structural steel beams framing into girders in Figure 11–1, we found the setback dimension that determined the overall length of the beam by taking one-half of the web thickness of the girder plus ⅟₁₆, a dimension commonly referred to as the C-dimension. The C-dimension is mentioned again here because it figures into some of the framing conditions depicted in Figure 11–5. With this in mind, the end-to-end or overall beam dimensions for the structural steel framing conditions shown in Figure 11–5 are are follows:

① **Beam to Girder Webs.** Minus C distance at each end for one, two, or three bays. If this condition repeats for more than three bays, use minus C distance at one end and one-half of girder web at the other.

② **Beam to Column Webs.** Minus one-half of column web at each end for one or two bays. Minus one-half of column web at one end and one-half of column web plus ⅟₁₆ inch at other for more than three bays.

③ **Beam to Column Flanges.** Minus one-half of column depth at each end for one or two bays. Minus one-half of column depth at one end and one-half column depth plus ⅟₁₆ inch at other end for three or more bays.

④ **Beam to Column Flange at One End and Foundation Wall at Other End.** Minus one-half of column depth when one end is wall bearing. Note: This was the example illustrated in Figure 11–4.

⑤ **Beam to Column Web at One End and Foundation Wall at Other End.** Minus one-half of column web when one end is wall bearing.

⑥ **Beam to Column Web at One End and Column Flange at Other End.** Minus one-half of column web and beam depth for one or two bays. Minus one-half of column depth or beam web at one end, and one-half of column web or beam depth plus ⅟₁₆ inch at other end for three or more bays.

Notice in the preceding examples that the C-dimension of ⅟₁₆ inch does not apply for beam-to-column connections when one end of the beam is wall bearing as illustrated in examples ④ and ⑤ in Figure 11–5, and only applies with beam-to-column connections when the same situation is repeated for three or more bays. That is because anchor bolt holes through column baseplates, beam bearing plates, or the bottom of beam flanges are oversized and thus permit adjustments as required when the connection is used for only one or two bays. However, after three consecutive bays, the C-dimension allowance is required as specified to facilitate field erection of the structural steel.

Figure 11–6 illustrates shop detail drawings for three structural steel beams. These examples have been selected to show the various methods of dimensioning beams. Two of the beams, 4B1 and 4B2, are simply supported, and the third beam, 4B3, is an overhanging beam. Although overhanging

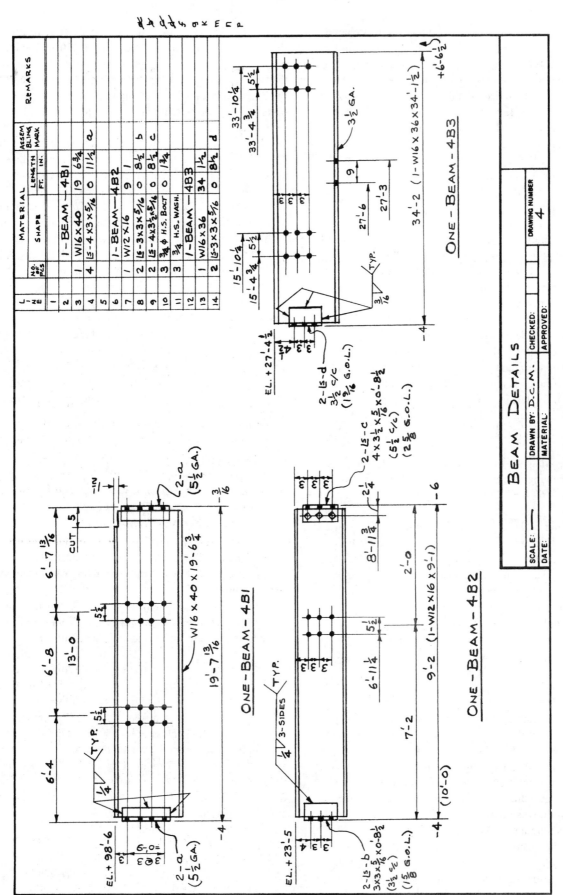

Figure 11-6   Beam details

beams will be discussed in more depth later in this chapter, this example is included to introduce a particular dimensioning concept. We will begin by examining the shop detail drawing for beam 4B1.

Notice first that the overall dimension of 19′–7¹³⁄₁₆ shows a –4 setback on the left end and a –³⁄₁₆ on the right end. Notice also that the top flange on the right side has a 1½ × 5 cope. This indicates that the beam could be fastened to the flange face of a W8 column on the left and the web of a girder on the right, and that the grid distance from the center of the column to the center of the girder is:

$$4 + 19′–7³⁄₁₆ + ³⁄₁₆ = 20′–0$$

A very important feature is the top dimension string, which locates points along the beam where open holes have been punched or drilled through the web for connections to intermediate beams. The dimensions given (6′–4 and 6′–8) are to the center of each connection. The 5½-inch distance between the vertical rows of holes is assumed to be 2¾ inches on either side of the center line of the connection. Notice that the center of the first connection is 6′–4 from the left end of the beam, and the next 6′–8 beyond that, but the second row of holes is also located by a running dimension of 13′–0 from the left end. This dimension is important for the shop, which locates all the holes from the left end of the beam. However, the top dimension string (6′–4, 6′–8, etc.) is more important for the person in the drafting room who will check this detail against the erection plan and the engineering design drawing framing plan. The checker will also verify that 6′–4 + 6′–8 = 13′–0. In other words, the numbers must add up.

Notice that no hole sizes are shown, indicating that all of the holes through both the beam web and connection angles are the standard ¹³⁄₁₆ diameter for ¾-diameter high-strength bolts. The beam itself is called out to be 19′–6¾ in length, which will result in approximately ½ inch of setback on either side from the face of the outstanding legs of the connection angles. Top-of-steel elevation, weld symbol, and framing angles are shown as previously discussed, and the material required to fabricate this beam is called for in the material list on the upper right side of the sheet.

Beam 4B2 is dimensioned somewhat differently than beam 4B1. First notice that the overall dimension of 9′–2 back-to-back of framing angles shows setbacks of –4 on the left end and –6 on the right, which might indicate that this beam connects to the flanges of a W8 column on the left side and a W10 column on the right. This also might mean a 10′–0 center-to-center of column grid dimension. Notice that the 10′–0 grid dimension is shown below and to the left of the overall dimension of 9′–2. The beam size (W12 × 16) and the beam length (9′–1) are shown just to the right of the 9′–2 overall dimension.

It is important to note that the holes through the web for an intermediate beam connection are dimensioned slightly differently than was illustrated on beam 4B1. In this case, the center of the connection location, 7′–2 from the left end, is similar to 4B1, but a 6′–11¼ running dimension locates the first vertical row of connection holes from the left end of the beam, as does the 8′–11¾ running dimension for the web bolts or shop bolts required for the framing angles at the right end of the beam.

Notice also that the framing angles are called for differently on beam 4B2 than on 4B1. First, the angle size is shown on the detail. Also, the 3½ and 5½ GA. distances are shown as 3½ and 5½ c/c (center-to-center of holes). The 1⅝ G.O.L. and 2⅝ G.O.L. stand for 1⅝ or 2⅝ gage outstanding leg. *Gage Outstanding Leg* means that the holes are located 1⅝ or 2⅝ from the back side of the web leg of the angle, and when the angles are in place on either side of the ¼-inch-thick beam web, the resulting c/c or gage distance between the vertical rows of holes will be 3½ and 5½, respectively.

Beam 4B3 is dimensioned differently from both beams 4B1 and 4B2. Notice that all the connecting holes through both the beam web and the bottom flange are located from the left end of the beam by running dimensions. This dimensioning method is clearly helpful for the fabricating shop where automatic hole-punching and drilling equipment must be pre-set to locate holes from one end as the beam moves along the beam line. The holes through the bottom flange indicate a condition where the beam extends over the cap plate of a column, which could be a W-shape, structural tube, or steel pipe column (see Figure 7–18). The example in the shop detail drawing for beam 4B3 might indicate a situation where the beam would connect to the flange face of a W8 column and then pass over the top of a pipe column and overhang 6′–6½ beyond the center of the pipe column. In that case, a +6′–6½ would be shown just to the right of the 34′–2 overall dimension because that is how far the beam extends beyond the center line of the supporting pipe column.

Beams 5B1 and 5B2 in Figure 11–7 illustrate the very common practice of combining details. This simply means drawing one beam and using that detail for two, three, or even four beams. Beams 5B1 and 5B2 are both simple beams with the same sized framing angles on either end, and both have two rows of open holes through their webs for connections to intermediate beams. But there the similarity ends. The beams are of different sizes and lengths, and the spacing of the connection holes for the intermediate beams is completely different. However, the structural detailer provided the required information to fabricate both beams on the same detail, thus saving a substantial amount of drafting time and money. Structural steel detailers are always expected to look for opportunities to combine details and save expenses for their employers.

The detail of beam 5B3 illustrates a beam requiring copes at each end that are so deep it becomes necessary to cut off the bottom flange of the beam flush so the required

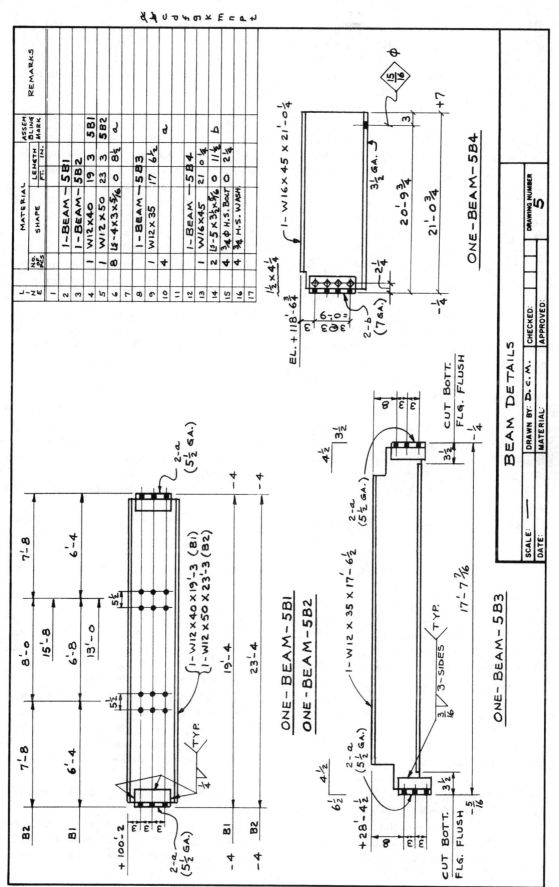

Figure 11-7   Beam details

3-row framing angles can be welded to the beam web. Notice that the cope marks indicate a 6½-inch vertical and a 4½-inch horizontal cut on the left end of the beam and a 3½-inch vertical and 4½-inch horizontal cut at the right end of the beam. Also, the framing angles at either end of beam 5B3 are exactly the same as those used for beams 5B1 and 5B2, thus they can use the same assembling mark, *a*. However, while the angles called for on the material list for beam 5B3 are angles *a*, the actual angle size is not listed because it is the same as the size already called for with beams 5B1 and 5B2.

Beam 5B4 is a shop detail drawing of a beam that might be connected into a girder web on the left end and set onto a foundation wall on the right. Notice first that arrowheads are not placed at the ends of dimension lines; rather, short *tick marks,* or "slash marks," are used. This practice has become much more prevalent in the industry in recent years since the advent of computer-aided drafting because tick marks are the default (most commonly used symbol) for a great deal of the structural software on the market. Also see on beam 5B4 the ◇$\frac{15}{16}$◇ diameter holes called for through the bottom flange of the beam. This indicates the beam will probably not require a beam bearing plate and the anchor bolts will extend upward through the bottom flange of the beam itself.

At the left end of beam 5B4, the required cope at the top flange is shown by a different cope mark symbol from the one used on beam 5B3. Also notice that framing angles *b* for beam 5B4 are 5 × 3½ × ⁵⁄₁₆ angles, indicating a connection similar to the one in Figure 11–3 where the beams fasten into the girder from both sides.

Beam detail 6B1 in Figure 11–8 illustrates an extremely common situation in structural steel framing systems—beams framing into column webs. In Figure 11–5 and on the structural steel framing plans in Chapter 6 (Figures 6–20, 6–21, 6–22, etc.), beams were shown framing into the webs of structural steel columns. In actual practice, it is probably true that more beams frame into the webs of columns than into the column flanges. This method helps stabilize the weak axis of the column, requires fewer field bolts, and many times reduces the eccentricity of the load. Also, it works well if the drafter is framing a W14 × 30 beam into the web of a W12 × 50 column because the flange width of a W14 × 30 is 6¾ inches, and the T-distance of a W12 × 50 is 9½ inches.

However, what if a W24 × 68 beam with a flange width of 9 inches is framing into the web of a W10 × 54 column? The W10 × 54 column is 10 inches deep, its flanges are ⅝-inch thick, and its T-distance is 7⅝ inches. Obviously, because of the width of its flanges, the W24 × 68 will not fit between the flanges of the W10 × 54 column. The top and bottom flanges of the W24 × 68 will have to be cut to a narrower width. But what width should the beam flanges be cut

down to and how far back from the end of the beam should they be cut? The AISC manual, in a table entitled Flange Cuts for Column Web Connections (reproduced here as Table 11–1), shows the width and length to which beam flanges should be cut for a wide variety of beam-to-column web connections.

Notice first on Table 11–1 that a circle with sizes of various bolt heads or rivets is given in each of the four corner areas of the column. Sizes indicated are for ¾, ⅞, 1, and 1⅛-inch-diameter bolts and rivets, although the use of rivets is extremely rare in present-day steel construction. Because ¾ = inch-diameter high-strength bolts are the most common size for commercial and light industrial construction, we will take our data for the width of the cut from the lower right-hand corner of the table.

Table 11–1 shows beams coming into the column web from two sides. The dimensions are given from the center line of the beam web to the outside edge of the beam flange after it has been cut to clear the bolts as required. These dimensions are listed for various sizes of columns (not beams), from a W14 × 730 down to a W8 × 24.

For purposes of illustration, we will assume the W24 × 68 beam shown as beam 6B1 in Figure 11–8 is framing into the web of a W10 × 54 column. Again, the only dimensions we are concerned about are those below the center line of the beam in the lower right-hand quadrant of the column in Table 11–1. Here we can see that any beam framing into the web of a W10 × 54 column should have its flanges trimmed to a maximum of 3¼ inches out from the center line of the beam web. Thus, 3¼ inches either side of the web of the W24 × 68 means its top and bottom flanges should be cut down to a width of 3¼ × 2 = 6½ inches. Thus, the flanges of the W24 × 68 must be cut from 9 inches down to 6½ inches to fit inside the flanges of the W10 × 54 structural steel column. On detail 6B1, notes at each end of the beam indicate that the top and bottom flanges in the shaded areas should be cut down to a width of 6½ inches. The next task for the structural detailer is to determine how far back from the end of the beam to make the flange cuts.

The required length of flange cuts for various-sized columns, not beams, is found by referring to the dimensions and data at the bottom of the table of Flange Cuts for Column Web Connections. Reading down the table reveals that a W10 × 54 column requires a 6-inch-long cut from the center line of the column. Most fabricating shops use the dimension given on the table directly on the shop detail, ignoring the fact that the clearance will be increased by half the thickness of the column web. Notice how the required length of flange cut is shown on the shop detail drawing for beam 6B1.

Detail 6B2 in Figure 11–8 is an example of a C-shape (American Standard Channel Shape) used as a beam. Notice first that the principal view in elevation shows the channel as

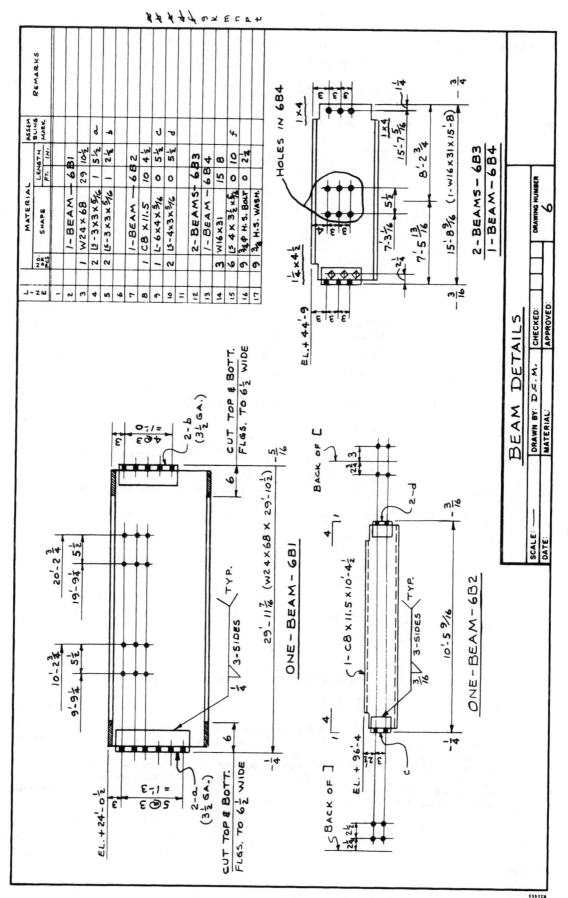

Figure 11–8   Beam details

Table 11–1
Field Erection Clearances and Flange Cuts for Column Web Connectoins

RIVETS AND THREADED FASTENERS
Field erection clearances

RIVET CLEARANCE—W COLUMNS

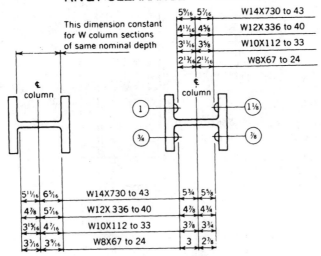

5⁹⁄₁₆	5⁷⁄₁₆	W14X730 to 43
4¹¹⁄₁₆	4⅝	W12X336 to 40
3¹¹⁄₁₆	3⅝	W10X112 to 33
2¹³⁄₁₆	2¹¹⁄₁₆	W8X67 to 24

This dimension constant for W column sections of same nominal depth

5¹¹⁄₁₆	6⁵⁄₁₆	W14X730 to 43
4⅞	5⁷⁄₁₆	W12X 336 to 40
3¹⁵⁄₁₆	4⁷⁄₁₆	W10X112 to 33
3³⁄₁₆	3⁹⁄₁₆	W8X67 to 24

5¾	5⅝	
4⅞	4¾	
3⅞	3¾	
3	2⅞	

FLANGE CUTS FOR COLUMN WEB CONNECTIONS

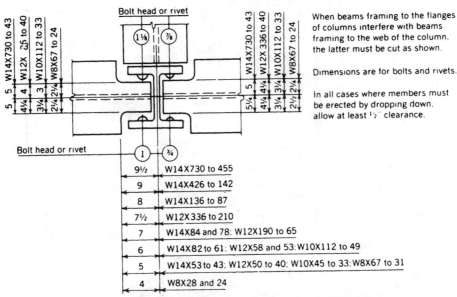

Bolt head or rivet

Bolt head or rivet

When beams framing to the flanges of columns interfere with beams framing to the web of the column. the latter must be cut as shown.

Dimensions are for bolts and rivets.

In all cases where members must be erected by dropping down. allow at least ½" clearance.

9½	W14X730 to 455
9	W14X426 to 142
8	W14X136 to 87
7½	W12X336 to 210
7	W14X84 and 78; W12X190 to 65
6	W14X82 to 61; W12X58 and 53;W10X112 to 49
5	W14X53 to 43; W12X50 to 40; W10X45 to 33;W8X67 to 31
4	W8X28 and 24

Notes:
1. Information shown on these clearance diagrams applies to both the old and new WF and W series. Maximum clearances are shown to accommodate the slight differences in dimensions.
2. Values shown for clearances over rivet heads are applicable when applied to bolt heads, but not to the nut and stick-through. See Table of Assembling Clearances for high-strength bolt clearance dimensions.
3. Based on Table of Dimensions of Structural Rivets.

(Courtesy of the American Institution of Steel Construction Inc.)

it would be seen from the back side. This standard procedure in shop detail drawings immediately identifies the beam as a channel rather than a W-shape. Also notice that the dimensions locating the holes through the outstanding legs of the framing angles are referenced from the back side of the channel, not from the center line of a web with W-shape beams. Because the connection on the left end is one-sided, requiring two vertical rows of bolts, the angle must be a 6 × 4, with the 6-inch leg being the outstanding leg to provide room for the bolts.

The detail for beams 6B3 and 6B4 shows the framing angles on the left end shop-bolted to the web of the W16 × 31 beam with staggered holes as shown in Figure 11–3*B* (see also Figure 9–9*B*). The copes and web connection holes at the right end of beams 6B3 and 6B4 would be for a single-plate connection similar to that shown in Figure 7–37. Also notice the note that the open holes through the web for a 3-row connection near the center of the beam are for beam B4 only. This is another example of how the structural steel detail drafter can save costly drafting time by combining details.

Detail 7B1 in Figure 11–9 shows how a section view is required to clarify a feature that cannot be clearly represented on the principal view of a beam detail. In this case, Section A-A indicates that the flange cut on the bottom flange of the W12 × 30 beam is to be made only on one side of the beam's web.

Details 7B2 and 7B3 are examples of how simple beam detailing, fabrication, and erection can be. Detail 7B2 is the shop detail for the W18 × 46 beam shown in Figure 7–29, assuming the grid distance between ST 6 × 6 columns is 27′–8. Detail 7B3 illustrates the shop detail for the W18 × 46 beam shown in Figure 7–25. This is a seated beam connection and assumes the grid distance between the W8 × 31 columns on either end of the beam is 26′–0.

Notice that the engineering design drawing detail in Figure 7–25 calls for the mounting holes through the bottom flange of the beam to be located 2¾ inches from the back of the seat angle and 3 inches apart, or 3 GA. on the bottom flange. If a detail were shown with these dimensions on the engineering drawing detail, this is what the steel detailer in the fabricator's office would call for on the shop fabrication detail.

Very likely, neither of these two dimensions is what the steel detailer would have selected if the dimensions had not been shown on the engineering design drawing detail. First, as previously discussed, the steel detailer would most likely have selected a 3½-inch gage distance on the bottom flange because the flanges of a W18 × 46 are 6 inches wide. Also, the steel detailer would have looked in the AISC manual table entitled Usual Gages for Angles, Inches and selected a

G.O.L. of 2 inches for the 4-inch outstanding leg of the seat angle as shown in Table 6–1.

However, the AISC does permit other gages to meet specific requirements subject to clearance and edge distance limitations. This is mentioned here to point out that the competent structural steel detailer will always check to see that design details conform with AISC recommendations, while at the same time striving to detail, as much as possible, exactly what the structural engineering design drafter has called for on the design plans and details.

Overhanging Beams

Chapter 7 discussed the economic advantage of using overhanging beams in structural steel construction. You may recall that the negative bending moments produced by the overhang offset the positive bending moments developed by the load between the supporting columns. Thus, the value of the maximum moment is less than it would otherwise be, resulting in lighter-weight, and thus more economical, beams. Figure 11–10*A* is an elevation view of this situation, which is widely used for structural steel roof framing systems.

Notice in Figure 11–10*A* that grid lines ⑦ to ⑩ show standard steel pipe columns spaced at 20′–0 center-to-center. The W16 × 31 beam supported at grid lines ⑦ and 8 is a double overhanging beam, extending to splice plate connections 4′–0 beyond the supports. The W16 × 31 beam supported by columns at grid lines ⑨ and ⑩ overhangs on one side only, 4′–0 to the left of grid line ⑨. Grid line ⑩ can be assumed to be at the end of the building, which is why the beam would not overhang at that point. The W12 × 26 beam between grid lines ⑧ and ⑨ is actually a simply supported beam whose reactions constitute a concentrated load at the ends of the W16 × 31 overhanging beams. The connection between the two W16 × 31s and the W12 × 26 is made with splice plates. Figure 11–10*B* shows that the center of the splice plate connections is 4′–0 beyond the column supports, with a ½-inch space between the beams. Thus, the end of the W16 × 31 is 3′–11¾ beyond the supporting column (or +3′–11¾), and the end of the W12 × 26 is 4′0¼ short of (or –4′–0¼) the supporting columns at grid lines ⑧ and ⑨. Keeping this in mind, we will now examine Figure 11–11, which illustrates how the beams in Figure 11–10 might be detailed.

Notice first in Figure 11–11 that beam 8B1 (the W12 × 26) is detailed with the required splice plates as part of the detail. The splice plates at either end of the W12 × 26 are shown as bolted to the web of the beam with ¾-inch-diameter high-strength bolts. These bolts are called out in the material list in the upper right-hand corner of the detail

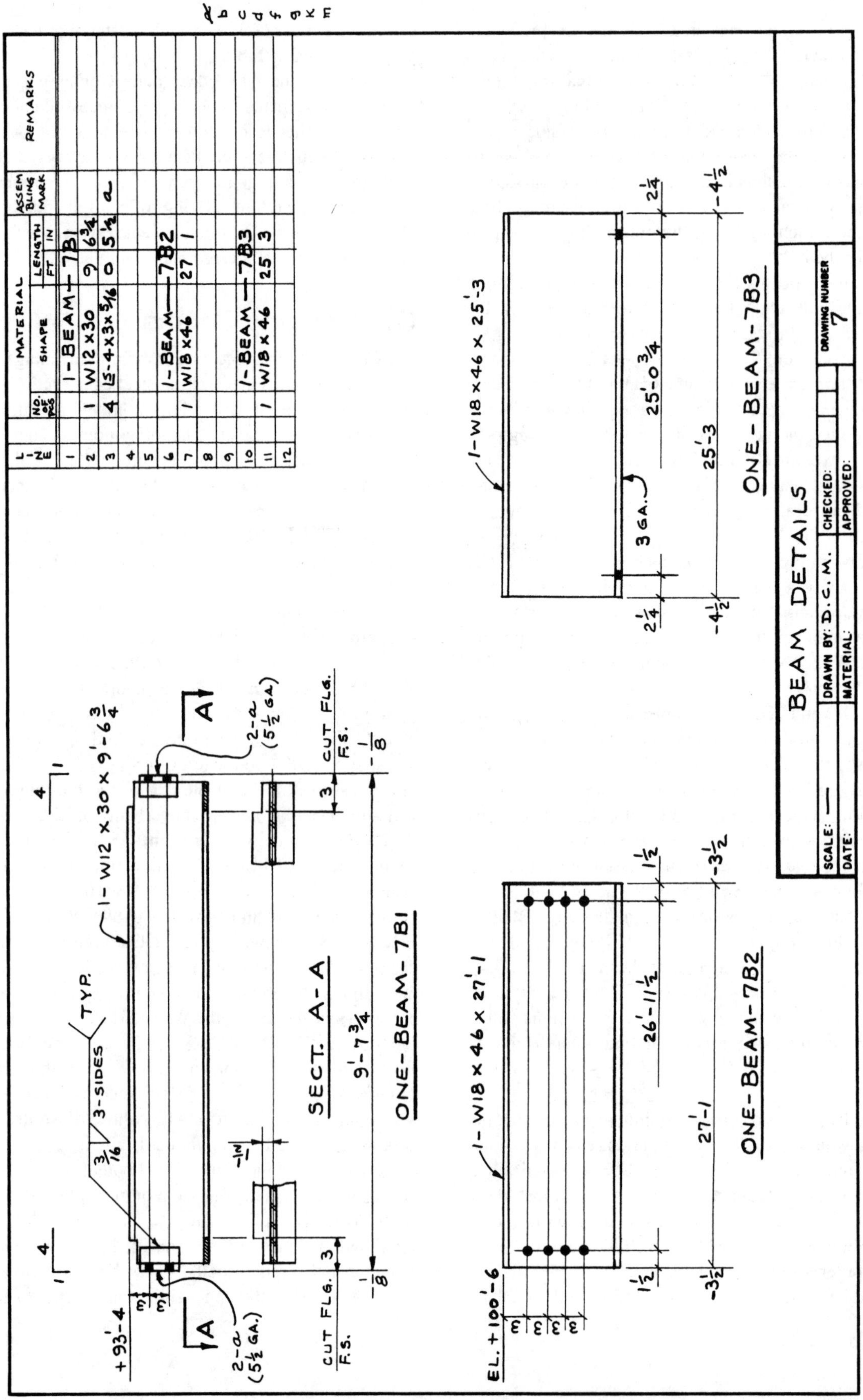

Figure 11-9 Beam details

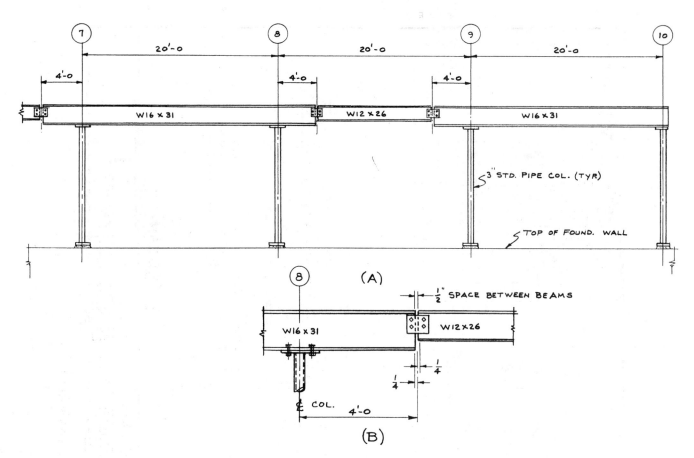

Figure 11–10   Overhanging beam detail

sheet. For beams requiring splice plate connections, many structural fabricators prefer to specify that the splice plates be bolted in place to the beam in the fabricating shop as shown for beam 8B1. This eliminates the need for the ironworkers to look for the splice plates at the job site. Notice also the –4′–0¼ dimension just beyond the overall beam dimension of 11′–11½:

$$11'-11\tfrac{1}{2} + 4'-0\tfrac{1}{4} + 4'-0\tfrac{1}{4} = 20'-0$$
(the distance between grid lines ⑧ and ⑨)

Both of the W16 × 31 beams, 8B2 and 8B3, are very easy to detail and fabricate. All that is required is to show the overall beam length, locate the holes through the bottom flanges to match the holes through the column cap plates, and locate the required holes through the webs of the beams to make the splice plate connection. Notice the +3′–11¼ overhang on one side of beam 8B2 and on both sides of beam 8B3. The right end of beam 8B2, which will connect to the column cap plate at grid line ⑩ , would not under

normal circumstances need to extend more than 2 or 3 inches beyond the center line of the column.

Figure 11–12 illustrates a slightly different situation and another way to detail overhanging beams. All of the beams are the same size, which is not unusual because they are most surely lighter, and thus more economical, than if they had been designed as simply supported beams. In this example, the splice plates are detailed separately and identified with an M number. Many structural steel fabrication shops use the M classification for miscellaneous steel, a category that can include a wide variety of structural steel components. (A brief discussion and some examples of miscellaneous steel will conclude this chapter.)

Figure 11–12 shows that beams 9B1 and 9B2 have ¾-inch-thick web stiffener plates welded in place on either side of their webs at the supports. Web stiffeners were discussed in Chapter 7 and illustrated in Figures 7–17B, 7–18, and 7–19. When shown on the engineering design drawing details, these web stiffeners are called for on the shop

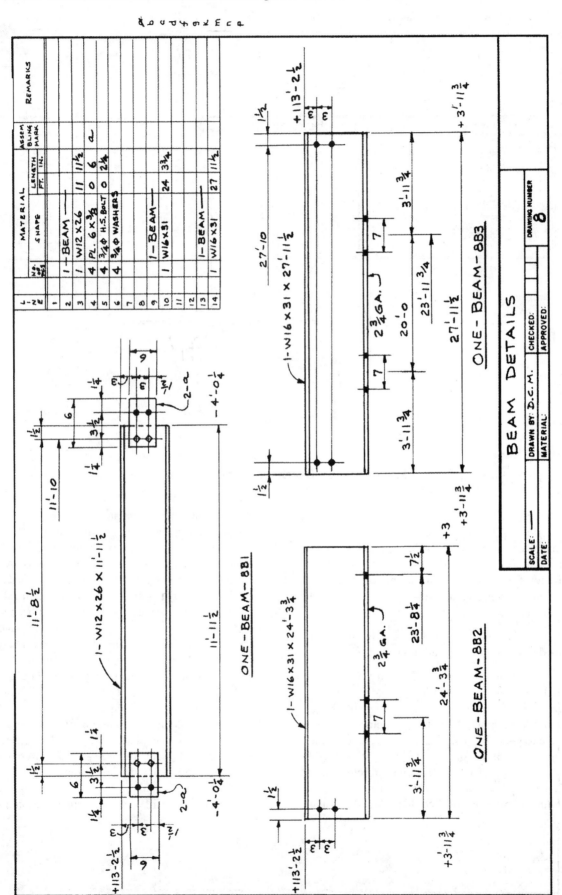

Figure 11–11 Beam details

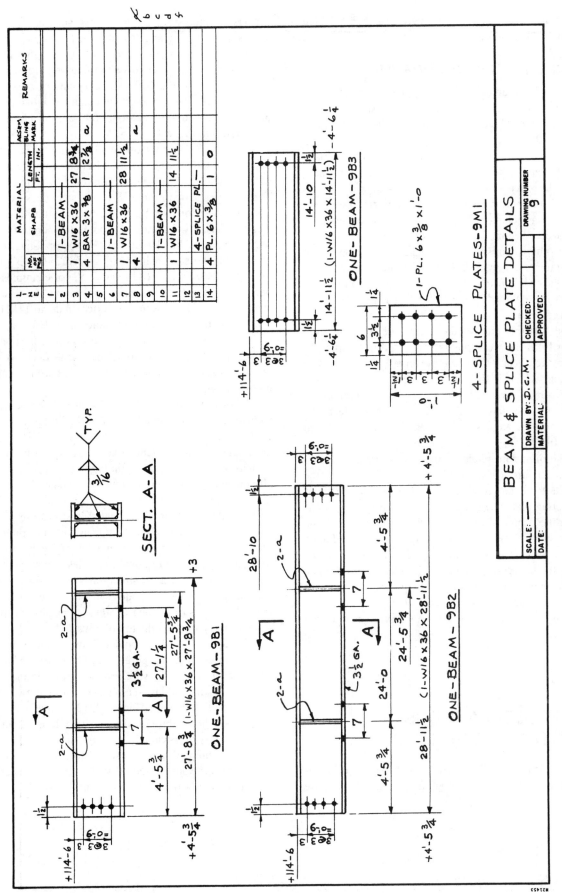

Figure 11-12 Beam and splice plate details

fabrication drawings as shown in Figure 11–12 and welded in place in the shop before the beam goes out to the job site.

Sloped and Skewed Beams

Most of the structural steel components that comprise the support frame of a commercial or industrial building consist of horizontal beams with vertical webs connecting at right angles (90 degrees) to the vertical webs of supporting girders or the faces or webs of supporting columns. This is called *rectangular framing*. However, as discussed in Chapter 7 and illustrated in Figures 7–33, 7–34, and 7–40, it is not unusual for some parts of the structural support system to require non-rectangular framing—that is, connections made between beams and their supporting members at something other than a 90-degree angle. The examples of non-rectangular framing discussed in Chapter 7 were sloped beams and skewed beams. Beam 10B1 in Figure 11–13 is a shop detail for a sloped beam. Beam 10B2 is a shop detail for a skewed beam.

Notice first that both of these beam details dimension everything from the work points (discussed in Chapter 7), and that in each detail, the angle of slope or skew is shown by the pitch, which is the ratio between the rise and run. While the run is always shown as 12 inches on structural drawings, the rise varies, depending on how steep the designer wants the angle of slope. The ⅛ to 12 slope shown on beam 10B1 is very common for roof systems because it is about the minimum pitch required to ensure good water drainage, even on so-called "flat" roofs. Notice that the work points for beam 10B2 are 3/16 inch beyond the face of the connecting plate. This is because the work point is figured from the center of the web of the girder to which the beam is connected, and in this case, the 3/16 is assumed to be one-half of the girder web thickness. Notice also how the flanges of the beam must be cut back to avoid interference with the flanges of the girder. (Another example of this situation is illustrated in Figure 7–40.)

This discussion is intended as only a brief introduction to non-rectangular framing. More in-depth coverage must be left to an advanced course in structural steel design or detail drafting.

11.4 MISCELLANEOUS STRUCTURAL STEEL

In addition to the main structural steel framework, steel-frame commercial or industrial buildings frequently require various secondary steel components, such as deck support angles, joist bearing angles and plates, beam bearing plates, lintels, roof frames, and steel ladders. The structural steel detail drafter often prepares shop drawings for these components, which are called secondary and/or miscellaneous structural steel. This chapter will conclude with a discussion of some of the important structural components that many times fall into that category.

Figure 11–14 is a shop detail drawing of several items that are usually classified as miscellaneous steel. Details 11M1 and 11M2 are shop details of deck support angles. These angles support the ends of the steel deck at the junction of a floor or roof framing system and a vertical wall. They are usually fastened to a poured concrete foundation wall or masonry block wall with ½-inch or ⅝-inch-diameter concrete anchors or anchor bolts spaced 2'–0 to 3'–0 center-to-center. (See Figures 6–14 and 7–52 for examples of how deck support angles are used to support the end of the steel deck for a floor system.) Notice on detail 11M1 in Figure 11–14 that the mounting holes called for are $\langle \tfrac{9}{16} \rangle \phi$, which means they must be called out on the detail. Deck support angle 11M2 requires $\langle \tfrac{13}{16} \rangle \phi$ holes, so in keeping with standard practice, the hole sizes are not called out on the drawing.

Notice also on detail 11M2 that a 1-inch-deep by 7-inch-long cut has been taken out of the bottom of the vertical leg of the angle. This type of cut is often called for if the deck support angle extends over the top of a beam flange, such as the W18 × 46 shown in Figure 7–52. The cut is usually made large enough to allow for ½ inch clearance around the top flange of the beam.

Details 11M3 and 11M4 show joist bearing plates and angles used when steel joists are supported at poured concrete or concrete masonry unit walls. (See Figures 6–15, 6–16, and 7–53 for additional examples of how these angles and plates are shown on structural steel design drawings.) Joist support angles and plates can be detailed as shown for individual joists or as long continuous angles and plates to support many joists at the top of a wall.

Detail 11M5 is the shop detail for a beam bearing plate like the one shown in Figure 7–52. The bearing plate would be set in place on the wall, leveled, grouted, and secured. Then the beam would be set in place by the ironworkers and field-welded as shown in Figure 7–52. Notice that since the anchor bolt holes through the plate are larger than 13/16″ in diameter, the size is called out on the detail.

In commercial and industrial steel and masonry construction, *lintels* are structural steel members placed above door and window openings to support the weight of the wall

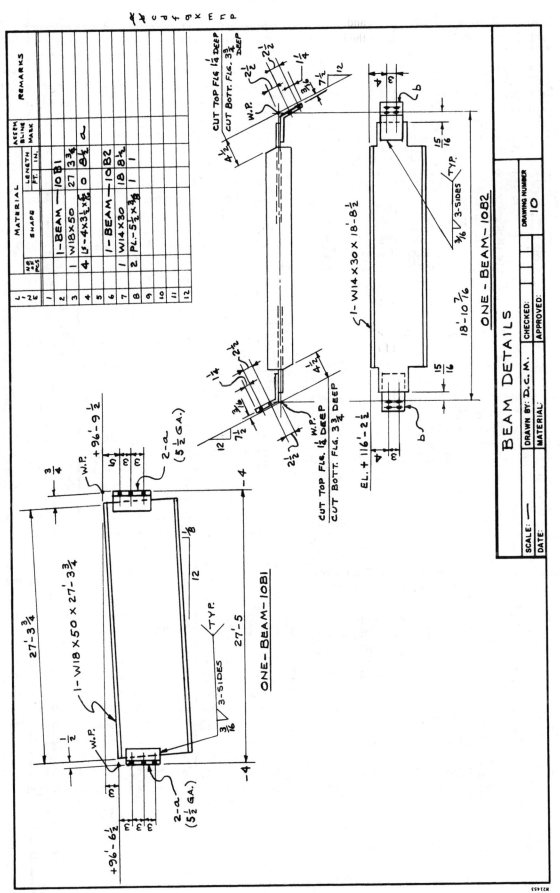

Figure 11–13 Sloped and skewed beam details

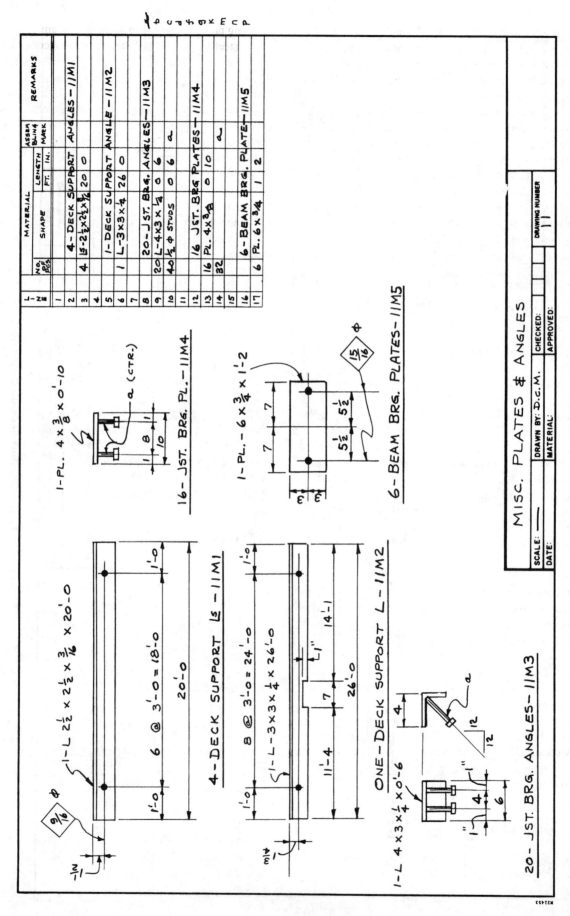

Figure 11–14 Miscellaneous plate and angle details

above the door or window. Lintels are usually very simple to detail because only the lintel shape and any required welds need to be shown. Lintels usually bear a minimum of 4 inches beyond the masonry opening on either side for small window or door openings and up to 1′–0 or more for long openings. It is common to detail lintels with a single view as shown by the examples in Figure 11–15, with the required length specified in the material list. Notice the variety of welds called for on the lintel details in Figure 11–15.

Details 13M1 and 13M2 in Figure 11–16 illustrate joist bearing angles similar to those shown in Figure 11–14 except that these are long angles, which can be set in place side-by-side to provide a continuous strip of bearing angle along the entire length of a wall. This example also shows how the structural detailer combines details to show two different joist bearing angles on the same detail.

Another structural component found in virtually any type of structural steel building is the roof frame. A typical shop fabrication detail for a roof frame is shown in detail 13RF1 in Figure 11–16.

Roof frames are supports under the area of a roof through which an opening must be cut. Almost any flat-roofed commercial or industrial building has several of these openings to support the steel deck around openings for roof drains, vent pipes, and supply and exhaust ducts from the building's heating and ventilating systems. The roof frame is generally a welded angle frame, similar to the example in detail 13RF1, which supports the steel deck near the opening.

On most roof framing plans, roof frames are shown only in their approximate locations and are identified either by an *M* or *RF*. Also, while the inside size of the opening might be given, the location of the roof duct opening is not because that is best left up to the general contractor. After the contractor determines where the duct or vent pipe will come up through the roof, he informs the fabricator, who then fabricates the roof frame. Figure 11–17A illustrates how a roof frame might be shown on the roof framing plan. Figure 11–17B shows how the roof frame sets on a joist or beam.

Detail 13LA1 in Figure 11–16 is a typical shop detail for a small steel ladder. This type of ladder, commonly seen in commercial and industrial building construction, is fastened to the floor with concrete anchors through angles *c* and to a wall through bent bars *f*. The design of such ladders is rigidly controlled by OSHA requirements, which both the structural design and detail drafter must be aware of as they prepare the design and shop detail drawings.

11.5   SUMMARY

This has been one of the longest and most important chapters in *Structural Steel Drafting*. The discussion of

detailing procedures for structural steel beams and miscellaneous steel has introduced many basic concepts relevant to detailing structural steel for virtually any size project. Both this chapter and Chapter 10 have demonstrated the extent to which the structural steel detailer must first analyze the engineering design drawings of a building project and then carefully and methodically prepare the shop detail drawings. Each coping cut, flange cut, beam length, hole size, and weld symbol must be exactly right and clearly shown on the shop drawings. Until the steel detailer has memorized many of the standards through repetitive use, the AISC *Manual of Steel Construction* must be consulted often for information on flange widths and thicknesses, beam web thicknesses, recommended angle gages, flange cuts, and of course, connections. The material presented here has barely scratched the surface of the situations encountered in a steel detailer's day-to-day work, but the information in both Chapter 10 and 11 has hopefully provided a practical introduction to structural steel detailing.

However, perhaps some of the most important lessons of Chapters 10 and 11 do not involve the actual procedures for detailing beams, columns, and miscellaneous steel. Rather, by providing numerous examples of the degree of detail and visualization with which a structural drafter must work and think every day, these chapters may now have given new meaning to the material presented back in Chapter 3. It might be worthwhile to review sections 3.3 (Desirable Characteristics in a Structural Drafter) and 3.4 (Essential Skills for a Structural Drafter) in light of the background provided by Chapters 10 and 11.

STUDY QUESTIONS

1. For beams framing into girders with framing angles, the setback dimension is commonly referred to as the C-dimension. What is the C-dimension?

2. What is the k-distance of a W-shape steel beam?

3. Many structural fabricators prefer to show required coping cuts on their shop detail drawings with a cope mark or symbol. Show two examples of cope marks.

4. When detailing structural steel beams, the steel detailer often shows running dimensions. These dimensions are always taken from what end of the beam?

5. On structural steel beam details, any open hole for which no size is indicated is assumed to be what diameter?

6. Why should structural steel detailers always look for opportunities to combine details?

7. What is the recommended length of flange cut for a W16 × 50 beam framing into the web of a W8 × 31 column?

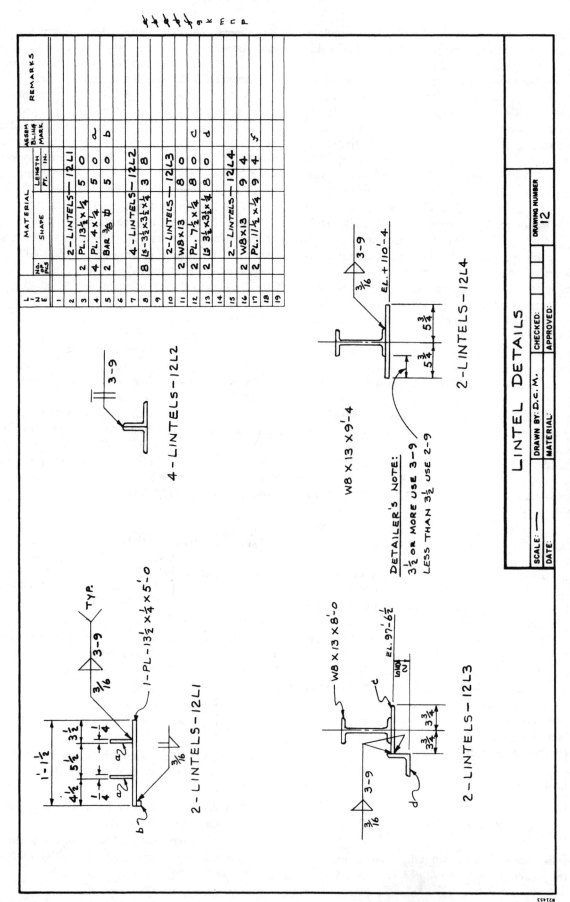

Figure 11–15   Structural steel lintel details

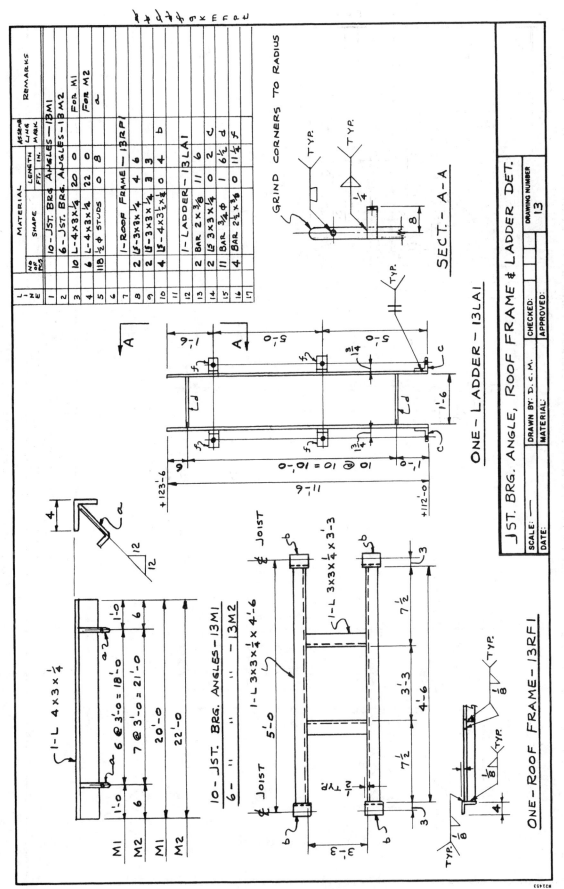

Figure 11–16   Miscellaneous structural steel details

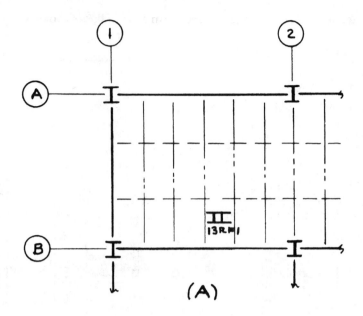

(A)

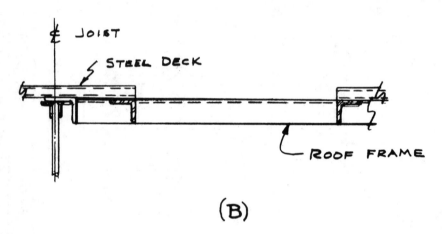

(B)

Figure 11–17 Structural steel roof frame

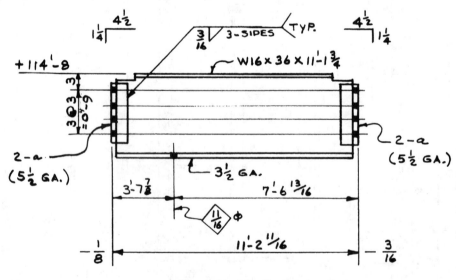

Figure 11–18 Beam detail

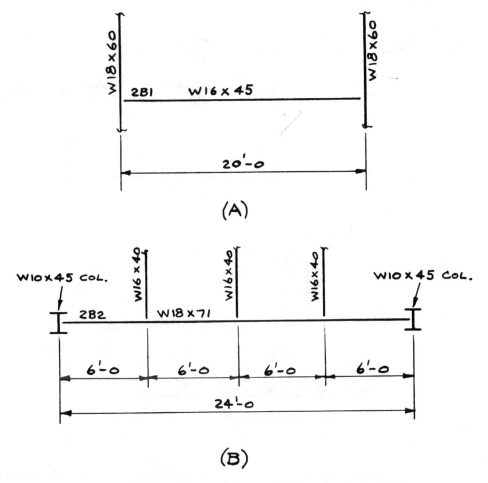

Figure 11–19   Beams for student drafting activity 1

8. The top and bottom flanges of a W16 × 50 beam framing into a W8 × 31 column web should be trimmed down to what width if ¾-inch-diameter high-strength bolts are to be used for all beam-to-column connections?

9. What is meant by G.O.L., and what is the recommended G.O.L. for a 3 × 3 structural steel angle?

10. When detailing sloped and/or skewed beams, the structural detailer will locate almost all dimensions from the _____ .

Note: The following questions refer to Figure 11–18.

11. What does the +114′–8 in the upper left-hand corner of the detail indicate?

12. What do the 2¼ GA. on the lower flange and the 5½ GA. on the connection angles indicate?

13. What do the –⅛ and –³⁄₁₆ at either end of the overall dimension indicate?

14. Is the overall dimension an end-to-end of beam or an end-to-end of angle measurement?

15. How deep and how long are the cutout, or cope, dimensions on the upper flange of the beam?

16. What is the distance down from the top of the beam to the first hole on the connecting angles?

17. How many rows of holes are on the connecting angles?

18. What size holes are on the outstanding legs of the connecting angles?

19. What size are the holes through the bottom flange of the beam?

20. The connecting angles are shop-welded to the web of the beam. What size and type of weld are specified?

STUDENT ACTIVITY

1. Using either manual or CAD drafting technique, make a shop detail drawing of the beams marked 2B1 and 2B2 in Figure 11–19 on a 12″ × 18″ sheet. Show framing angle and connections, with the framing

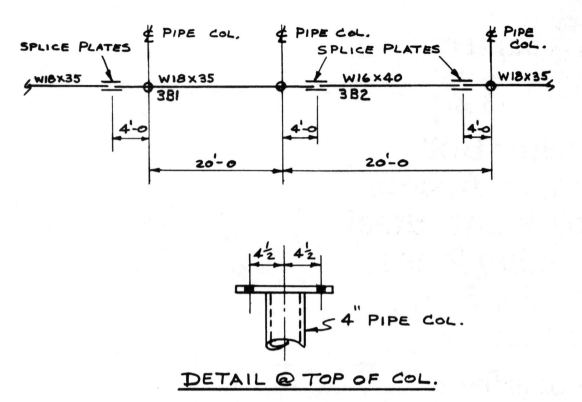

Figure 11–20 Beams for student drafting activity 2

angles for beam 2B1 shop- and field-bolted and the framing angles for beam 2B2 shop-welded and field-bolted. All connections are to be 4-row connections. Make a material list in the upper right corner of the sheet. Draw the beams to a scale of 1″ = 1′–0″ in depth, but the length dimension need not be to scale. Following the examples in this chapter, make all object lines heavy and dark, all extension and dimension lines thin but sharp and easy to read, and all lettering dark, easy to read, and about ⅛″ high, except

for titles, which can be ¼″ high. Be especially careful not to crowd dimensions.

2. Detail beams 3B1 and 3B2 in Figure 11–20 on a 12″ × 18″ sheet. Draw to a depth scale of 1″ = 1′–0″ and show splice plate connections as 4-row connections. Make a material list in the upper right corner of the sheet and follow the directions given in Activity 1 concerning linework and lettering technique.

chapter 12

Anchor Bolt Details, Anchor Bolt Plans, Steel Erection Plans, and the Field Bolt List

OBJECTIVES

Upon completion of this chapter, you should be able to:

- prepare anchor bolt details for a commercial or industrial building.
- prepare anchor bolt plans for a commercial or industrial building.
- prepare structural steel erection plans for a commercial or industrial building.
- prepare a field bolt list for a commercial or industrial building.

12.1   INTRODUCTION

Chapters 10 and 11 introduced the various methods of detailing structural steel columns, beams, and miscellaneous steel for shop fabrication. However, the structural detailer's responsibility extends beyond the preparation of shop fabrication drawings. On a typical project, the structural detailer must also prepare the *erection drawings,* the drawings used by the general contractor and the ironworkers to install the structural steel at the job site. Additionally, the structural detailer provides the *field bolt list,* which shows the high-strength bolts required to erect the structural steel.

The erection drawings normally consist of anchor bolt plans, floor framing erection plans, and roof framing erection plans. Erection plans show the ironworkers and the general contractor where and how to install each anchor bolt, beam, column, support angle, or roof frame. In actual practice, the anchor bolt plans and anchor bolt detail drawings are usually the first drawings prepared by the structural steel

detailer because the anchor bolts must be ready for installation at the job site before the footings are poured. The floor and roof erection plans are usually prepared as the structural steel components are being detailed. The field bolt list is compiled after the erection plans and shop fabrication drawings are finished. Accurate anchor bolt plans, erection plans, and field bolt lists are extremely important to the successful completion of the structural steel portion of any commercial or industrial building.

12.2   ANCHOR BOLT DETAILS

As discussed in previous chapters, the design of steel-frame structures generally assumes that beams and/or steel joists support floor and roof loads, girders support the reactions of beams, and columns support loads induced by the reactions of girders and/or beams. These column loads, including lateral wind loads, are then transmitted down to

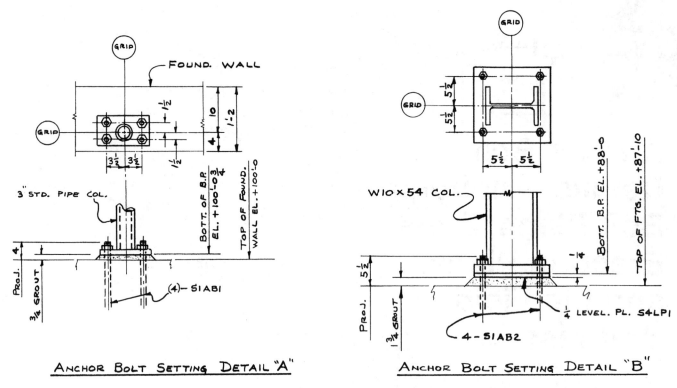

ANCHOR BOLT SETTING DETAIL "A"

ANCHOR BOLT SETTING DETAIL "B"

Figure 12–1 **Anchor bolt setting details**

the footings and foundation walls. Obviously, the baseplates of these columns must be strongly anchored to their supporting footings and walls, not only because of the vertical loads, but also because of other factors, such as lateral stresses caused by wind against exterior wall columns or shock loads if a column should be bumped. These lateral loads cause an overturning effect on the column, which must be resisted by the anchor bolts. The overturning action tends to tip the column over, which in turn, pulls the anchor bolts out of the footing or foundation wall. Thus, the two primary considerations when designing anchor bolts are:

1. The tensile strength of the bolt to resist tension caused by the overturning moment, and
2. The bond strength between the anchor bolt and the concrete in which it is placed.

Bond strength is developed by determining a length of required embedded area (surface area) of the bolt into the concrete so the bond developed between the concrete and the steel bolt will prevent the bolt from pulling out if the expected tensile load is applied. Most anchor bolts used in building construction are also hooked at one end to further increase their anchorage into the footing or foundation wall and to prevent turning.

In most structural engineering offices, anchor bolt design is either done by established office standards or calculated for specific situations by experienced engineers or designers. (An example of an office standard might be the requirement that a ¾-inch-diameter bolt be embedded a minimum of x inches into the concrete footing or foundation wall to develop bond equal to its allowable stress.) The structural steel detailer is primarily concerned with either drawing details of the anchor bolts so they can be made in the shop or making anchor bolt plans to show exactly how every anchor bolt should be placed into a footing or wall. Before specifically discussing anchor bolts and anchor bolt plans, we will review a few typical details illustrating how anchor bolts are used in building construction.

The plan and elevation view in Figure 12–1 of anchor bolt setting detail "A" illustrates a detail for the 3-inch standard pipe column shown in Figure 7–4. Notice that, while the two details are very similar in many ways, the emphasis is completely different. For example, the column baseplate detail in Figure 7–4, which would have been drawn by the structural drafter in the design office as part of the structural engineering design drawings, locates the column on the grid system, shows the required baseplate size and thickness, calls for a fillet weld all around to connect the pipe to the baseplate, and specifies two ¾-inch-diameter anchor bolts.

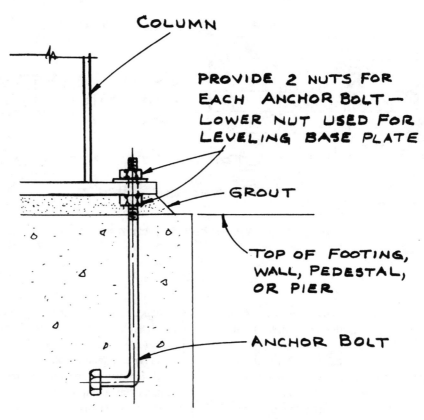

COLUMN

PROVIDE 2 NUTS FOR
EACH ANCHOR BOLT —
LOWER NUT USED FOR
LEVELING BASE PLATE

GROUT

TOP OF FOOTING,
WALL, PEDESTAL,
OR PIER

ANCHOR BOLT

Figure 12–2 Anchor bolt setting detail

By contrast, anchor bolt setting detail "A" in Figure 12–1 gives no information at all about the baseplate or the type of weld required to connect the steel pipe to the baseplate. This detail is only meant to show the building contractor how and where to set the anchor bolts before the concrete foundation wall is poured. The anchor bolts are located on the grid system, the required 4-inch projection of the anchor bolt above the top of the foundation wall is shown, and the four anchor bolts are identified as anchor bolts S1AB1.

Because anchor bolts do not require complex fabrication details for the shop and are needed at the job site very early, many fabricators prefer to have them detailed on small 8½ × 11-inch sheets, often referred to as "S" sheets. Thus, if the anchor bolts required for anchor bolt setting detail "A" were the first anchor bolts detailed on an S-sheet, they would be identified in most structural steel fabrication shops as S1AB1.

Smaller fabricated columns, such as 3-inch pipe columns, are taken out to the job site where the footings and foundation walls have already been poured. With the anchor bolts sticking up out of the concrete in the proper location and at the proper projection, each column can simply be set

down over the anchor bolts, leveled up with shims, tightened into place, and grouted with a non-shrinking grout. Sometimes the baseplate and column can be leveled by using two nuts with each bolt. One nut, which goes under the bottom of the plate, is used for leveling the plate and column. The other nut goes above the plate and is used to tighten the baseplate and column in place. Figure 12–2 illustrates how this would look.

Anchor bolt setting detail "B" in Figure 12–1 illustrates the W10 × 54 column shown in Figure 7–6. This detail is very much like the one previously discussed with two exceptions. First, because the W10 × 54 is a much heavier column that could well extend up through two or even three floor levels, it could be very cumbersome, and thus time-consuming and expensive, to set in place and level up. Therefore, setting plates or leveling plates are often positioned first and leveled and grouted before the column is set in place over the anchor bolts. Also, because the baseplate for the W10 × 54 is so much larger than the baseplate for the 3-inch steel pipe column, more room is required to solidly pack the non-shrinking grout between the underside of the leveling plate and the top of the footing. That explains the ¾-inch grout for setting detail "A" and the 1¾-inch grout for setting detail "B."

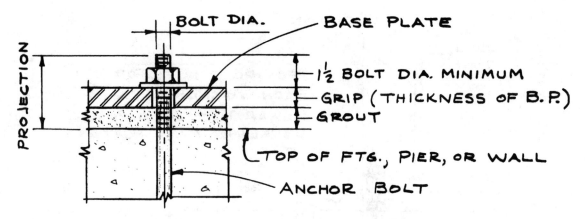

Figure 12–3   Anchor bolt projection detail

When preparing shop detail drawings for the anchor bolts themselves, the structural detailer is concerned with two factors: (1) the length the anchor bolt must project above the top of the footing, pedestal, pier, or foundation wall, and (2) the length the bolt must be embedded into the concrete so that bond stresses can be developed to equal the tensile strength of the bolt, thus keeping the bolt from pulling out of the concrete.

Calculating the projection above the top of the footing or foundation wall is very simple. The structural detailer just adds the thickness of the grout, plus the thickness of the baseplate, plus a minimum of 1½ times the diameter of the bolt. In actual practice, twice the diameter of the bolt is probably more common, which means the projection above the top of the column baseplate for ¾-inch diameter bolts will be 1½ inches. This will ensure enough thread above the baseplate for a nut and washer. Figure 12–3 illustrates how the projection is calculated.

Determining the length of embedment of the anchor bolt into the concrete is accomplished in structural offices either by office standards based upon the diameter of the bolt or by calculation by experienced engineers or designers. Again, the larger the bolt diameter, the greater the expected stresses are likely to be, and thus the deeper the anchor bolt must be embedded into the concrete. Figure 12–4, taken from an actual project, illustrates various recommended embedments for several sizes and types of anchor bolts.

The most common type of anchor bolt used in construction is a version of the hooked bolt shown as Type 2 in Figure 12–4. That type of anchor bolt is usually made from a steel rod or machine bolt of either standard ASTM A307 or a higher-strength steel, with a "hook" formed on the bottom by simply bending a straight machine bolt or round

steel shape. For the vast majority of anchor bolts, which are ¾-inch to 1-inch in diameter, the hook at the bottom is from 2 inches to 6 inches long.

The type of anchor bolt shown as Type 4 in Figure 12–4 is used mainly to fasten mechanical equipment such as pumps or air compressors to their foundation piers. The pipe sleeve is usually two to three times the diameter of the bolt. The bolt is designed so that the pipe sleeve does not extend above the top surface of the pier foundation, which should have a rough top surface to provide a good hold for the grout.

Figure 12–5 illustrates an anchor bolt detail, unlike those in Figure 12–4, that is also commonly used in commercial and industrial construction. This is an *anchor bolt set*, in which a set of bolts is fabricated together in the shop at the center-to-center distance to which they will be set in place at the job site. The anchor bolt set illustrated in Figure 12–5 was actually installed into a reinforced concrete pedestal similar to the one shown in Figure 7–2 to fasten down the heavy steel columns of a support frame used to mount some large mechanical equipment. In this installation, the 1¼-inch diameter anchor bolts were embedded almost 2½ feet into the reinforced concrete pedestal.

Figure 12–6 illustrates details as they might be prepared by the structural detailer for the anchor bolts shown in Figure 12–1. Washers are called for because, while anchor bolts will be furnished with hex. nuts, many fabricators do not furnish washers unless they are specifically called for on the detail. Also, the length of the hook is not shown, although obviously the length of machine bolt called for, bent as shown on the detail, will provide an approximate 3-inch-long hook on S1AB1 and a 4-inch-long hook on S1AB2.

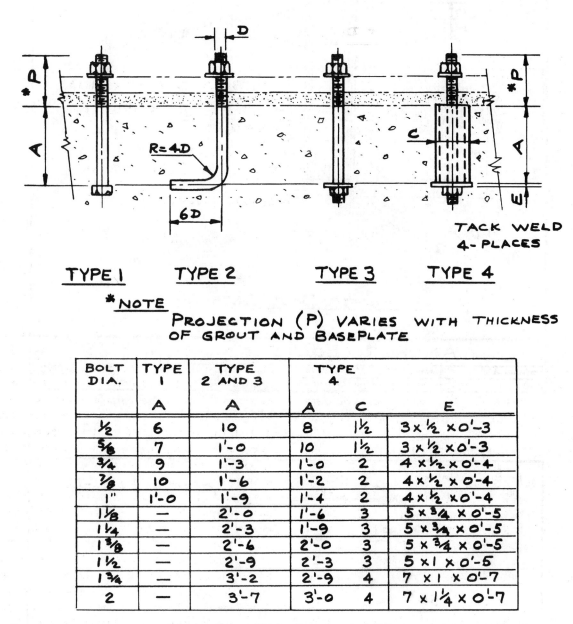

BOLT DIA.	TYPE 1	TYPE 2 AND 3	TYPE 4		
	A	A	A	C	E
½	6	10	8	1½	3 x ½ x 0'-3
⅝	7	1'-0	10	1½	3 x ½ x 0'-3
¾	9	1'-3	1'-0	2	4 x ½ x 0'-4
⅞	10	1'-6	1'-2	2	4 x ½ x 0'-4
1"	1'-0	1'-9	1'-4	2	4 x ½ x 0'-4
1⅛	—	2'-0	1'-6	3	5 x ¾ x 0'-5
1¼	—	2'-3	1'-9	3	5 x ¾ x 0'-5
1⅜	—	2'-6	2'-0	3	5 x ¾ x 0'-5
1½	—	2'-9	2'-3	3	5 x 1 x 0'-5
1¾	—	3'-2	2'-9	4	7 x 1 x 0'-7
2	—	3'-7	3'-0	4	7 x 1¼ x 0'-7

Figure 12–4   Typical anchor bolt types

12.3   ANCHOR BOLT PLANS

The *anchor bolt plan,* generally the first drawing prepared on most structural steel jobs, is usually not difficult to draw. It is simply a location plan or placing plan (not necessarily drawn to scale) that shows the building contractor the exact location and identification of every anchor bolt as well as specific anchor bolt setting details similar to those in Figure 12–1. Anchor bolt plans must show each of these items very clearly, because, if the structural steel columns

arrive at the job site and the anchor bolts already set in place do not match up with the mounting holes in the column baseplates, extremely costly delays will occur literally before the project even gets off the ground.

Figure 12–7 shows a very small and simple anchor bolt plan. Taken from an actual building project, it represents a row of structural tube columns at the end of a loading dock. Study it. Does it show the anchor bolt location, anchor bolt identification, and anchor bolt setting detail (including the required anchor bolt projection) previously discussed?

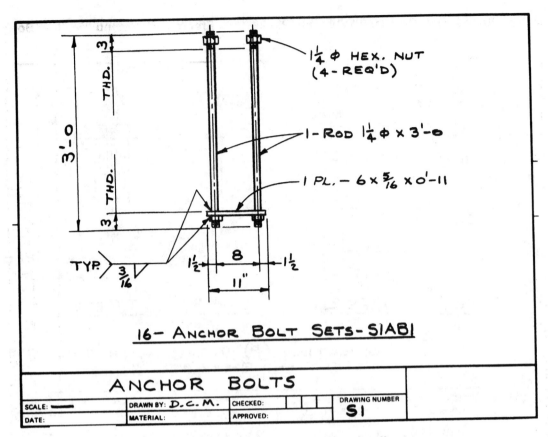

Figure 12–5 Anchor bolt set detail

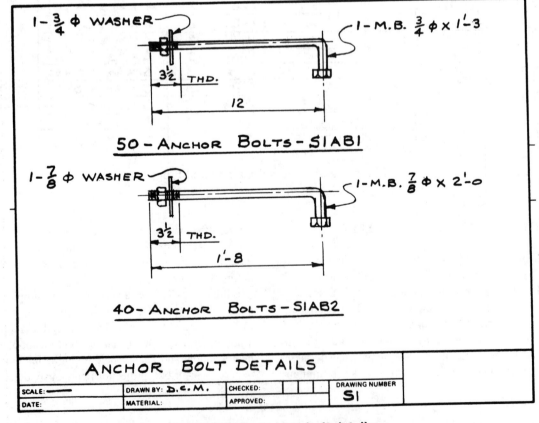

Figure 12–6 Anchor bolt details

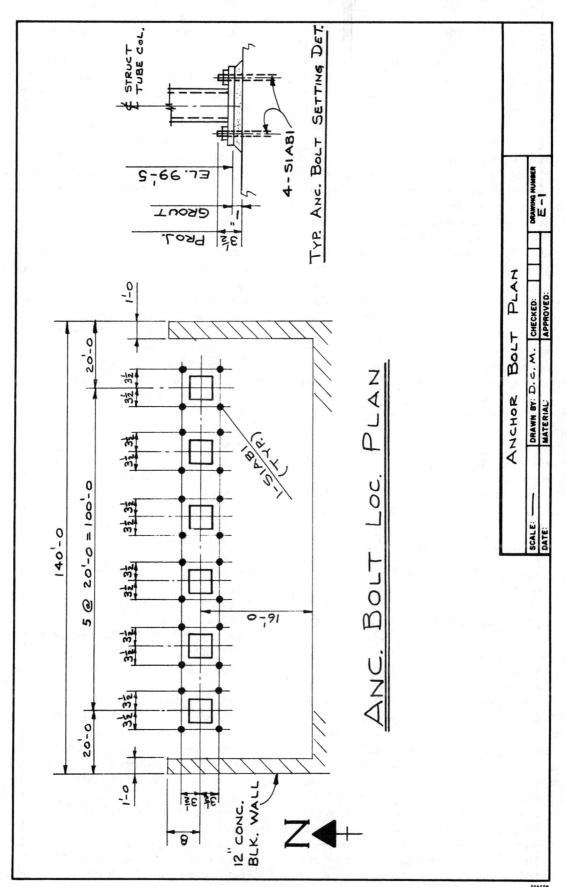

Figure 12-7   Anchor bolt plan

Figure 12–8 is an anchor bolt plan for a small commercial building such as a real estate office, insurance office, or branch bank. Notice first that this is an "E" drawing, or *erection drawing,"* thus no material list appears in the upper right-hand corner as did with the shop detail drawings previously examined. Also, this drawing has an "E" number rather than the simple sheet numbers seen on previous detail sheets.

The anchor bolt plan on the left side of the sheet is a plan view showing the outline of the building, grid system, location of anchor bolts for interior wide-flange columns on pad footings, exterior pipe columns at the top of the foundation wall, and beam pockets in the foundation wall. Notice that the width of the foundation wall has been exaggerated to help clarify anchor bolt settings for the pipe columns and beam pockets on the wall. Column sizes, bottom-of-baseplate elevations, and references to various anchor bolt setting details are also shown on the anchor bolt plan. However, the actual location and identification of the anchor bolts are not shown.

The anchor bolt setting details clearly illustrate bottom-of-baseplate elevations, exact anchor bolt locations, anchor bolt identification and projection, and any required setting or leveling plates. Taken together, the anchor bolt plan and anchor bolt setting details should show the building contractor precisely how to set the anchor bolts so that they can be properly installed well before the structural steel arrives at the job site. The structural steel detailer must always make sure the anchor bolt plan and setting details are simple, clear, and above all, accurate to the greatest degree possible.

12.4   STEEL ERECTION PLANS

The *steel erection plan,* sometimes incorrectly referred to as a framing plan, consists of a line diagram of the structural steel for each floor and the roof of a steel-frame building structure. Steel erection plans show the locations of all structural steel framing members, including beams, columns, and deck support angles. Erection plans are usually started immediately after the anchor bolt plan is complete, but they cannot be finished until all the shop detail drawings are done. Sometimes elevations of the structural steel are drawn when needed to show the location of girts, struts, and similar pieces, especially on industrial buildings.

Structural steel erection plans are gradually completed as the detailing work progresses. Usually, in the beginning, the structural detailer simply copies the design floor plans, roof plan, etc., showing all the structural steel beams, columns, deck support angles, and other parts without identifying any of them. Then, as the shop details are drawn for the individual components, the detailer transfers the identification or detail number for each beam, column, or other structural part to the proper location on the steel erection plan. It is very important that this be done often while the structural component parts are being detailed. The structural steel detailer who waits until he or she has detailed several sheets of structural steel, thinking to transfer them all to the erection plan at one time, could end up making costly mistakes. The erection plan must be done carefully and correctly because this is the plan the ironworkers will use for reference when they erect the building's structural steel. The elevations of floor and roof framing members (top of structural steel) may be given as either a benchmark elevation for various structural members or as an elevation below the finished floor level unless most of the members are at the same elevation, in which case a general note may list the elevation of certain components for that particular floor or roof. On smaller buildings, the sizes of beams (i.e., $W18 \times 50$, $W21 \times 44$, etc.) are not usually shown. However, this information may be given on erection plans for large buildings because if a large amount of steel is used in a structure, showing the beam sizes on the steel erection plans will help the ironworkers locate various pieces at the job site.

The most important item on the erection plan is the identification mark assigned to each individual component. For example, columns may be labeled 1C1, 1C2, and 1C3, while beams are called 2B1, 2B2, and 2B3. This identification mark shows the ironworkers exactly where each part fits into the structural steel framework. Again, the ironworkers will assemble the support frame *exactly* the way the pieces are shown and numbered on the steel erection plan. The design plans (those prepared by the structural design firm) will usually be used only for reference if the ironworkers run into a problem.

When doing an erection plan layout, the structural detailer should follow a logical sequence as much as possible on the framing plan. Often structural steel detailers start detailing the steel in the upper left-hand corner of the sheet and progress down and to the right across the building. The beams that tie into each other should be clustered together in the details and on the erection plans. That is, the structural steel detailer should not locate beam 1B1 in the upper right-hand corner of a building, 1B2 in the lower left, and 1B3 in the lower right.

Also, if several steel beams are identical and interchangeable in the structure, the same number should be given to all those pieces. The number or detail mark is usually painted in bright paint near one end of a component such as a beam on the same relative end of the beam that it is shown on the steel erection plan. In other words, the ironworkers will install the beam with the 3B1, 4B6, or whatever in the same location as the detailer has it shown on the structural steel erection plan. This is why the detailer should not show a beam number such as 4B2 in the center of the beam on the steel erection plan. This would confuse the ironworkers as to how to orient the beam.

Figure 12–9 shows the first-floor steel erection plan for the building in Figure 12–8. Notice how the lines representing structural steel beams, columns, and deck support angles

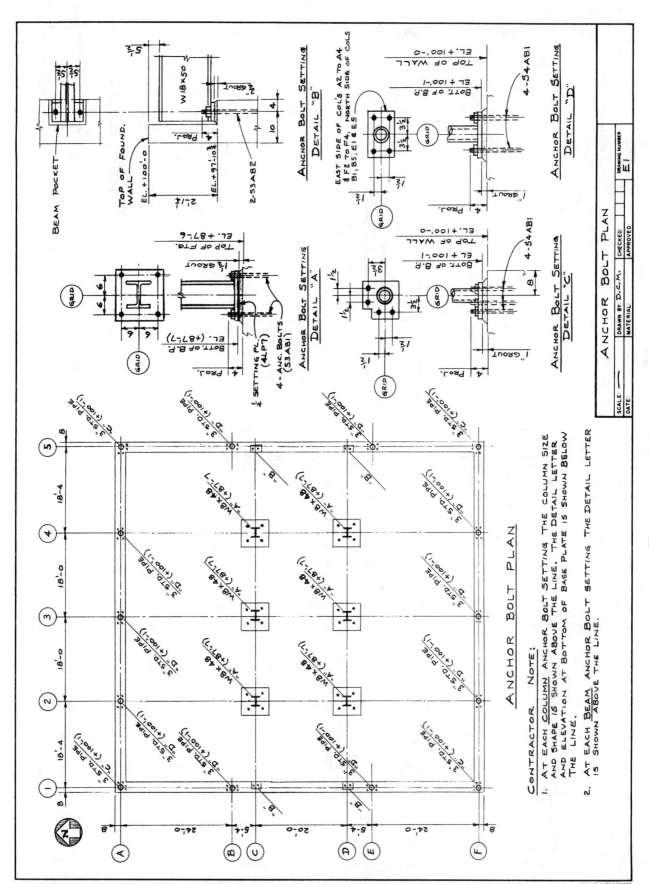

Figure 12–8   Anchor bolt plan

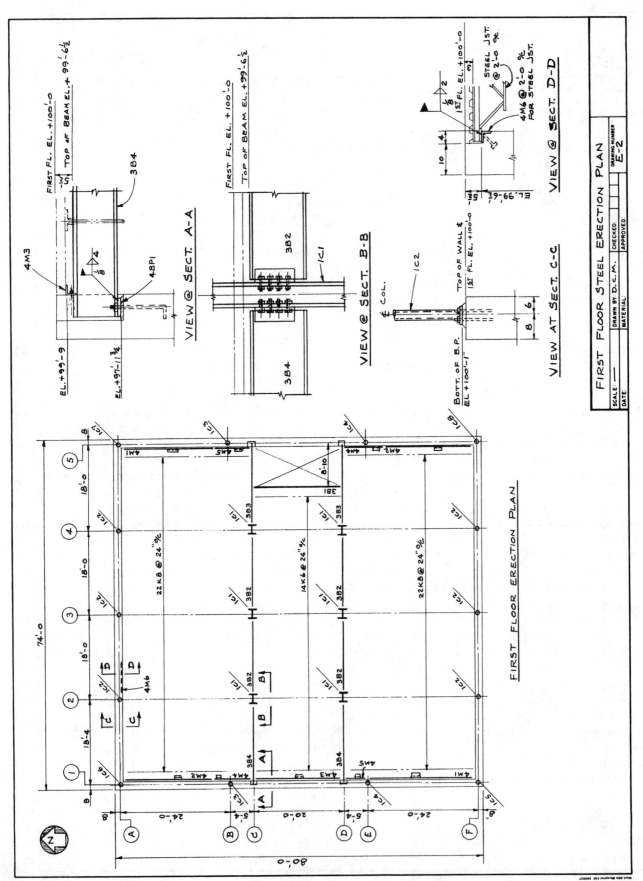

Figure 12-9   Structural Steel erection plan

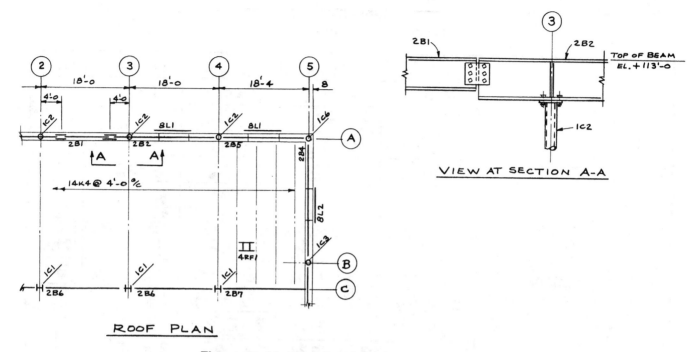

Figure 12–10 Structural steel roof erection plan

are bold and dark, while the lines for the building outline and grid are distinctly thinner. Also see how the beam and support angle numbers (i.e., 3B2, 3B3, 3B4, and 4M1, 4M2, 4M3) are located toward one end of the beams and angles. This is how the ironworkers will install them.

Section views showing top-of-steel elevations, top-of-floor and wall elevations, field weld symbols, beams, columns, and various support angles, all identified by their detail marks, are included as well. This erection plan also shows the open-web steel joists, although steel joists are often not shown on erection plans. The most important information on a steel erection plan is that concerning the placement of the main structural steel members.

Steel erection plans for roof systems are very similar to those for floor systems. Figure 12–10 illustrates a partial roof steel erection plan for the building shown in Figures 12–8 and 12–9. Notice on this plan that a roof frame is shown and identified in its approximate location. The actual location of roof frames is usually left up to the mechanical contractor. Also notice how lintels above window openings are shown on the roof erection plan. That is, the window openings are shown on the building outline, while the lintels are indicated in plan view just outside the window openings and identified as 8L1 and 8L2.

12.5 THE FIELD BOLT LIST

After the structural steel detailer has completed all the anchor bolt plans, anchor bolt details, shop fabrication draw-

ings, and required steel erection plans, the project is almost finished as far as he or she is concerned. Almost, but not quite. One final, very important task remains: preparing the field bolt list.

The *field bolt list* is simply a list of all the high-strength bolts, standard bolts, or other fasteners that must be shipped out to the job site so the ironworkers can erect the structural steel. The preparation of this list requires the structural steel detailer to go back through all of the shop detail and steel erection drawings he or she has already prepared and very carefully determine the size, length, and quantity of field bolts needed to fasten together the components of the steel framework at the job site.

To assist with this task, structural fabrication offices have reference tables that tell the structural detailer the length of bolt required to make a connection, depending on the diameter of the bolt and the length of the grip. The *grip* is simply the sum of the thicknesses of all the parts to be fastened together. Table 12–1 is an example of a reference table showing the recommended grip for high-strength bolts from ⅝ to 1 inch in diameter and up to 4 inches in length. For example, Table 12–1 indicates that structural steel parts with a grip of 1¼ inches fastened together with ¾-inch-diameter high-strength bolts would require a bolt length of 2½ inches.

Figures 12–11A and 12–11B illustrate two examples of the type of connections the structural steel detailer examines when preparing the field bolt list. Figure 12–11A shows a W16 × 40 beam connected through its bottom flange to a ½-inch-thick cap plate at the top of a 4 × 4 structural tube

Table 12–1
Lengths for High-Strength Bolts

GRIP				LENGTH
5/8 φ	3/4 φ	7/8 φ	1 φ	
1/4 – 7/16	1/8 – 5/16	—	—	1 1/2
1/2 – 11/16	3/8 – 9/16	1/4 – 7/16	1/8 – 5/16	1 3/4
3/4 – 15/16	5/8 – 13/16	1/2 – 11/16	3/8 – 7/16	2
1 – 1 3/16	7/8 – 1 1/16	3/4 – 15/16	5/8 – 13/16	2 1/4
1 1/4 – 1 7/16	1 1/8 – 1 5/16	1 – 1 3/16	7/8 – 1 1/16	2 1/2
1 1/2 – 1 11/16	1 3/8 – 1 9/16	1 1/4 – 1 7/16	1 1/8 – 1 5/16	2 3/4
1 3/4 – 1 15/16	1 5/8 – 1 13/16	1 1/2 – 1 11/16	1 3/8 – 1 9/16	3
2 – 2 3/16	1 7/8 – 2 1/16	1 3/4 – 1 15/16	1 5/8 – 1 13/16	3 1/4
2 1/4 – 2 7/16	2 1/8 – 2 5/16	2 – 2 3/16	1 7/8 – 2 1/16	3 1/2
2 1/2 – 2 11/16	2 3/8 – 2 9/16	2 1/4 – 2 7/16	2 1/8 – 2 5/16	3 3/4
2 3/4 – 2 15/16	2 5/8 – 2 13/16	2 1/2 – 2 11/16	2 3/8 – 2 9/16	4

LENGTHS FOR H.S. BOLTS

SCALE:	DRAWN BY: D.C.M.	CHECKED:	DRAWING NUMBER
DATE:	MATERIAL:	APPROVED:	

column. Looking in Part 1 of the AISC manual, the structural detailer finds that the flange thickness of a W16 × 40 beam is ½ inch. Thus, the grip for the high-strength bolts needed for the connection would be: ½″ + ½″ = 1″.

Referring to Table 12–1, the steel detailer sees that ¾-inch diameter bolts making a connection with a 1-inch grip require a bolt length of 2¼ inches. If the angle thickness connecting the framing angles to the flange face of the W10 × 49 column in Figure 12–11*B* is ⁵⁄₁₆ inch, the detailer looks in Part 1 of the AISC manual, sees that the flange thickness of a W10 × 49 column is ⁹⁄₁₆ inch, and determines that the grip would be: ⁵⁄₁₆″ + ⁹⁄₁₆″ = ⅞″. Table 12–1 would then tell the detailer that the eight ¾-inch-diameter high-strength field bolts for this connection should

also be 2¼ inches in length. Using the process just described, the steel detailer selects all the field bolts for the project.

After calculating the required field bolts for all of the connections for a project, the steel detailer prepares a field bolt list like the one shown in Figure 12–12.

As previously explained, the field bolt list ensures that the proper sizes, lengths, and quantities of bolts are shipped to the project site so that the ironworkers can assemble the structural steel. The field bolt list also indicates where the various lengths and sizes of bolts are to be used. For example, Figure 12–12 indicates that 140 field bolts, 1¼ inches long, are to be shipped to the Hollister Industries office project to connect beams 2B1, 2B2, 2B3, and 2B4 to beams

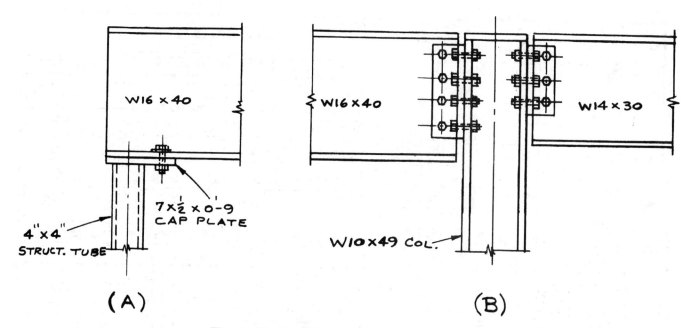

Figure 12–11   Beam-to-column connections

2B5, 2B6, 3B1, and 3B2. The bolts are to be ¾-inch in diameter. They are to be hex. head (H) bolts with hex. nuts (H) and furnished with one round (1R) washer per bolt. They are to be high-strength (H.S.) bolts. Also be shipped with the field bolts are (50) ½-inch-diameter "Kwik Bolt" concrete anchors, which might be used to fasten deck support angles such as angle 4M3 in Section A-A of Figure 12–9 to a foundation wall. The field bolt list also shows the thickness of the material connected, the sum of which adds up to the total grip upon which the length of the bolt is calculated from Table 12–1.

Notice also on the field bolt list that the quantity of bolts shipped is approximately 5 percent more than the actual number the steel detailer thinks will be required. This is why the detailer called for (148) ¾-inch-diameter × 1¼-inch-long bolts to be shipped, even though he or she had calculated 140 as the actual requirement. Upon completion of the field bolt list, the structural steel detailer's contribution to the project is complete, and he or she is ready for another assignment.

12.6   SUMMARY

This chapter, along with Chapters 10 and 11, has provided a fairly comprehensive introduction to the world of the structural steel detailer. The discussion has been designed around the types of experiences the structural drafter could expect to encounter working in a typical structural steel fabricator's drafting office. Topics have included the preparation of structural steel shop fabrication drawings for a variety of beams, columns, and miscellaneous structural steel, as well as structural steel erection drawings, anchor bolt plans and details, and the field bolt list. These are the types of assignments the entry-level structural steel drafter would carry out on almost any project. Special emphasis has been placed on the relationship between the structural design and fabrication drawings. If the student now feels better acquainted and more comfortable with structural steel fabrication drawings, steel erection plans, anchor bolt plans and details, and field bolt lists, the objectives of this unit have been accomplished.

STUDY QUESTIONS

1. What is the difference between shop fabrication drawings and erection drawings?
2. Name three types of erection drawings.
3. On a typical building project, what drawings are usually prepared first by the structural steel detailer?
4. On a typical building project, what is the last task usually performed by the structural steel detailer?
5. What are the two primary considerations in designing anchor bolts?

NO. PCS.	φ	BOLTS LGTH	HD	NUT	WRS	TYPE	TOT. GRIP	THICKNESS OF MATERIAL CONNECTED			LOCATION
140	3/4	1 3/4	H	H	IR	H.S.	9/16	3/8	3/16		2B1, 2B2, 2B3, 2B4 TO 2B5, 2B6, 3B1 & 3B2
56	3/4	2	H	H	IR	H.S.	3/4	3/8	3/8		1C1, 1C2, 1C3, 1C4 TO 3B3, 3B4, 3B5 & 3B6, 4C1, 4C2 TO 5B1, 5B2, 5B3 & 5B4
40	3/4	2 1/4	H	H	IR	H.S.	1"	3/8	3/8	1/4	6B6, 6B7, 6B8 TO 1C5, 1C6
24	3/4	2 1/2	H	H	IR	H.S.	1 1/8	3/4	3/8		6B1, 6B2, 6B3, 6B4 & 6B5 TO 1C7, 1C8 & 1C9
50	1/2	3	"KWIK" ANCH.				3/8	3/8			4M1, 4M2, 4M3, 4M4 TO CONCRETE WALL

ACTUAL REQ'D	NO. SHIPPED	φ	BOLTS LGTH	HD.	NUT	TYPE
140	148	3/4	1 3/4	H	H	H.S.
56	60	3/4	2	H	H	H.S.
40	43	3/4	2 1/4	H.	H.	H.S.
24	26	3/4	2 1/2	H.	H.	H.S.

FURNISH:

274 - 3/4 φ H.S. WASHERS
50 - 1/2 φ × 3" LG. "KWIK BOLT" CONC. ANCHORS

SHIP TO:
HOLLISTER INDUSTRIES OFFICE PROJ.
EAU CLAIRE, WISCONSIN

FIELD BOLT LIST

SCALE: ▬	DRAWN BY: D.C.M.	CHECKED:			DRAWING NUMBER S4	
DATE:	MATERIAL:	APPROVED:				

Figure 12–12   A field bolt list

6. What are the three most important items that must be shown on anchor bolt plans and details?

7. Sometimes anchor bolt plans call for two nuts on each anchor bolt. Why?

8. How does the structural steel detailer calculate the required projection of an anchor bolt above a footing or foundation wall?

9. When preparing an anchor bolt plan, why does the structural detailer often exaggerate the width of the foundation wall?

10. What is shown on the steel erection plan?

11. Why is the steel erection plan important?

12. Structural steel erection plans usually include section views. What is shown on the section views?

13. What is a field bolt list?

14. What is meant by the word *grip*?

15. When ordering bolts on the field bolt list, the structural steel detailer usually orders about _____ percent more than the actual number he or she has calculated are required.

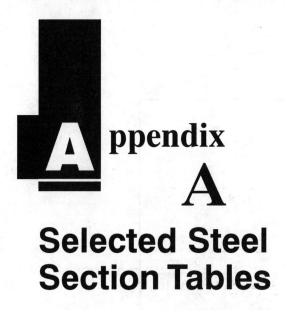

Appendix A

A

Selected Steel Section Tables

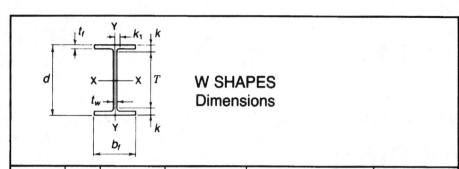

W SHAPES
Dimensions

Desig-nation	Area A	Depth d		Web Thickness t_w		$\frac{t_w}{2}$	Flange Width b_f		Flange Thickness t_f		Distance T	k	k_1
	In.²	In.		In.		In.	In.		In.		In.	In.	In.
W 24×492ᵃ	144.0	29.65	29⅝	1.970	2	1	14.115	14⅛	3.540	3⁹⁄₁₆	21	4⁵⁄₁₆	1⁹⁄₁₆
×450ᵃ	132.0	29.09	29⅛	1.810	1¹³⁄₁₆	¹⁵⁄₁₆	13.955	14	3.270	3¼	21	4¹⁄₁₆	1½
×408ᵃ	119.0	28.54	28½	1.650	1⅝	¹³⁄₁₆	13.800	13¾	2.990	3	21	3¾	1⅜
×370ᵃ	108.0	27.99	28	1.520	1½	¾	13.660	13⅝	2.720	2¾	21	3½	1⁵⁄₁₆
×335ᵃ	98.4	27.52	27½	1.380	1⅜	¹¹⁄₁₆	13.520	13½	2.480	2½	21	3¼	1¼
×306ᵃ	89.8	27.13	27⅛	1.260	1¼	⅝	13.405	13⅜	2.280	2¼	21	3¹⁄₁₆	1³⁄₁₆
×279ᵃ	82.0	26.73	26¾	1.160	1³⁄₁₆	⅝	13.305	13¼	2.090	2¹⁄₁₆	21	2⅞	1⅛
×250ᵃ	73.5	26.34	26⅜	1.040	1¹⁄₁₆	⁹⁄₁₆	13.185	13⅛	1.890	1⅞	21	2¹¹⁄₁₆	1⅛
×229	67.2	26.02	26	0.960	1	½	13.110	13⅛	1.730	1¾	21	2½	1
×207	60.7	25.71	25¾	0.870	⅞	⁷⁄₁₆	13.010	13	1.570	1⁹⁄₁₆	21	2⅜	1
×192	56.3	25.47	25½	0.810	¹³⁄₁₆	⁷⁄₁₆	12.950	13	1.460	1⁷⁄₁₆	21	2¼	1
×176	51.7	25.24	25¼	0.750	¾	⅜	12.890	12⅞	1.340	1⁵⁄₁₆	21	2⅛	¹⁵⁄₁₆
×162	47.7	25.00	25	0.705	¹¹⁄₁₆	⅜	12.955	13	1.220	1¼	21	2	1¹⁄₁₆
×146	43.0	24.74	24¾	0.650	⅝	⁵⁄₁₆	12.900	12⅞	1.090	1¹⁄₁₆	21	1⅞	1¹⁄₁₆
×131	38.5	24.48	24½	0.605	⅝	⁵⁄₁₆	12.855	12⅞	0.960	¹⁵⁄₁₆	21	1¾	1¹⁄₁₆
×117	34.4	24.26	24¼	0.550	⁹⁄₁₆	⁵⁄₁₆	12.800	12¾	0.850	⅞	21	1⅝	1
×104	30.6	24.06	24	0.500	½	¼	12.750	12¾	0.750	¾	21	1½	1
W 24×103ᵇ	30.3	24.53	24½	0.550	⁹⁄₁₆	⁵⁄₁₆	9.000	9	0.980	1	21	1¾	¹³⁄₁₆
× 94	27.7	24.31	24¼	0.515	½	¼	9.065	9⅛	0.875	⅞	21	1⅝ *	1
× 84	24.7	24.10	24⅛	0.470	½	¼	9.020	9	0.770	¾	21	1⁹⁄₁₆	¹⁵⁄₁₆
× 76	22.4	23.92	23⅞	0.440	⁷⁄₁₆	¼	8.990	9	0.680	¹¹⁄₁₆	21	1⁷⁄₁₆	¹⁵⁄₁₆
× 68	20.1	23.73	23¾	0.415	⁷⁄₁₆	¼	8.965	9	0.585	⁹⁄₁₆	21	1⅜	¹⁵⁄₁₆
W 24× 62	18.2	23.74	23¾	0.430	⁷⁄₁₆	¼	7.040	7	0.590	⁹⁄₁₆	21	1⅜	¹⁵⁄₁₆
× 55	16.2	23.57	23⅝	0.395	⅜	³⁄₁₆	7.005	7	0.505	½	21	1⁵⁄₁₆	¹⁵⁄₁₆

ᵃFor application refer to Notes in Table 2.
ᵇHeavier shapes in this series are available from some producers.

American Institute of Steel Construction, Inc.

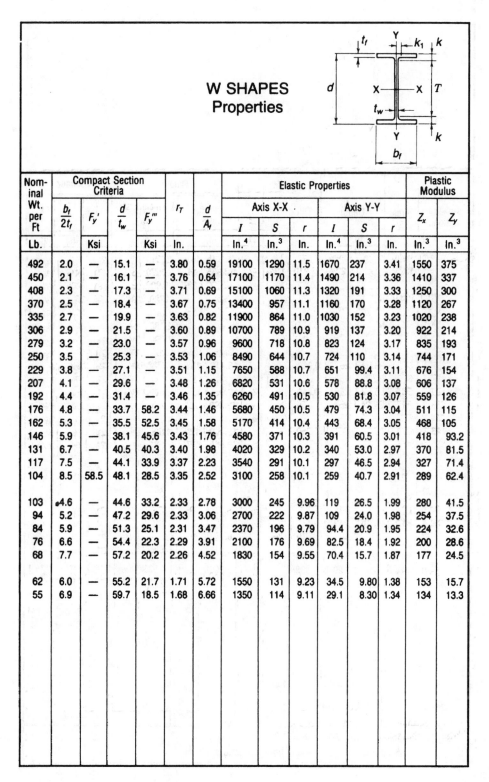

**W SHAPES
Properties**

Nom- inal Wt. per Ft	Compact Section Criteria				r_T	$\dfrac{d}{A_f}$	Elastic Properties						Plastic Modulus	
	$\dfrac{b_f}{2t_f}$	F_y'	$\dfrac{d}{t_w}$	F_y'''			Axis X-X			Axis Y-Y			Z_x	Z_y
							I	S	r	I	S	r		
Lb.		Ksi		Ksi	In.		In.⁴	In.³	In.	In.⁴	In.³	In.	In.³	In.³
492	2.0	—	15.1	—	3.80	0.59	19100	1290	11.5	1670	237	3.41	1550	375
450	2.1	—	16.1	—	3.76	0.64	17100	1170	11.4	1490	214	3.36	1410	337
408	2.3	—	17.3	—	3.71	0.69	15100	1060	11.3	1320	191	3.33	1250	300
370	2.5	—	18.4	—	3.67	0.75	13400	957	11.1	1160	170	3.28	1120	267
335	2.7	—	19.9	—	3.63	0.82	11900	864	11.0	1030	152	3.23	1020	238
306	2.9	—	21.5	—	3.60	0.89	10700	789	10.9	919	137	3.20	922	214
279	3.2	—	23.0	—	3.57	0.96	9600	718	10.8	823	124	3.17	835	193
250	3.5	—	25.3	—	3.53	1.06	8490	644	10.7	724	110	3.14	744	171
229	3.8	—	27.1	—	3.51	1.15	7650	588	10.7	651	99.4	3.11	676	154
207	4.1	—	29.6	—	3.48	1.26	6820	531	10.6	578	88.8	3.08	606	137
192	4.4	—	31.4	—	3.46	1.35	6260	491	10.5	530	81.8	3.07	559	126
176	4.8	—	33.7	58.2	3.44	1.46	5680	450	10.5	479	74.3	3.04	511	115
162	5.3	—	35.5	52.5	3.45	1.58	5170	414	10.4	443	68.4	3.05	468	105
146	5.9	—	38.1	45.6	3.43	1.76	4580	371	10.3	391	60.5	3.01	418	93.2
131	6.7	—	40.5	40.3	3.40	1.98	4020	329	10.2	340	53.0	2.97	370	81.5
117	7.5	—	44.1	33.9	3.37	2.23	3540	291	10.1	297	46.5	2.94	327	71.4
104	8.5	58.5	48.1	28.5	3.35	2.52	3100	258	10.1	259	40.7	2.91	289	62.4
103	4.6	—	44.6	33.2	2.33	2.78	3000	245	9.96	119	26.5	1.99	280	41.5
94	5.2	—	47.2	29.6	2.33	3.06	2700	222	9.87	109	24.0	1.98	254	37.5
84	5.9	—	51.3	25.1	2.31	3.47	2370	196	9.79	94.4	20.9	1.95	224	32.6
76	6.6	—	54.4	22.3	2.29	3.91	2100	176	9.69	82.5	18.4	1.92	200	28.6
68	7.7	—	57.2	20.2	2.26	4.52	1830	154	9.55	70.4	15.7	1.87	177	24.5
62	6.0	—	55.2	21.7	1.71	5.72	1550	131	9.23	34.5	9.80	1.38	153	15.7
55	6.9	—	59.7	18.5	1.68	6.66	1350	114	9.11	29.1	8.30	1.34	134	13.3

American Institute of Steel Construction, Inc.

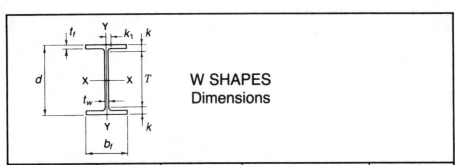

W SHAPES
Dimensions

Desig-nation	Area A (In.²)	Depth d (In.)		Web Thickness tw (In.)		tw/2 (In.)	Flange Width bf (In.)		Flange Thickness tf (In.)		Distance T (In.)	k (In.)	k₁ (In.)
W 21×402ᵃ	118.0	26.02	26	1.730	1 3/4	7/8	13.405	13 3/8	3.130	3 1/8	18 1/4	3 7/8	1 7/16
×364ᵃ	107.0	25.47	25 1/2	1.590	1 9/16	13/16	13.265	13 1/4	2.850	2 7/8	18 1/4	3 5/8	1 3/8
×333ᵃ	97.9	25.00	25	1.460	1 7/16	3/4	13.130	13 1/8	2.620	2 5/8	18 1/4	3 3/8	1 5/16
×300ᵃ	88.2	24.53	24 1/2	1.320	1 5/16	11/16	12.990	13	2.380	2 3/8	18 1/4	3 1/8	1 1/4
×275ᵃ	80.8	24.13	24 1/8	1.220	1 1/4	5/8	12.890	12 7/8	2.190	2 3/16	18 1/4	3	1 3/16
×248ᵃ	72.8	23.74	23 3/4	1.100	1 1/8	9/16	12.775	12 3/4	1.990	2	18 1/4	2 3/4	1 1/8
×223	65.4	23.35	23 3/8	1.000	1	1/2	12.675	12 5/8	1.790	1 13/16	18 1/4	2 9/16	1 1/16
×201	59.2	23.03	23	0.910	15/16	1/2	12.575	12 5/8	1.630	1 5/8	18 1/4	2 3/8	1
×182	53.6	22.72	22 3/4	0.830	13/16	7/16	12.500	12 1/2	1.480	1 1/2	18 1/4	2 1/4	1
×166	48.8	22.48	22 1/2	0.750	3/4	3/8	12.420	12 3/8	1.360	1 3/8	18 1/4	2 1/8	15/16
×147	43.2	22.06	22	0.720	3/4	3/8	12.510	12 1/2	1.150	1 1/8	18 1/4	1 7/8	1 1/16
×132	38.8	21.83	21 7/8	0.650	5/8	5/16	12.440	12 1/2	1.035	1 1/16	18 1/4	1 13/16	1
×122	35.9	21.68	21 5/8	0.600	5/8	5/16	12.390	12 3/8	0.960	15/16	18 1/4	1 11/16	1
×111	32.7	21.51	21 1/2	0.550	9/16	5/16	12.340	12 3/8	0.875	7/8	18 1/4	1 5/8	15/16
×101	29.8	21.36	21 3/8	0.500	1/2	1/4	12.290	12 1/4	0.800	13/16	18 1/4	1 9/16	15/16
W 21× 93	27.3	21.62	21 5/8	0.580	9/16	5/16	8.420	8 3/8	0.930	15/16	18 1/4	1 11/16	1
× 83	24.3	21.43	21 3/8	0.515	1/2	1/4	8.355	8 3/8	0.835	13/16	18 1/4	1 9/16	15/16
× 73	21.5	21.24	21 1/4	0.455	7/16	1/4	8.295	8 1/4	0.740	3/4	18 1/4	1 1/2	15/16
× 68	20.0	21.13	21 1/8	0.430	7/16	1/4	8.270	8 1/4	0.685	11/16	18 1/4	1 7/16	7/8
× 62	18.3	20.99	21	0.400	3/8	3/16	8.240	8 1/4	0.615	5/8	18 1/4	1 3/8	7/8
W 21× 57	16.7	21.06	21	0.405	3/8	3/16	6.555	6 1/2	0.650	5/8	18 1/4	1 3/8	7/8
× 50	14.7	20.83	20 7/8	0.380	3/8	3/16	6.530	6 1/2	0.535	9/16	18 1/4	1 5/16	7/8
× 44	13.0	20.66	20 5/8	0.350	3/8	3/16	6.500	6 1/2	0.450	7/16	18 1/4	1 3/16	7/8

ᵃFor application refer to Notes in Table 2.
Shapes in shaded rows are not available from domestic producers.

American Institute of Steel Construction, Inc.

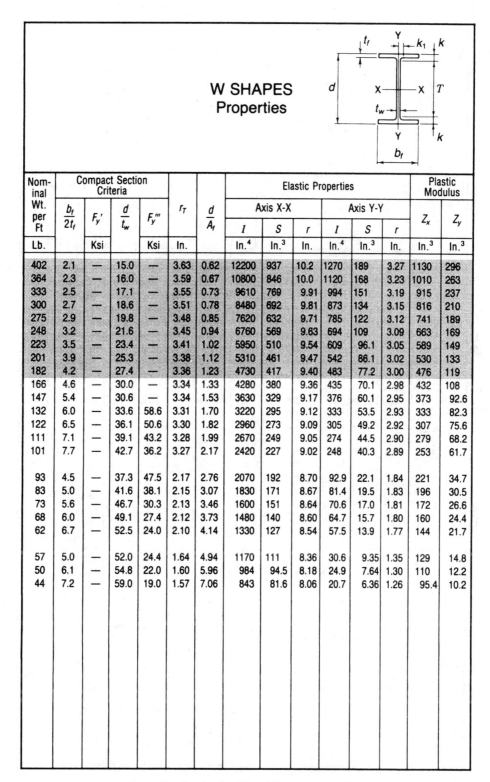

Nom-inal Wt. per Ft	Compact Section Criteria				r_T	$\dfrac{d}{A_f}$	Elastic Properties						Plastic Modulus	
	$\dfrac{b_f}{2t_f}$	F_y'	$\dfrac{d}{t_w}$	F_y'''			Axis X-X			Axis Y-Y			Z_x	Z_y
							I	S	r	I	S	r		
Lb.		Ksi		Ksi	In.		In.⁴	In.³	In.	In.⁴	In.³	In.	In.³	In.³
402	2.1	—	15.0	—	3.63	0.62	12200	937	10.2	1270	189	3.27	1130	296
364	2.3	—	16.0	—	3.59	0.67	10800	846	10.0	1120	168	3.23	1010	263
333	2.5	—	17.1	—	3.55	0.73	9610	769	9.91	994	151	3.19	915	237
300	2.7	—	18.6	—	3.51	0.78	8480	692	9.81	873	134	3.15	816	210
275	2.9	—	19.8	—	3.48	0.85	7620	632	9.71	785	122	3.12	741	189
248	3.2	—	21.6	—	3.45	0.94	6760	569	9.63	694	109	3.09	663	169
223	3.5	—	23.4	—	3.41	1.02	5950	510	9.54	609	96.1	3.05	589	149
201	3.9	—	25.3	—	3.38	1.12	5310	461	9.47	542	86.1	3.02	530	133
182	4.2	—	27.4	—	3.36	1.23	4730	417	9.40	483	77.2	3.00	476	119
166	4.6	—	30.0	—	3.34	1.33	4280	380	9.36	435	70.1	2.98	432	108
147	5.4	—	30.6	—	3.34	1.53	3630	329	9.17	376	60.1	2.95	373	92.6
132	6.0	—	33.6	58.6	3.31	1.70	3220	295	9.12	333	53.5	2.93	333	82.3
122	6.5	—	36.1	50.6	3.30	1.82	2960	273	9.09	305	49.2	2.92	307	75.6
111	7.1	—	39.1	43.2	3.28	1.99	2670	249	9.05	274	44.5	2.90	279	68.2
101	7.7	—	42.7	36.2	3.27	2.17	2420	227	9.02	248	40.3	2.89	253	61.7
93	4.5	—	37.3	47.5	2.17	2.76	2070	192	8.70	92.9	22.1	1.84	221	34.7
83	5.0	—	41.6	38.1	2.15	3.07	1830	171	8.67	81.4	19.5	1.83	196	30.5
73	5.6	—	46.7	30.3	2.13	3.46	1600	151	8.64	70.6	17.0	1.81	172	26.6
68	6.0	—	49.1	27.4	2.12	3.73	1480	140	8.60	64.7	15.7	1.80	160	24.4
62	6.7	—	52.5	24.0	2.10	4.14	1330	127	8.54	57.5	13.9	1.77	144	21.7
57	5.0	—	52.0	24.4	1.64	4.94	1170	111	8.36	30.6	9.35	1.35	129	14.8
50	6.1	—	54.8	22.0	1.60	5.96	984	94.5	8.18	24.9	7.64	1.30	110	12.2
44	7.2	—	59.0	19.0	1.57	7.06	843	81.6	8.06	20.7	6.36	1.26	95.4	10.2

American Institute of Steel Construction, Inc.

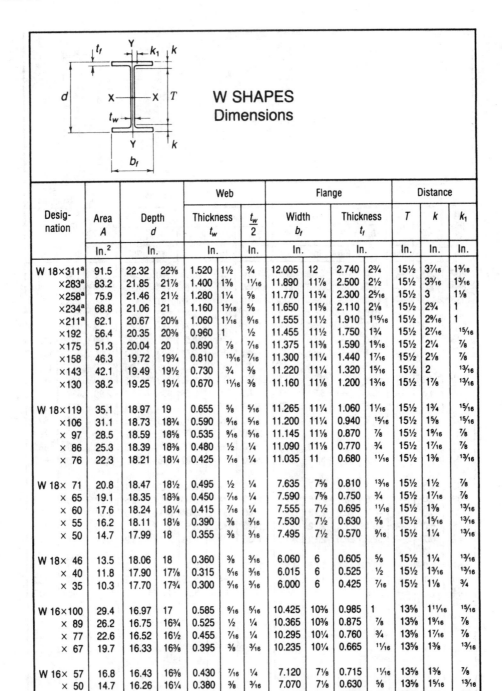

W SHAPES
Dimensions

Desig- nation	Area A	Depth d		Web Thickness t_w		$\dfrac{t_w}{2}$	Flange Width b_f		Flange Thickness t_f		Distance T	k	k_1
	In.²	In.		In.		In.	In.		In.		In.	In.	In.
W 18×311[a]	91.5	22.32	22⅜	1.520	1½	¾	12.005	12	2.740	2¾	15½	3⁷⁄₁₆	1³⁄₁₆
×283[a]	83.2	21.85	21⅞	1.400	1⅜	¹¹⁄₁₆	11.890	11⅞	2.500	2½	15½	3³⁄₁₆	1³⁄₁₆
×258[a]	75.9	21.46	21½	1.280	1¼	⅝	11.770	11¾	2.300	2⁵⁄₁₆	15½	3	1⅛
×234[a]	68.8	21.06	21	1.160	1³⁄₁₆	⅝	11.650	11⅝	2.110	2⅛	15½	2¾	1
×211[a]	62.1	20.67	20⅝	1.060	1¹⁄₁₆	⁹⁄₁₆	11.555	11½	1.910	1¹⁵⁄₁₆	15½	2⁹⁄₁₆	1
×192	56.4	20.35	20⅜	0.960	1	½	11.455	11½	1.750	1¾	15½	2⁷⁄₁₆	¹⁵⁄₁₆
×175	51.3	20.04	20	0.890	⅞	⁷⁄₁₆	11.375	11⅜	1.590	1⁹⁄₁₆	15½	2¼	⅞
×158	46.3	19.72	19¾	0.810	¹³⁄₁₆	⁷⁄₁₆	11.300	11¼	1.440	1⁷⁄₁₆	15½	2⅛	⅞
×143	42.1	19.49	19½	0.730	¾	⅜	11.220	11¼	1.320	1⁵⁄₁₆	15½	2	¹³⁄₁₆
×130	38.2	19.25	19¼	0.670	¹¹⁄₁₆	⅜	11.160	11⅛	1.200	1³⁄₁₆	15½	1⅞	¹³⁄₁₆
W 18×119	35.1	18.97	19	0.655	⅝	⁵⁄₁₆	11.265	11¼	1.060	1¹⁄₁₆	15½	1¾	¹⁵⁄₁₆
×106	31.1	18.73	18¾	0.590	⁹⁄₁₆	⁵⁄₁₆	11.200	11¼	0.940	¹⁵⁄₁₆	15½	1⅝	¹⁵⁄₁₆
× 97	28.5	18.59	18⅝	0.535	⁹⁄₁₆	⁵⁄₁₆	11.145	11⅛	0.870	⅞	15½	1⁹⁄₁₆	⅞
× 86	25.3	18.39	18⅜	0.480	½	¼	11.090	11⅛	0.770	¾	15½	1⁷⁄₁₆	⅞
× 76	22.3	18.21	18¼	0.425	⁷⁄₁₆	¼	11.035	11	0.680	¹¹⁄₁₆	15½	1⅜	¹³⁄₁₆
W 18× 71	20.8	18.47	18½	0.495	½	¼	7.635	7⅝	0.810	¹³⁄₁₆	15½	1½	⅞
× 65	19.1	18.35	18⅜	0.450	⁷⁄₁₆	¼	7.590	7⅝	0.750	¾	15½	1⁷⁄₁₆	⅞
× 60	17.6	18.24	18¼	0.415	⁷⁄₁₆	¼	7.555	7½	0.695	¹¹⁄₁₆	15½	1⅜	¹³⁄₁₆
× 55	16.2	18.11	18⅛	0.390	⅜	³⁄₁₆	7.530	7½	0.630	⅝	15½	1⁵⁄₁₆	¹³⁄₁₆
× 50	14.7	17.99	18	0.355	⅜	³⁄₁₆	7.495	7½	0.570	⁹⁄₁₆	15½	1¼	¹³⁄₁₆
W 18× 46	13.5	18.06	18	0.360	⅜	³⁄₁₆	6.060	6	0.605	⅝	15½	1¼	¹³⁄₁₆
× 40	11.8	17.90	17⅞	0.315	⁵⁄₁₆	³⁄₁₆	6.015	6	0.525	½	15½	1³⁄₁₆	¹³⁄₁₆
× 35	10.3	17.70	17¾	0.300	⁵⁄₁₆	³⁄₁₆	6.000	6	0.425	⁷⁄₁₆	15½	1⅛	¾
W 16×100	29.4	16.97	17	0.585	⁹⁄₁₆	⁵⁄₁₆	10.425	10⅜	0.985	1	13⅝	1¹¹⁄₁₆	¹⁵⁄₁₆
× 89	26.2	16.75	16¾	0.525	½	¼	10.365	10⅜	0.875	⅞	13⅝	1⁹⁄₁₆	⅞
× 77	22.6	16.52	16½	0.455	⁷⁄₁₆	¼	10.295	10¼	0.760	¾	13⅝	1⁷⁄₁₆	⅞
× 67	19.7	16.33	16⅜	0.395	⅜	³⁄₁₆	10.235	10¼	0.665	¹¹⁄₁₆	13⅝	1⅜	¹³⁄₁₆
W 16× 57	16.8	16.43	16⅜	0.430	⁷⁄₁₆	¼	7.120	7⅛	0.715	¹¹⁄₁₆	13⅝	1⅜	⅞
× 50	14.7	16.26	16¼	0.380	⅜	³⁄₁₆	7.070	7⅛	0.630	⅝	13⅝	1⁵⁄₁₆	¹³⁄₁₆
× 45	13.3	16.13	16⅛	0.345	⅜	³⁄₁₆	7.035	7	0.565	⁹⁄₁₆	13⅝	1¼	¹³⁄₁₆
× 40	11.8	16.01	16	0.305	⁵⁄₁₆	³⁄₁₆	6.995	7	0.505	½	13⅝	1³⁄₁₆	¹³⁄₁₆
× 36	10.6	15.86	15⅞	0.295	⁵⁄₁₆	³⁄₁₆	6.985	7	0.430	⁷⁄₁₆	13⅝	1⅛	¾

American Institute of Steel Construction, Inc.

W SHAPES
Properties

Nominal Wt. per Ft	Compact Section Criteria				r_T	$\dfrac{d}{A_f}$	Elastic Properties						Plastic Modulus	
	$\dfrac{b_f}{2t_f}$	F_y'	$\dfrac{d}{t_w}$	F_y'''			Axis X-X			Axis Y-Y			Z_x	Z_y
							I	S	r	I	S	r		
Lb.		Ksi		Ksi	In.		In.^{4}	In.^{3}	In.	In.^{4}	In.^{3}	In.	In.^{3}	In.^{3}
311	2.2	—	14.7	—	3.26	0.68	6960	624	8.72	795	132	2.95	753	207
283	2.4	—	15.6	—	3.23	0.74	6160	564	8.61	704	118	2.91	676	185
258	2.6	—	16.8	—	3.19	0.79	5510	514	8.53	628	107	2.88	611	166
234	2.8	—	18.2	—	3.16	0.86	4900	466	8.44	558	95.8	2.85	549	149
211	3.0	—	19.5	—	3.13	0.94	4330	419	8.35	493	85.3	2.82	490	132
192	3.3	—	21.2	—	3.10	1.02	3870	380	8.28	440	76.8	2.79	442	119
175	3.6	—	22.5	—	3.07	1.11	3450	344	8.20	391	68.8	2.76	398	106
158	3.9	—	24.3	—	3.05	1.21	3060	310	8.12	347	61.4	2.74	356	94.8
143	4.2	—	26.7	—	3.03	1.32	2750	282	8.09	311	55.5	2.72	322	85.4
130	4.6	—	28.7	—	3.01	1.44	2460	256	8.03	278	49.9	2.70	291	76.7
119	5.3	—	29.0	—	3.02	1.59	2190	231	7.90	253	44.9	2.69	261	69.1
106	6.0	—	31.7	—	3.00	1.78	1910	204	7.84	220	39.4	2.66	230	60.5
97	6.4	—	34.7	54.7	2.99	1.92	1750	188	7.82	201	36.1	2.65	211	55.3
86	7.2	—	38.3	45.0	2.97	2.15	1530	166	7.77	175	31.6	2.63	186	48.4
76	8.1	64.2	42.8	36.0	2.95	2.43	1330	146	7.73	152	27.6	2.61	163	42.2
71	4.7	—	37.3	47.4	1.98	2.99	1170	127	7.50	60.3	15.8	1.70	145	24.7
65	5.1	—	40.8	39.7	1.97	3.22	1070	117	7.49	54.8	14.4	1.69	133	22.5
60	5.4	—	44.0	34.2	1.96	3.47	984	108	7.47	50.1	13.3	1.69	123	20.6
55	6.0	—	46.4	30.6	1.95	3.82	890	98.3	7.41	44.9	11.9	1.67	112	18.5
50	6.6	—	50.7	25.7	1.94	4.21	800	88.9	7.38	40.1	10.7	1.65	101	16.6
46	5.0	—	50.2	26.2	1.54	4.93	712	78.8	7.25	22.5	7.43	1.29	90.7	11.7
40	5.7	—	56.8	20.5	1.52	5.67	612	68.4	7.21	19.1	6.35	1.27	78.4	9.95
35	7.1	—	59.0	19.0	1.49	6.94	510	57.6	7.04	15.3	5.12	1.22	66.5	8.06
100	5.3	—	29.0	—	2.81	1.65	1490	175	7.10	186	35.7	2.51	198	54.9
89	5.9	—	31.9	64.9	2.79	1.85	1300	155	7.05	163	31.4	2.49	175	48.1
77	6.8	—	36.3	50.1	2.77	2.11	1110	134	7.00	138	26.9	2.47	150	41.1
67	7.7	—	41.3	38.6	2.75	2.40	954	117	6.96	119	23.2	2.46	130	35.5
57	5.0	—	38.2	45.2	1.86	3.23	758	92.2	6.72	43.1	12.1	1.60	105	18.9
50	5.6	—	42.8	36.1	1.84	3.65	659	81.0	6.68	37.2	10.5	1.59	92.0	16.3
45	6.2	—	46.8	30.2	1.83	4.06	586	72.7	6.65	32.8	9.34	1.57	82.3	14.5
40	6.9	—	52.5	24.0	1.82	4.53	518	64.7	6.63	28.9	8.25	1.57	72.9	12.7
36	8.1	64.0	53.8	22.9	1.79	5.28	448	56.5	6.51	24.5	7.00	1.52	64.0	10.8

American Institute of Steel Construction, Inc.

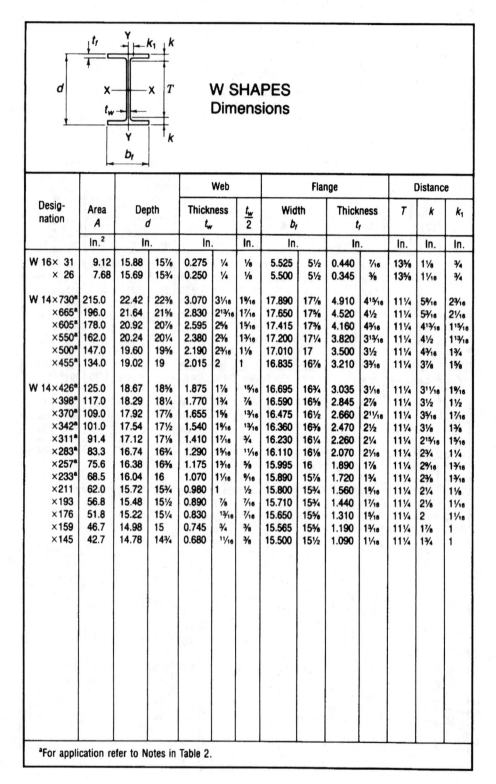

W SHAPES
Dimensions

Desig-nation	Area A	Depth d	Web Thickness t_w		$\frac{t_w}{2}$	Flange Width b_f		Thickness t_f		Distance T	k	k_1
	In.²	In.	In.		In.	In.		In.		In.	In.	In.
W 16× 31	9.12	15.88	15⅞	0.275	¼	⅛	5.525	5½	0.440	⁷⁄₁₆	13⅝	1⅛	¾
× 26	7.68	15.69	15¾	0.250	¼	⅛	5.500	5½	0.345	⅜	13⅝	1¹⁄₁₆	¾
W 14×730ª	215.0	22.42	22⅜	3.070	3¹⁄₁₆	1⁹⁄₁₆	17.890	17⅞	4.910	4¹⁵⁄₁₆	11¼	5⁹⁄₁₆	2⁵⁄₁₆
×665ª	196.0	21.64	21⅝	2.830	2¹³⁄₁₆	1⁷⁄₁₆	17.650	17⅝	4.520	4½	11¼	5⁵⁄₁₆	2¹⁄₁₆
×605ª	178.0	20.92	20⅞	2.595	2⅝	1⁵⁄₁₆	17.415	17⅜	4.160	4³⁄₁₆	11¼	4¹³⁄₁₆	1¹⁵⁄₁₆
×550ª	162.0	20.24	20¼	2.380	2⅜	1³⁄₁₆	17.200	17¼	3.820	3¹³⁄₁₆	11¼	4½	1¹³⁄₁₆
×500ª	147.0	19.60	19⅝	2.190	2³⁄₁₆	1⅛	17.010	17	3.500	3½	11¼	4³⁄₁₆	1¾
×455ª	134.0	19.02	19	2.015	2	1	16.835	16⅞	3.210	3³⁄₁₆	11¼	3⅞	1⅝
W 14×426ª	125.0	18.67	18⅝	1.875	1⅞	¹⁵⁄₁₆	16.695	16¾	3.035	3¹⁄₁₆	11¼	3¹¹⁄₁₆	1⁹⁄₁₆
×398ª	117.0	18.29	18¼	1.770	1¾	⅞	16.590	16⅝	2.845	2⅞	11¼	3½	1½
×370ª	109.0	17.92	17⅞	1.655	1⅝	¹³⁄₁₆	16.475	16½	2.660	2¹¹⁄₁₆	11¼	3⁵⁄₁₆	1⁷⁄₁₆
×342ª	101.0	17.54	17½	1.540	1⁹⁄₁₆	¹³⁄₁₆	16.360	16⅜	2.470	2½	11¼	3⅛	1⅜
×311ª	91.4	17.12	17⅛	1.410	1⁷⁄₁₆	¾	16.230	16¼	2.260	2¼	11¼	2¹⁵⁄₁₆	1⁵⁄₁₆
×283ª	83.3	16.74	16¾	1.290	1⁵⁄₁₆	¹¹⁄₁₆	16.110	16⅛	2.070	2¹⁄₁₆	11¼	2¾	1¼
×257ª	75.6	16.38	16⅜	1.175	1³⁄₁₆	⅝	15.995	16	1.890	1⅞	11¼	2⁹⁄₁₆	1³⁄₁₆
×233ª	68.5	16.04	16	1.070	1¹⁄₁₆	⁹⁄₁₆	15.890	15⅞	1.720	1¾	11¼	2⅜	1³⁄₁₆
×211	62.0	15.72	15¾	0.980	1	½	15.800	15¾	1.560	1⁹⁄₁₆	11¼	2¼	1⅛
×193	56.8	15.48	15½	0.890	⅞	⁷⁄₁₆	15.710	15¾	1.440	1⁷⁄₁₆	11¼	2⅛	1¹⁄₁₆
×176	51.8	15.22	15¼	0.830	¹³⁄₁₆	⁷⁄₁₆	15.650	15⅝	1.310	1⁵⁄₁₆	11¼	2	1¹⁄₁₆
×159	46.7	14.98	15	0.745	¾	⅜	15.565	15⅝	1.190	1³⁄₁₆	11¼	1⅞	1
×145	42.7	14.78	14¾	0.680	¹¹⁄₁₆	⅜	15.500	15½	1.090	1¹⁄₁₆	11¼	1¾	1

ªFor application refer to Notes in Table 2.

American Institute of Steel Construction, Inc.

W SHAPES
Properties

Nom- inal Wt. per Ft	Compact Section Criteria				r_T	$\dfrac{d}{A_f}$	Elastic Properties						Plastic Modulus	
	$\dfrac{b_f}{2t_f}$	F_y'	$\dfrac{d}{t_w}$	F_y'''			Axis X-X			Axis Y-Y			Z_x	Z_y
							I	S	r	I	S	r		
Lb.		Ksi		Ksi	In.		In.⁴	In.³	In.	In.⁴	In.³	In.	In.³	In.³
31	6.3	—	57.7	19.8	1.39	6.53	375	47.2	6.41	12.4	4.49	1.17	54.0	7.03
26	8.0	—	62.8	16.8	1.36	8.27	301	38.4	6.26	9.59	3.49	1.12	44.2	5.48
730	1.8	—	7.3	—	4.99	0.25	14300	1280	8.17	4720	527	4.69	1660	816
665	2.0	—	7.6	—	4.92	0.27	12400	1150	7.98	4170	472	4.62	1480	730
605	2.1	—	8.1	—	4.85	0.29	10800	1040	7.80	3680	423	4.55	1320	652
550	2.3	—	8.5	—	4.79	0.31	9430	931	7.63	3250	378	4.49	1180	583
500	2.4	—	8.9	—	4.73	0.33	8210	838	7.48	2880	339	4.43	1050	522
455	2.6	—	9.4	—	4.68	0.35	7190	756	7.33	2560	304	4.38	936	468
426	2.8	—	10.0	—	4.64	0.37	6600	707	7.26	2360	283	4.34	869	434
398	2.9	—	10.3	—	4.61	0.39	6000	656	7.16	2170	262	4.31	801	402
370	3.1	—	10.8	—	4.57	0.41	5440	607	7.07	1990	241	4.27	736	370
342	3.3	—	11.4	—	4.54	0.43	4900	559	6.98	1810	221	4.24	672	338
311	3.6	—	12.1	—	4.50	0.47	4330	506	6.88	1610	199	4.20	603	304
283	3.9	—	13.0	—	4.46	0.50	3840	459	6.79	1440	179	4.17	542	274
257	4.2	—	13.9	—	4.43	0.54	3400	415	6.71	1290	161	4.13	487	246
233	4.6	—	15.0	—	4.40	0.59	3010	375	6.63	1150	145	4.10	436	221
211	5.1	—	16.0	—	4.37	0.64	2660	338	6.55	1030	130	4.07	390	198
193	5.5	—	17.4	—	4.35	0.68	2400	310	6.50	931	119	4.05	355	180
176	6.0	—	18.3	—	4.32	0.74	2140	281	6.43	838	107	4.02	320	163
159	6.5	—	20.1	—	4.30	0.81	1900	254	6.38	748	96.2	4.00	287	146
145	7.1	—	21.7	—	4.28	0.88	1710	232	6.33	677	87.3	3.98	260	133

American Institute of Steel Construction, Inc.

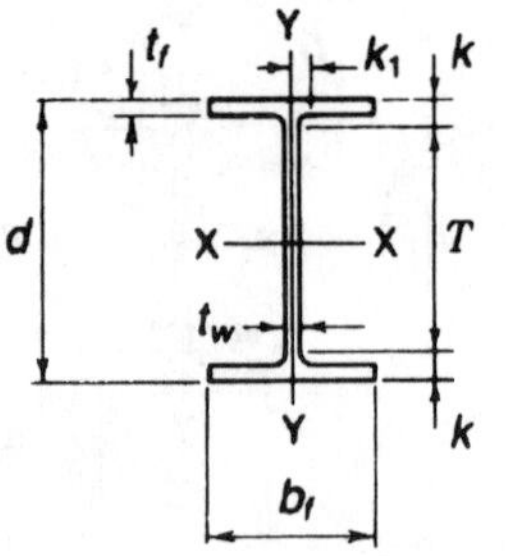

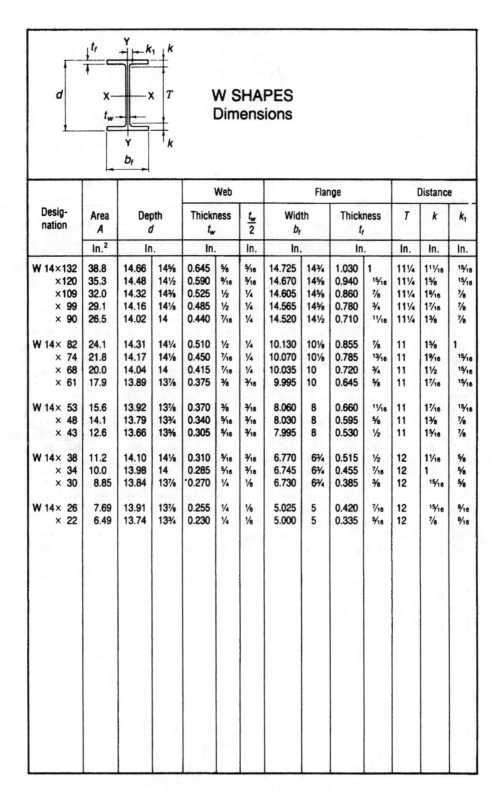

W SHAPES
Dimensions

Designation	Area A	Depth d		Web Thickness t_w		Web $\frac{t_w}{2}$	Flange Width b_f		Flange Thickness t_f		Distance T	Distance k	Distance k_1
	In.^{2}	In.		In.		In.	In.		In.		In.	In.	In.
W 14×132	38.8	14.66	14 5/8	0.645	5/8	5/16	14.725	14 3/4	1.030	1	11 1/4	1 11/16	15/16
×120	35.3	14.48	14 1/2	0.590	9/16	5/16	14.670	14 5/8	0.940	15/16	11 1/4	1 5/8	15/16
×109	32.0	14.32	14 3/8	0.525	1/2	1/4	14.605	14 5/8	0.860	7/8	11 1/4	1 9/16	7/8
× 99	29.1	14.16	14 1/8	0.485	1/2	1/4	14.565	14 5/8	0.780	3/4	11 1/4	1 7/16	7/8
× 90	26.5	14.02	14	0.440	7/16	1/4	14.520	14 1/2	0.710	11/16	11 1/4	1 3/8	7/8
W 14× 82	24.1	14.31	14 1/4	0.510	1/2	1/4	10.130	10 1/8	0.855	7/8	11	1 5/8	1
× 74	21.8	14.17	14 1/8	0.450	7/16	1/4	10.070	10 1/8	0.785	13/16	11	1 9/16	15/16
× 68	20.0	14.04	14	0.415	7/16	1/4	10.035	10	0.720	3/4	11	1 1/2	15/16
× 61	17.9	13.89	13 7/8	0.375	3/8	3/16	9.995	10	0.645	5/8	11	1 7/16	15/16
W 14× 53	15.6	13.92	13 7/8	0.370	3/8	3/16	8.060	8	0.660	11/16	11	1 7/16	15/16
× 48	14.1	13.79	13 3/4	0.340	5/16	3/16	8.030	8	0.595	5/8	11	1 3/8	7/8
× 43	12.6	13.66	13 5/8	0.305	5/16	3/16	7.995	8	0.530	1/2	11	1 5/16	7/8
W 14× 38	11.2	14.10	14 1/8	0.310	5/16	3/16	6.770	6 3/4	0.515	1/2	12	1 1/16	5/8
× 34	10.0	13.98	14	0.285	5/16	3/16	6.745	6 3/4	0.455	7/16	12	1	5/8
× 30	8.85	13.84	13 7/8	·0.270	1/4	1/8	6.730	6 3/4	0.385	3/8	12	15/16	5/8
W 14× 26	7.69	13.91	13 7/8	0.255	1/4	1/8	5.025	5	0.420	7/16	12	15/16	9/16
× 22	6.49	13.74	13 3/4	0.230	1/4	1/8	5.000	5	0.335	5/16	12	7/8	9/16

American Institute of Steel Construction, Inc.

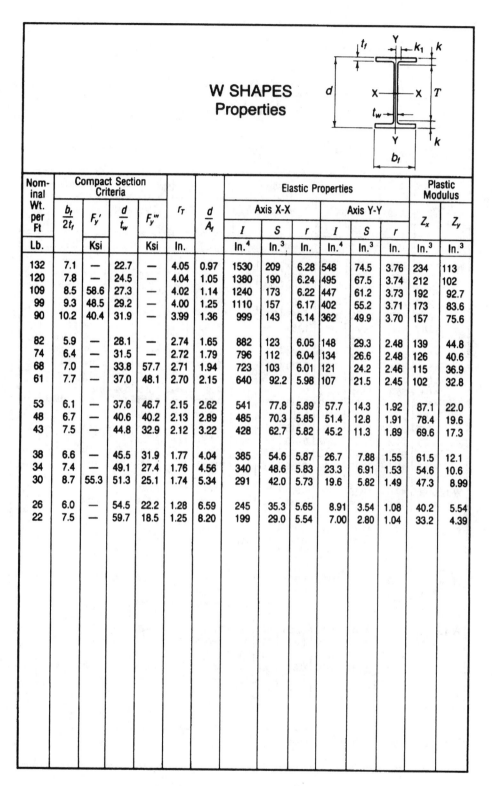

Nom-inal Wt. per Ft	Compact Section Criteria				r_T	$\dfrac{d}{A_f}$	Elastic Properties						Plastic Modulus	
	$\dfrac{b_f}{2t_f}$	F_y'	$\dfrac{d}{t_w}$	F_y'''			Axis X-X			Axis Y-Y			Z_x	Z_y
							I	S	r	I	S	r		
Lb.		Ksi		Ksi	In.		In.⁴	In.³	In.	In.⁴	In.³	In.	In.³	In.³
132	7.1	—	22.7	—	4.05	0.97	1530	209	6.28	548	74.5	3.76	234	113
120	7.8	—	24.5	—	4.04	1.05	1380	190	6.24	495	67.5	3.74	212	102
109	8.5	58.6	27.3	—	4.02	1.14	1240	173	6.22	447	61.2	3.73	192	92.7
99	9.3	48.5	29.2	—	4.00	1.25	1110	157	6.17	402	55.2	3.71	173	83.6
90	10.2	40.4	31.9	—	3.99	1.36	999	143	6.14	362	49.9	3.70	157	75.6
82	5.9	—	28.1	—	2.74	1.65	882	123	6.05	148	29.3	2.48	139	44.8
74	6.4	—	31.5	—	2.72	1.79	796	112	6.04	134	26.6	2.48	126	40.6
68	7.0	—	33.8	57.7	2.71	1.94	723	103	6.01	121	24.2	2.46	115	36.9
61	7.7	—	37.0	48.1	2.70	2.15	640	92.2	5.98	107	21.5	2.45	102	32.8
53	6.1	—	37.6	46.7	2.15	2.62	541	77.8	5.89	57.7	14.3	1.92	87.1	22.0
48	6.7	—	40.6	40.2	2.13	2.89	485	70.3	5.85	51.4	12.8	1.91	78.4	19.6
43	7.5	—	44.8	32.9	2.12	3.22	428	62.7	5.82	45.2	11.3	1.89	69.6	17.3
38	6.6	—	45.5	31.9	1.77	4.04	385	54.6	5.87	26.7	7.88	1.55	61.5	12.1
34	7.4	—	49.1	27.4	1.76	4.56	340	48.6	5.83	23.3	6.91	1.53	54.6	10.6
30	8.7	55.3	51.3	25.1	1.74	5.34	291	42.0	5.73	19.6	5.82	1.49	47.3	8.99
26	6.0	—	54.5	22.2	1.28	6.59	245	35.3	5.65	8.91	3.54	1.08	40.2	5.54
22	7.5	—	59.7	18.5	1.25	8.20	199	29.0	5.54	7.00	2.80	1.04	33.2	4.39

American Institute of Steel Construction, Inc.

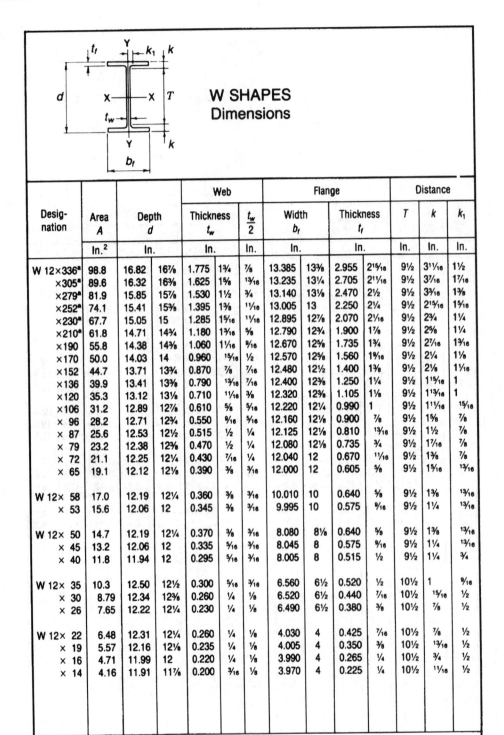

W SHAPES
Dimensions

Desig-nation	Area A	Depth d		Web Thickness t_w		$\frac{t_w}{2}$	Flange Width b_f		Flange Thickness t_f		T	k	k_1
	In.²	In.		In.		In.	In.		In.		In.	In.	In.
W 12×336ª	98.8	16.82	16 7/8	1.775	1 3/4	7/8	13.385	13 3/8	2.955	2 15/16	9 1/2	3 11/16	1 1/2
×305ª	89.6	16.32	16 3/8	1.625	1 5/8	13/16	13.235	13 1/4	2.705	2 11/16	9 1/2	3 7/16	1 7/16
×279ª	81.9	15.85	15 7/8	1.530	1 1/2	3/4	13.140	13 1/8	2.470	2 1/2	9 1/2	3 3/16	1 3/8
×252ª	74.1	15.41	15 3/8	1.395	1 3/8	11/16	13.005	13	2.250	2 1/4	9 1/2	2 15/16	1 5/16
×230ª	67.7	15.05	15	1.285	1 5/16	11/16	12.895	12 7/8	2.070	2 1/16	9 1/2	2 3/4	1 1/4
×210ª	61.8	14.71	14 3/4	1.180	1 3/16	5/8	12.790	12 3/4	1.900	1 7/8	9 1/2	2 5/8	1 1/4
×190	55.8	14.38	14 3/8	1.060	1 1/16	9/16	12.670	12 5/8	1.735	1 3/4	9 1/2	2 7/16	1 3/16
×170	50.0	14.03	14	0.960	15/16	1/2	12.570	12 5/8	1.560	1 9/16	9 1/2	2 1/4	1 1/8
×152	44.7	13.71	13 3/4	0.870	7/8	7/16	12.480	12 1/2	1.400	1 3/8	9 1/2	2 1/8	1 1/16
×136	39.9	13.41	13 3/8	0.790	13/16	7/16	12.400	12 3/8	1.250	1 1/4	9 1/2	1 15/16	1
×120	35.3	13.12	13 1/8	0.710	11/16	3/8	12.320	12 3/8	1.105	1 1/8	9 1/2	1 13/16	1
×106	31.2	12.89	12 7/8	0.610	5/8	5/16	12.220	12 1/4	0.990	1	9 1/2	1 11/16	15/16
× 96	28.2	12.71	12 3/4	0.550	9/16	5/16	12.160	12 1/8	0.900	7/8	9 1/2	1 5/8	7/8
× 87	25.6	12.53	12 1/2	0.515	1/2	1/4	12.125	12 1/8	0.810	13/16	9 1/2	1 1/2	7/8
× 79	23.2	12.38	12 3/8	0.470	1/2	1/4	12.080	12 1/8	0.735	3/4	9 1/2	1 7/16	7/8
× 72	21.1	12.25	12 1/4	0.430	7/16	1/4	12.040	12	0.670	11/16	9 1/2	1 3/8	7/8
× 65	19.1	12.12	12 1/8	0.390	3/8	3/16	12.000	12	0.605	5/8	9 1/2	1 5/16	13/16
W 12× 58	17.0	12.19	12 1/4	0.360	3/8	3/16	10.010	10	0.640	5/8	9 1/2	1 3/8	13/16
× 53	15.6	12.06	12	0.345	3/8	3/16	9.995	10	0.575	9/16	9 1/2	1 1/4	13/16
W 12× 50	14.7	12.19	12 1/4	0.370	3/8	3/16	8.080	8 1/8	0.640	5/8	9 1/2	1 3/8	13/16
× 45	13.2	12.06	12	0.335	5/16	3/16	8.045	8	0.575	9/16	9 1/2	1 1/4	13/16
× 40	11.8	11.94	12	0.295	5/16	3/16	8.005	8	0.515	1/2	9 1/2	1 1/4	3/4
W 12× 35	10.3	12.50	12 1/2	0.300	5/16	3/16	6.560	6 1/2	0.520	1/2	10 1/2	1	9/16
× 30	8.79	12.34	12 3/8	0.260	1/4	1/8	6.520	6 1/2	0.440	7/16	10 1/2	15/16	1/2
× 26	7.65	12.22	12 1/4	0.230	1/4	1/8	6.490	6 1/2	0.380	3/8	10 1/2	7/8	1/2
W 12× 22	6.48	12.31	12 1/4	0.260	1/4	1/8	4.030	4	0.425	7/16	10 1/2	7/8	1/2
× 19	5.57	12.16	12 1/8	0.235	1/4	1/8	4.005	4	0.350	3/8	10 1/2	13/16	1/2
× 16	4.71	11.99	12	0.220	1/4	1/8	3.990	4	0.265	1/4	10 1/2	3/4	1/2
× 14	4.16	11.91	11 7/8	0.200	3/16	1/8	3.970	4	0.225	1/4	10 1/2	11/16	1/2

ªFor application refer to Notes in Table 2.

American Institute of Steel Construction, Inc.

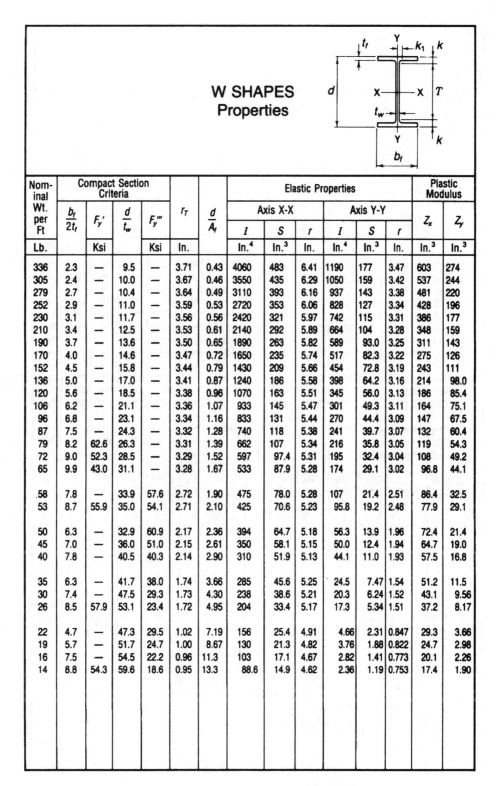

Nom-inal Wt. per Ft	Compact Section Criteria				r_T	$\dfrac{d}{A_f}$	Elastic Properties						Plastic Modulus	
	$\dfrac{b_f}{2t_f}$	F_y'	$\dfrac{d}{t_w}$	F_y'''			Axis X-X			Axis Y-Y			Z_x	Z_y
							I	S	r	I	S	r		
Lb.		Ksi		Ksi	In.		In.^{4}	In.^{3}	In.	In.^{4}	In.^{3}	In.	In.^{3}	In.^{3}
336	2.3	—	9.5	—	3.71	0.43	4060	483	6.41	1190	177	3.47	603	274
305	2.4	—	10.0	—	3.67	0.46	3550	435	6.29	1050	159	3.42	537	244
279	2.7	—	10.4	—	3.64	0.49	3110	393	6.16	937	143	3.38	481	220
252	2.9	—	11.0	—	3.59	0.53	2720	353	6.06	828	127	3.34	428	196
230	3.1	—	11.7	—	3.56	0.56	2420	321	5.97	742	115	3.31	386	177
210	3.4	—	12.5	—	3.53	0.61	2140	292	5.89	664	104	3.28	348	159
190	3.7	—	13.6	—	3.50	0.65	1890	263	5.82	589	93.0	3.25	311	143
170	4.0	—	14.6	—	3.47	0.72	1650	235	5.74	517	82.3	3.22	275	126
152	4.5	—	15.8	—	3.44	0.79	1430	209	5.66	454	72.8	3.19	243	111
136	5.0	—	17.0	—	3.41	0.87	1240	186	5.58	398	64.2	3.16	214	98.0
120	5.6	—	18.5	—	3.38	0.96	1070	163	5.51	345	56.0	3.13	186	85.4
106	6.2	—	21.1	—	3.36	1.07	933	145	5.47	301	49.3	3.11	164	75.1
96	6.8	—	23.1	—	3.34	1.16	833	131	5.44	270	44.4	3.09	147	67.5
87	7.5	—	24.3	—	3.32	1.28	740	118	5.38	241	39.7	3.07	132	60.4
79	8.2	62.6	26.3	—	3.31	1.39	662	107	5.34	216	35.8	3.05	119	54.3
72	9.0	52.3	28.5	—	3.29	1.52	597	97.4	5.31	195	32.4	3.04	108	49.2
65	9.9	43.0	31.1	—	3.28	1.67	533	87.9	5.28	174	29.1	3.02	96.8	44.1
58	7.8	—	33.9	57.6	2.72	1.90	475	78.0	5.28	107	21.4	2.51	86.4	32.5
53	8.7	55.9	35.0	54.1	2.71	2.10	425	70.6	5.23	95.8	19.2	2.48	77.9	29.1
50	6.3	—	32.9	60.9	2.17	2.36	394	64.7	5.18	56.3	13.9	1.96	72.4	21.4
45	7.0	—	36.0	51.0	2.15	2.61	350	58.1	5.15	50.0	12.4	1.94	64.7	19.0
40	7.8	—	40.5	40.3	2.14	2.90	310	51.9	5.13	44.1	11.0	1.93	57.5	16.8
35	6.3	—	41.7	38.0	1.74	3.66	285	45.6	5.25	24.5	7.47	1.54	51.2	11.5
30	7.4	—	47.5	29.3	1.73	4.30	238	38.6	5.21	20.3	6.24	1.52	43.1	9.56
26	8.5	57.9	53.1	23.4	1.72	4.95	204	33.4	5.17	17.3	5.34	1.51	37.2	8.17
22	4.7	—	47.3	29.5	1.02	7.19	156	25.4	4.91	4.66	2.31	0.847	29.3	3.66
19	5.7	—	51.7	24.7	1.00	8.67	130	21.3	4.82	3.76	1.88	0.822	24.7	2.98
16	7.5	—	54.5	22.2	0.96	11.3	103	17.1	4.67	2.82	1.41	0.773	20.1	2.26
14	8.8	54.3	59.6	18.6	0.95	13.3	88.6	14.9	4.62	2.36	1.19	0.753	17.4	1.90

American Institute of Steel Construction, Inc.

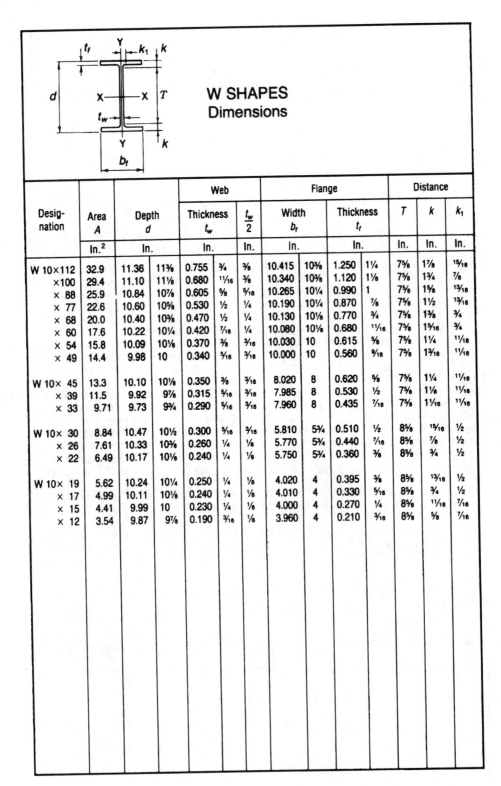

W SHAPES
Dimensions

Desig-nation	Area A (In.²)	Depth d (In.)		Web Thickness t_w (In.)		$\frac{t_w}{2}$ (In.)	Flange Width b_f (In.)		Flange Thickness t_f (In.)		Distance T (In.)	Distance k (In.)	Distance k_1 (In.)
W 10×112	32.9	11.36	11⅜	0.755	¾	⅜	10.415	10⅜	1.250	1¼	7⅝	1⅞	15⁄16
×100	29.4	11.10	11⅛	0.680	11⁄16	⅜	10.340	10⅜	1.120	1⅛	7⅝	1¾	⅞
× 88	25.9	10.84	10⅞	0.605	⅝	5⁄16	10.265	10¼	0.990	1	7⅝	1⅝	13⁄16
× 77	22.6	10.60	10⅝	0.530	½	¼	10.190	10¼	0.870	⅞	7⅝	1½	13⁄16
× 68	20.0	10.40	10⅜	0.470	½	¼	10.130	10⅛	0.770	¾	7⅝	1⅜	¾
× 60	17.6	10.22	10¼	0.420	7⁄16	¼	10.080	10⅛	0.680	11⁄16	7⅝	1 5⁄16	¾
× 54	15.8	10.09	10⅛	0.370	⅜	3⁄16	10.030	10	0.615	⅝	7⅝	1¼	11⁄16
× 49	14.4	9.98	10	0.340	5⁄16	3⁄16	10.000	10	0.560	9⁄16	7⅝	1 3⁄16	11⁄16
W 10× 45	13.3	10.10	10⅛	0.350	⅜	3⁄16	8.020	8	0.620	⅝	7⅝	1¼	11⁄16
× 39	11.5	9.92	9⅞	0.315	5⁄16	3⁄16	7.985	8	0.530	½	7⅝	1⅛	11⁄16
× 33	9.71	9.73	9¾	0.290	5⁄16	3⁄16	7.960	8	0.435	7⁄16	7⅝	1 1⁄16	11⁄16
W 10× 30	8.84	10.47	10½	0.300	5⁄16	3⁄16	5.810	5¾	0.510	½	8⅝	15⁄16	½
× 26	7.61	10.33	10⅜	0.260	¼	⅛	5.770	5¾	0.440	7⁄16	8⅝	⅞	½
× 22	6.49	10.17	10⅛	0.240	¼	⅛	5.750	5¾	0.360	⅜	8⅝	¾	½
W 10× 19	5.62	10.24	10¼	0.250	¼	⅛	4.020	4	0.395	⅜	8⅝	13⁄16	½
× 17	4.99	10.11	10⅛	0.240	¼	⅛	4.010	4	0.330	5⁄16	8⅝	¾	½
× 15	4.41	9.99	10	0.230	¼	⅛	4.000	4	0.270	¼	8⅝	11⁄16	7⁄16
× 12	3.54	9.87	9⅞	0.190	3⁄16	⅛	3.960	4	0.210	3⁄16	8⅝	⅝	7⁄16

American Institute of Steel Construction, Inc.

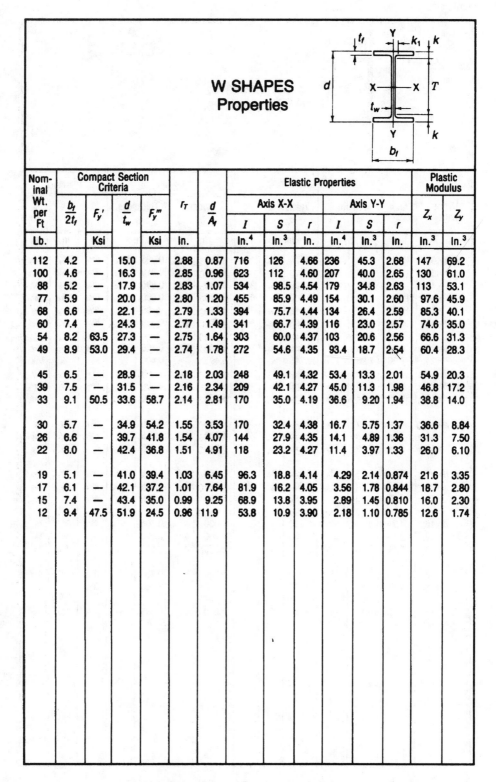

W SHAPES
Properties

Nom-inal Wt. per Ft	Compact Section Criteria				r_T	$\dfrac{d}{A_f}$	Elastic Properties						Plastic Modulus	
	$\dfrac{b_f}{2t_f}$	F_y'	$\dfrac{d}{t_w}$	F_y^m			Axis X-X			Axis Y-Y			Z_x	Z_y
							I	S	r	I	S	r		
Lb.		Ksi		Ksi	In.		In.⁴	In.³	In.	In.⁴	In.³	In.	In.³	In.³
112	4.2	—	15.0	—	2.88	0.87	716	126	4.66	236	45.3	2.68	147	69.2
100	4.6	—	16.3	—	2.85	0.96	623	112	4.60	207	40.0	2.65	130	61.0
88	5.2	—	17.9	—	2.83	1.07	534	98.5	4.54	179	34.8	2.63	113	53.1
77	5.9	—	20.0	—	2.80	1.20	455	85.9	4.49	154	30.1	2.60	97.6	45.9
68	6.6	—	22.1	—	2.79	1.33	394	75.7	4.44	134	26.4	2.59	85.3	40.1
60	7.4	—	24.3	—	2.77	1.49	341	66.7	4.39	116	23.0	2.57	74.6	35.0
54	8.2	63.5	27.3	—	2.75	1.64	303	60.0	4.37	103	20.6	2.56	66.6	31.3
49	8.9	53.0	29.4	—	2.74	1.78	272	54.6	4.35	93.4	18.7	2.54	60.4	28.3
45	6.5	—	28.9	—	2.18	2.03	248	49.1	4.32	53.4	13.3	2.01	54.9	20.3
39	7.5	—	31.5	—	2.16	2.34	209	42.1	4.27	45.0	11.3	1.98	46.8	17.2
33	9.1	50.5	33.6	58.7	2.14	2.81	170	35.0	4.19	36.6	9.20	1.94	38.8	14.0
30	5.7	—	34.9	54.2	1.55	3.53	170	32.4	4.38	16.7	5.75	1.37	36.6	8.84
26	6.6	—	39.7	41.8	1.54	4.07	144	27.9	4.35	14.1	4.89	1.36	31.3	7.50
22	8.0	—	42.4	36.8	1.51	4.91	118	23.2	4.27	11.4	3.97	1.33	26.0	6.10
19	5.1	—	41.0	39.4	1.03	6.45	96.3	18.8	4.14	4.29	2.14	0.874	21.6	3.35
17	6.1	—	42.1	37.2	1.01	7.64	81.9	16.2	4.05	3.56	1.78	0.844	18.7	2.80
15	7.4	—	43.4	35.0	0.99	9.25	68.9	13.8	3.95	2.89	1.45	0.810	16.0	2.30
12	9.4	47.5	51.9	24.5	0.96	11.9	53.8	10.9	3.90	2.18	1.10	0.785	12.6	1.74

American Institute of Steel Construction, Inc.

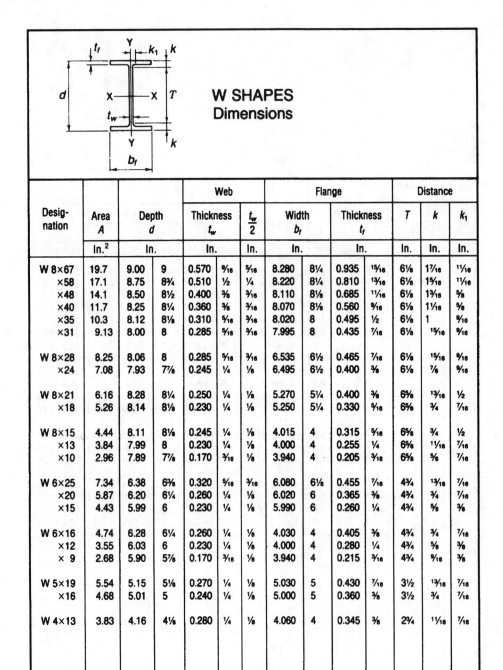

Desig-nation	Area A	Depth d		Web				Flange				Distance		
				Thickness t_w		$\frac{t_w}{2}$		Width b_f		Thickness t_f		T	k	k_1
	In.^{2}	In.		In.		In.		In.		In.		In.	In.	In.
W 8×67	19.7	9.00	9	0.570	9/16	5/16		8.280	8¼	0.935	15/16	6⅛	1 7/16	11/16
×58	17.1	8.75	8¾	0.510	½	¼		8.220	8¼	0.810	13/16	6⅛	1 5/16	11/16
×48	14.1	8.50	8½	0.400	⅜	3/16		8.110	8⅛	0.685	11/16	6⅛	1 3/16	⅝
×40	11.7	8.25	8¼	0.360	⅜	3/16		8.070	8⅛	0.560	9/16	6⅛	1 1/16	⅝
×35	10.3	8.12	8⅛	0.310	5/16	3/16		8.020	8	0.495	½	6⅛	1	9/16
×31	9.13	8.00	8	0.285	5/16	3/16		7.995	8	0.435	7/16	6⅛	15/16	9/16
W 8×28	8.25	8.06	8	0.285	5/16	3/16		6.535	6½	0.465	7/16	6⅛	15/16	9/16
×24	7.08	7.93	7⅞	0.245	¼	⅛		6.495	6½	0.400	⅜	6⅛	⅞	9/16
W 8×21	6.16	8.28	8¼	0.250	¼	⅛		5.270	5¼	0.400	⅜	6⅝	13/16	½
×18	5.26	8.14	8⅛	0.230	¼	⅛		5.250	5¼	0.330	5/16	6⅝	¾	7/16
W 8×15	4.44	8.11	8⅛	0.245	¼	⅛		4.015	4	0.315	5/16	6⅝	¾	½
×13	3.84	7.99	8	0.230	¼	⅛		4.000	4	0.255	¼	6⅝	11/16	7/16
×10	2.96	7.89	7⅞	0.170	3/16	⅛		3.940	4	0.205	3/16	6⅝	⅝	7/16
W 6×25	7.34	6.38	6⅜	0.320	5/16	3/16		6.080	6⅛	0.455	7/16	4¾	13/16	7/16
×20	5.87	6.20	6¼	0.260	¼	⅛		6.020	6	0.365	⅜	4¾	¾	7/16
×15	4.43	5.99	6	0.230	¼	⅛		5.990	6	0.260	¼	4¾	⅝	⅜
W 6×16	4.74	6.28	6¼	0.260	¼	⅛		4.030	4	0.405	⅜	4¾	¾	7/16
×12	3.55	6.03	6	0.230	¼	⅛		4.000	4	0.280	¼	4¾	⅝	⅜
× 9	2.68	5.90	5⅞	0.170	3/16	⅛		3.940	4	0.215	3/16	4¾	9/16	⅜
W 5×19	5.54	5.15	5⅛	0.270	¼	⅛		5.030	5	0.430	7/16	3½	13/16	7/16
×16	4.68	5.01	5	0.240	¼	⅛		5.000	5	0.360	⅜	3½	¾	7/16
W 4×13	3.83	4.16	4⅛	0.280	¼	⅛		4.060	4	0.345	⅜	2¾	11/16	7/16

American Institute of Steel Construction, Inc.

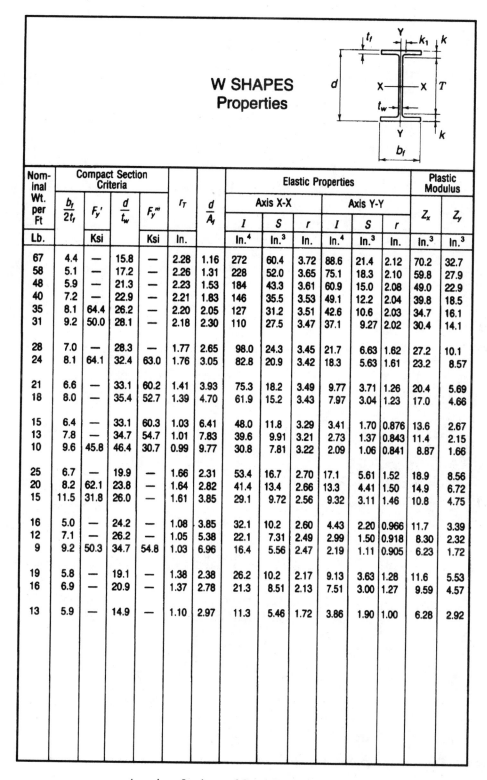

W SHAPES
Properties

Nom-inal Wt. per Ft	bf/2tf	Fy'	d/tw	Fy'''	rT	d/Af	Axis X-X I	Axis X-X S	Axis X-X r	Axis Y-Y I	Axis Y-Y S	Axis Y-Y r	Zx	Zy
Lb.		Ksi		Ksi	In.		In.⁴	In.³	In.	In.⁴	In.³	In.	In.³	In.³
67	4.4	—	15.8	—	2.28	1.16	272	60.4	3.72	88.6	21.4	2.12	70.2	32.7
58	5.1	—	17.2	—	2.26	1.31	228	52.0	3.65	75.1	18.3	2.10	59.8	27.9
48	5.9	—	21.3	—	2.23	1.53	184	43.3	3.61	60.9	15.0	2.08	49.0	22.9
40	7.2	—	22.9	—	2.21	1.83	146	35.5	3.53	49.1	12.2	2.04	39.8	18.5
35	8.1	64.4	26.2	—	2.20	2.05	127	31.2	3.51	42.6	10.6	2.03	34.7	16.1
31	9.2	50.0	28.1	—	2.18	2.30	110	27.5	3.47	37.1	9.27	2.02	30.4	14.1
28	7.0	—	28.3	—	1.77	2.65	98.0	24.3	3.45	21.7	6.63	1.62	27.2	10.1
24	8.1	64.1	32.4	63.0	1.76	3.05	82.8	20.9	3.42	18.3	5.63	1.61	23.2	8.57
21	6.6	—	33.1	60.2	1.41	3.93	75.3	18.2	3.49	9.77	3.71	1.26	20.4	5.69
18	8.0	—	35.4	52.7	1.39	4.70	61.9	15.2	3.43	7.97	3.04	1.23	17.0	4.66
15	6.4	—	33.1	60.3	1.03	6.41	48.0	11.8	3.29	3.41	1.70	0.876	13.6	2.67
13	7.8	—	34.7	54.7	1.01	7.83	39.6	9.91	3.21	2.73	1.37	0.843	11.4	2.15
10	9.6	45.8	46.4	30.7	0.99	9.77	30.8	7.81	3.22	2.09	1.06	0.841	8.87	1.66
25	6.7	—	19.9	—	1.66	2.31	53.4	16.7	2.70	17.1	5.61	1.52	18.9	8.56
20	8.2	62.1	23.8	—	1.64	2.82	41.4	13.4	2.66	13.3	4.41	1.50	14.9	6.72
15	11.5	31.8	26.0	—	1.61	3.85	29.1	9.72	2.56	9.32	3.11	1.46	10.8	4.75
16	5.0	—	24.2	—	1.08	3.85	32.1	10.2	2.60	4.43	2.20	0.966	11.7	3.39
12	7.1	—	26.2	—	1.05	5.38	22.1	7.31	2.49	2.99	1.50	0.918	8.30	2.32
9	9.2	50.3	34.7	54.8	1.03	6.96	16.4	5.56	2.47	2.19	1.11	0.905	6.23	1.72
19	5.8	—	19.1	—	1.38	2.38	26.2	10.2	2.17	9.13	3.63	1.28	11.6	5.53
16	6.9	—	20.9	—	1.37	2.78	21.3	8.51	2.13	7.51	3.00	1.27	9.59	4.57
13	5.9	—	14.9	—	1.10	2.97	11.3	5.46	1.72	3.86	1.90	1.00	6.28	2.92

American Institute of Steel Construction, Inc.

Index